Digital Humanities and Research Methods in Religious Studies

Introductions to Digital Humanities – Religion

Edited by
Claire Clivaz, Charles M. Ess, Gregory Price Grieve,
Kristian Petersen and Sally Promey

Volume 2

Digital Humanities and Research Methods in Religious Studies

―

An Introduction

Edited by
Christopher D. Cantwell and Kristian Petersen

DE GRUYTER

ISBN 978-3-11-057160-8
e-ISBN (PDF) 978-3-11-057302-2
e-ISBN (EPUB) 978-3-11-057194-3

Library of Congress Control Number: 2020944396

Bibliographic information published by the Deutsche Nationalbibliothek
The Deutsche Nationalbibliothek lists this publication in the Deutsche Nationalbibliografie; detailed bibliographic data are available on the Internet at http://dnb.dnb.de.

© 2021 Walter de Gruyter GmbH, Berlin/Boston
Cover image: Social network visualization. With friendly permission of Martin Grandjean.
Printing and binding: CPI books GmbH, Leck

www.degruyter.com

Table of Contents

Christopher D. Cantwell and Kristian Petersen
Digital Humanities and Religious Studies: A "Why" To Guide —— VII

Part I: **Texts**

Marcus Bingenheimer
Digital Tools for Buddhist Studies —— 3

Lincoln A. Mullen
The Making of *America's Public Bible*: Computational Text Analysis for Religious History —— 31

Frederik Elwert
Network Analysis of Religious Texts. Case Studies on Ancient Egyptian and Indian Religion —— 53

Rebecca Krawiec and Caroline T. Schroeder
Digital Approaches to Studying Authorial Style and Monastic Subjectivity in Early Christian Egypt —— 71

Part II: **Images**

Andrew Quintman and Kurtis R. Schaeffer
Synthesizing Image and Text in the Life of the Buddha —— 97

James S. Bielo & Claire Vaughn
Materializing the Bible: A Digital Scholarship Project from the Anthropology of Religion —— 119

Emily C. Floyd and Sally M. Promey
Collaboration and Access in the Study of Material and Visual Cultures of Religion —— 143

Erin Walcek Averett and Derek B. Counts
Scaling Religious Practice from Landscape to Artifact: Digital Approaches to Ancient Cyprus —— 167

Part III: **Places**

Abhishek Amar
Sacred Centers in India: Archiving Temples and Images of a Hindu City —— 199

J. E. E. Pettit, Fenggang Yang, Yuqian Huang
Developing a Database of Religions in Contemporary China —— 215

Louis Kaplan and Melissa Shiff
From Ararat to Kimberley: Activating Imaginary Jewish Homelands with Augmented and Virtual Reality —— 227

Caleb Elfenbein, Farah Bakaari, Julia Schafer
Mapping Anti-Muslim Hostility and its Effects —— 249

Part IV: **Issues**

Christopher R. Cotter and David G. Robertson
Critique and Community: Podcasting Religious Studies —— 273

S. Brent Plate
Public Pedagogy: MOOCs and Their Revolutionary Discontents —— 291

Wendi Bellar and Heidi A. Campbell
Building Social Sites of Collaborative Research: A Case Study of the Network for New Media, Religion, and Digital Culture Studies —— 303

Russell T. McCutcheon
Learning to Code: Digital Pebbles and Institutional Ripples —— 319

Index —— 339

Christopher D. Cantwell and Kristian Petersen
Digital Humanities and Religious Studies: A "Why" To Guide

In 2018, the American Academy of Religion hosted a special "wildcard" panel on digital research and teaching projects that encourage students to engage with their local communities. Titled "Teaching Local Religion with Digital Humanities: Objects, Methods, Pedagogies," the session featured ten educators who contributed to eight different projects. Some of the projects were based at major research universities and worked with communities across the country. Others were housed at small liberal arts colleges and focused on one place of worship. But almost all of the projects connected with the scholars' research agenda and employed the tools and methods of digital humanities (DH) in order to benefit from the work of student collaborators.[1]

Despite the panel's explicit focus on teaching, the question and answer session that followed forayed into other matters. The very first question from the audience asked how—or even if—such projects counted in tenure and promotion portfolios. Another wondered how the projects' directors received credit for their work. Finally, one audience members seemed to question the entire premise of the panel altogether, asking whether the time spent developing these projects would be better spent producing journal articles or monographs.

Such questions are important ones. They get at some of the tensions around integrating digital scholarship into the academic study of religion. But they also missed the point of the session. While the presenters were eager to talk about the pedagogical and intellectual benefits they had gained by employing digital methods, the audience was concerned with the professional issues that surround digital work.

---

[1] "Teaching Local Religion with Digital Humanities: Objects, Methods, Pedagogies," American Academy of Religion Annual Meeting, Denver, Colorado, 19 Nov. 2018. The projects featured at the roundtable were Emily Suzanne Clark, "Jesuit Missions on the Columbia Plateau"; Shana Sippy and Michael McNally, "Religious Diversity in Minnesota Initiative," https://religionsmn.carleton.edu/; Gale Kenny, "Religion in the Archive"; Rachel McBride Lindsey, "Arch City Religion," https://www.archcityreligion.org/; Christopher D. Cantwell, "Gathering Places: Religion and Community in Milwaukee," https://uwm.edu/gatheringplaces; Rachel Kranson, "Pittsburgh Torah Scrolls Project," https://pittsburghtorahscrolls.wordpress.com/; Amy DeRogatis and Isaac Weiner, "American Religious Sounds Project," https://religioussounds.osu.edu/; Jennifer S. Leath, "Black Religious Denver"; Cara Lea Burndige, "American Religion and Refugees in the Heartland."

It is a disconnect that also often defines conversations about digital humanities more broadly. From the moment scholars first took up the term "DH" in the early 2000s to describe work that either harnessed new technology for humanistic inquiry or studied new technologies humanistically, debates about DH tend to circulate around "how" questions. How is digital scholarship going to count? How is it going to be peer reviewed? How is it made?[2] Though, again, important, what can get lost in these more practical discussions is a sustained reflection on why one might want to take up digital methods in the first place. Why are an increasing number of scholars and educators finding digital humanities relevant to their work? What advantages have these methods presented to them? How have they benefited? And why might digital scholarship be vital for the study of religion's future?

This collection aims to take up this later set of questions. Rather than add to the already robust literature on best practices in the production, evaluation, and promotion of digital scholarship, it aims to focus on the ways scholars have found that digital methods enhance their research, teaching, and service to the profession. Gathering together the lead investigators of more than a dozen projects, the collection covers a wide swath of the digital work now being done in the study of religion. It considers why podcasting and social media platforms might be the future in promoting the public understanding of religion. It showcases why digital projects might be an invaluable resource in helping our students become critical producers of knowledge about religion. And it dwells upon why twenty-first century technology might be the most productive avenue for answering some of the questions that have animated our fields for almost a century. Though by no means a comprehensive survey of the ever-growing body of digital work taking place, the collection does hope to serve as a snapshot of the inspirations and revelations that have propelled digital scholarship's growth of late. In assembling the volume, we asked contributors to outline the journey of their inquiries and the logic behind the approaches they used. We hope that by

---

[2] On the history of digital humanities and the perennial debates that have defined the field, see Susan Hockey, "The History of Humanities Computing," in *A Companion to Digital Humanities*, Susan Schreibman, Ray Siemens, John Unsworth, eds. (Oxford: Blackwell, 2004), http://digital humanities.org:3030/companion/view?docId=blackwell/9781405103213/9781405103213.xml&chunk.id=ss1-2-1&toc.depth=1&toc.id=ss1-2-1&brand=9781405103213_brand; Matthew Kirschenbaum, "What Is Digital Humanities and What's It Doing in English Departments?" *ADE Bulletin* 150 (2010): 2–3; Matthew K. Gold, "The Digital Humanities Moment," in *Debates in Digital Humanities*, Gold, ed. (Minneapolis: University of Minnesota Press, 2012), http://dhdebates.gc.cuny.edu/debates; Christopher D. Cantwell and Hussein Rashid, *Religion, Media, and the Digital Turn* (New York: Social Science Research Council, 2015).

mapping the terrain in such a way—showcasing why digital tools aided in producing, managing, or teaching religious studies—the reader might come away with an alternative perspective on researching their own subjects or be inspired to try new methods that resonate with their materials.

Why Digital Humanities?

As the case studies presented here hopefully will demonstrate, there are as many reasons for drawing upon digital methods as there are digital projects. Scholars often find themselves drawn to digital methods because they help grapple with a question peculiar to their field, or because they open up new avenues for addressing perennial issues related to teaching, research, or service. But taken as a whole, the essays collected here do point to a series of overlapping inspirations that have propelled the field. For instance, several of the authors collected here turned to digital tools because analyzing a body of source material by traditional methods proved impractical. Lincoln Mullen's interest in documenting the most widely quoted passages of scripture throughout American history would have required him to read hundreds of millions of newspaper articles. Similarly, Frederick Elwert's research on the frequency with which certain characters or ideas appear in sacred texts across time, space, and tradition would have necessitated a lifetime of close reading and cataloging. These are issues that confront scholars across the academy today as the sheer volume of material or data related to our work grows exponentially. And by using textual and network analysis software, both Mullen and Elwert were able to process colossal quantities of data over a reasonable amount of time in order to contribute to debates in their field in new and vital ways.

The ability of digital research methods to facilitate the interpretation of data goes beyond just computational analysis, however. A number of the projects featured here have turned toward digital platforms in order to foster collaboration or consolidate resources around a particular kind of source material. Rebecca Krawiec and Caroline T. Schroeder's work on the Coptic SCRIPTORIUM, for example, not only uses text analysis software to analyze Coptic texts. It also takes advantage of open-source text encoding methods to create a shared database of Coptic items from multiple institutions across the globe. James Bielo and Claire Vaughan's piece on the Materializing the Bible project, meanwhile, similarly curates a collection of texts, images, and multimedia related to those instances when Christian communities across North America attempted to make Bible stories "real" through art, sculptures, plays, and other public performances. What's more, the expanse of Bielo and Vaughn's work is also made possible by the ability of many digital asset management systems, which allow them to host a variety of kinds of media in one place. Through their intentional curation, juxtaposition, and organization, Bielo and Vaughn, like Krawiec and Schroeder, open new avenues of inquiry and analysis.

In addition to creating new collections of source material, moreover, digital methods also allow scholars to reach and constitute new publics as well. Wendi Bellar and Heidi A. Campbell, for instance, argue in their contribution that social network platforms can actually facilitate the kind of interdisciplinarity the study of religion claims to foster. On the Network for New Media, Religion, and Digital Culture Studies they run, scholars from any discipline can create a profile in order to find a potential research partner or learn about new work that might have been published in another field. Christopher R. Cotter and David G. Robertson take this argument even further, claiming that new media outlets like podcasting not only connect scholars with each other, but can also connect scholars with the wider public. As founding contributors of the Religious Studies Project, Cotter and Robertson hope that their work will impact the public understanding of religion. The episodes and essays they host on their site are aimed at both scholars across the academy and lay readers who take an interest in the field.

As Cotter and Robertson's work suggests, one of the central promises of digital research methods are there ability to blur boundaries between previously separate categories or spheres of work. For the Religious Studies Project, this means collapsing the distance that exists between the academy and the wider public. With Coptic SCRIPTORIUM, the line where Krawiec and Schroeder's research begins and their service to the profession ends is blurred. Digital scholarship's multimodal nature is often what makes it incredibly difficult to fit into existing themes of promotion, peer review, and tenure. But for many of the contributors here, digital scholarship's genre-busting potential is also its greatest promise. Caleb Elfenbein's project documenting instances of anti-Muslim hostitily in the U.S., for instance, is a class assignment, research project, and public resource all at once. Students scour the media for reports of Islamaphobic or anti-Muslim hate crimes that official outlets fail to collect. This material then gets posted on an interactive map that journalists, activists, and policy makers can refer to. Abhishek Amar's work in developing virtual recreations of lost or inaccessible Hindu temples in India similarly builds upon the digital inclinations of twenty-first century students in order to make them partners in the research process. In a different vein, however, Louis Kaplan and and Melissa Shiff's project blurs the boundaries between scholarship and art, fact and speculation. Their work developing speculative recreations of failed efforts to create a Jewish homeland through virtual reality and other map-based software both documents these overlooked endeavors and interrogates the impulses that tie them together.

Each of these projects, in short, drew upon digital tools, platforms, or methods for a multitude of reasons. It is important to note, however, that despite this diversity all of the efforts documented here share a common understanding about the nature of DH work. Every project in its own way turned to new technologies because they allowed them to advance the study of religion's longstanding commitment to research, teaching, and public service. Emily Floyd and Sally Promey's essay on the Material and Visual Cultures of Religion (MAVCOR) perhaps documents this most clearly. MAVCOR began in part as an archive of sources related to religion's material and visual manifestations. But those connected with MAVCOR intended the archive to serve as a resource that could promote greater collaboration between art historians, visual studies scholars, and the study of religion. As part of its efforts, MAVCOR's team published an edited volume and established a new journal. Though it is published online in order to take advantage of the web's ability to publish images and other material at a lower cost than print publications, MAVCOR's journal functions, at its core, as an exemplar of best practices in scholarly publication.

The same is true for the other projects showcased here. The turn to the digital is in many ways a means to augment and enhance the work we do in analog. Quality digital work should never exist solely of its own accord. It should contribute to the study of religion's ongoing efforts in print, in the classroom, and in the public. The works gathered here exemplify this commitment to the field. Our hope is that their work can inspire others in their efforts as well.

What to Consider

While the reasons why scholars turn to digital work are diverse, the process by which projects came together, developed, and then launched is more uniform. The growth in digital scholarship in the last decades has provided a roadmap that many projects follow, and the contributions collected here are no different. From the essays readers may glean several themes that arise when using digital methods or producing digital scholarship. Taken together they draw attention to several things to consider when embarking on a digital humanities project. They include:

- *Do your homework.* Before starting a project do some exploratory research on the "how to" of DH. There is a growing body of resources, many of them open source and freely accessible, on how to employ certain methods, do certain types of analysis, and use specific tools.³ As you learn about new methods and tools some of what you'll encounter may seem beyond your abilities or patience. Don't get discouraged. Keep moving forward in your DH discovery and see where it brings you. Often you will not need to master lots of technical tools or languages in order to produce results that can advance your research. But it is important that you select the right tool for your project, and consulting the plethora of DH "how to" guides will help in that selection.

- *Collaboration is key.* Many projects will require the labor of several contributors, often those with very different skill sets. As with traditional research, collaboration may benefit from including various disciplinary specializations (i.e. different regional foci, expertise in varied traditions, or methodological approach, etc.). Digital projects, however, also often require collaboration with individuals from other fields altogether, including programmers, archivists, librarians, and students. Finding the right collaborators is key, and there are many outlets to do so. If you lack access to digital scholarship networks at the institution where you work, there are a number of DH blogs and professional associations you can connect with. Of particular interest will be the The Humanities and Technologies Camp (THAT Camp), which the American Academy of Religion and the Society for Biblical Literature sponsor at the start of their joint annual meetings.⁴

---

3 The literature and resources here are immense. For resources in the field see, The Digital Humanities Literacy Guidebook, (Cleveland: Carnegie Mellon University, 2018), https://cmu-lib.github.io/dhlg; Doing Digital Scholarship (New York: Social Science Research Council, 2018), https://labs.ssrc.org/dds/; Digital Humanities Now (Fairfax, VA: George Mason University, 2009), http://digitalhumanitiesnow.org/. Commonly cited introductory texts in the digital humanities also include Gold, ed., *Debates in Digital Humanities*; Daniel J. Cohen and Roy Rosenzweig, *Doing Digital History: A Guide to Gathering, Preserving, and Presenting the Past on the Web* (Philadelphia: University of Pennsylvania Press and Fairfax, VA: Center for History and New Media, 2005), http://chnm.gmu.edu/digitalhistory/; Anne Burdick, et al., *Digital_Humanities* (Cambridge, MA: MIT Press, 2016); Elieen Gardiner, *The Digital Humanities* (New York: Cambridge University Press, 2015); Claire Battershill and Shawna Ross, *Using Digital Humanities in the Classroom: A Practical Introduction for Teachers, Lecturers, and Students* (New York: Bloomsbury Academic, 2017); D. Berry, ed., *Understanding Digital Humanities* (New York: Palgrave MAcmillan, 2012); Geoffrey Rockwell and Stefan Sinclair, *Hermeneutica: Computer-Assisted Interpretation in the Humanities* (Cambridge, MA: MIT Press, 2016); Jentery Sayers, *Making Things and Drawing Boundaries: Experiments in the Digital Humanities* (Minneapolis: University of Minnesota Press, 2018); Cantwell and Rashid, *Religion, Media, and the Digital Turn.*
4 Other resources include the Alliance of Digital Humanities Organizations, http://adho.org/; MLA Commons, https://mla.hcommons.org/core/; Network Infrastructure for Nineteenth-Century

- *Think iteratively.* Traditional models of scholarly production based around print typically presume a project is finished upon publication. The appearance of a monograph or journal article represents years of labor. Digital scholarships, however, can—and often should—work in reverse. The launch of a digital project marks the beginning of the work to be done rather than its end. The projects featured here are growing and changing, not finished. What this means is that project directors have the ability to think about their work in phases or stages, allowing for both the project and the project team to grow and learn over time. If a project seems too daunting, consider breaking it into a series of phases that provide a set of benchmarks that can be set. The publics you envision or engage with can then provide multiple moments of connection and benefit. This iterative thinking also relates to scale, allowing a project to grow in scope over time.
- *Audience(s) is key.* A given project may address a number of constituencies. Who your intended audience is will shape the types of choices you make regarding the tools you employ, the user interface, privacy, accessibility, and phases. Projects might simultaneously be used by academic and general audiences, with their findings being used for scholarly research, activist outreach, and educational purposes all at the same time. These various components might develop over time but it is key to think about how your project will take shape as it moves beyond you computer and is taken up by others.
- *Is this ethical?* While digital tools and analysis may make new approaches available to researchers we should always think about the implications and social consequences of our research. Some projects may put religious actors, especially those from vulnerable communities, at risk. When designing projects one should consider: Am I endangering anyone by making this information available publicly? How can I protect my subjects? How might my data be used in the future? The ethical issues related to data and privacy may force you to imagine your project's objectives or methods in necessary ways.
- *Does this "count?"* For many scholars the question of institutional expectations may shape the types of scholarship they can pursue, especially in relation to tenure and promotion requirements. If one is not hired specifically to do digital humanities work, then a DH project may not fulfill narrow definitions of "scholarship." In these cases, academics often take an approach that places digital scholarship as an extra outcome in addition to more tra-

---

Electronice SCholarship (NINES), https://nines.org/; National Endowment for the Humanities' Office of Digital Humanities, https://www.neh.gov/divisions/odh.

ditional publishing methods, such as peer-reviewed journals and books. Part of the reason institutions don't recognize digital projects as fully as they might is because they may not be legible to reviewers using longstanding departmental standards. The American Academy of Religion (along with organizations such as the American Historical Association, Modern Language Association, and others) has produced "Guidelines for Evaluating Digital Scholarship," which are intended to aid evaluating committees in assessing digital scholarship. Part of the effort can be done by recognizing more traditional criterion for scholarship, such as grants, academic reviews, citations, or classroom use.

- *Think about sustainability.* While the ability to grow and evolve digital projects over time is one of the medium's benefits, this feature also raises the spectre of a lifetime of upkeep and maintenance. Therefore, it is incumbent to think about the lifetime of your project early on in order to identify both the kind and amount of resources you will need to maintain it. Not every project needs to involve a decade-long commitment to build a colossal archive that can call forth a new field of inquiry. Smaller scale projects can be completed in either a year or a semester's time and then left up with little change. Keeping projects online, however, often requires regular hosting and registration fees. So, accounting for when a project will sunset, and considering where the data from a project will live after the interface might break, are topics worthy of consideration early on.

While there are many issues to keep in mind when developing a digital project, the most important, perhaps, is managing expectations and keeping open the opportunity for surprise. Results derived from a DH project may not align with the initial expectations one may have going into a project. Alignment between the two shouldn't be the sole standard for measuring success. There are many unexpected elements that will arise when doing digital work but often these surprises will result in new opportunities. Unanticipated issues may be raised through the accumulation of valuable data, which may require learning how to use a new digital tool in order to analyze this data. Even grave challenges or out-right failures can lead to the next productive stage of a particular project. Luck (good or bad) can often be a factor in how a project takes shape or in the results it garners. What matters is that you thoughtfully and intentionally engage with these tools as you would any research method. And like other forms of inquiry and analysis, the best way to engage with them thoughtfully is by considering other examples. This is what we hope these essays provide.

A Guide to the Volume

As there are an infinite number of reasons scholars turn to DH and numerous ways these projects take shape, there are many ways to structure this volume as well. One version could have structured the collection by research method, dividing the essays by the kind of software or tool that they use. Another could have arranged the volume by activity type, grouping together those essays focused on research, teaching, or public engagement. We chose, however, to arrange the volume by source type, with sections on texts, images, places, and issues. Such an arrangement is in service of the fact that the kinds of sources scholars work with often dictate the tools or techniques used. Part I on "Texts," for instance, includes pieces on four projects whose data is primarily textual in nature. Marcus Bingenheimer's contribution focuses on the ways digitization and text encoding are changing the nature of research within Buddhist studies. The advent of computational methods have made the creation of encoded corpora valuable scholarly contributions to the field. Lincoln Mullen's essay, meanwhile, showcases the kind of public scholarship that can emerge from employing textual analysis. His America's Public Bible Project mined a massive collection of digitized American newspapers to pull out the frequencies of specific Bible quotes. In addition to advancing Mullen's own scholarship on the topic, the application of digital methods also resulted in the creation of a public-facing resource that allows anyone to chart the use of a particular passage of scripture over time. Similarly, Frederik Elwert's contribution documents how the digitization of textual material allows for the application of methods commonly associated with the study of institutions, organizations, or relationships. By tagging keywords, character, and parts of speech, Elwert's SeNeReKo project was able to chart the history of religious contact by the appearance of certain themes in ancient Egyptian and Buddhist texts. Like Bingenheimer, Mullen, and Elwert, Rebecca Krawiec and Caroline T. Schroeder's essay on the Coptic SCRIPTORIUM project underscores how the digitization and encoding of texts are vital scholarly contributions.

Building upon Part I's focus on texts, Part II on "Images" considers the digitization and interpretation of sources beyond the written word. Andrew Quintman and Kurtis R. Schaeffer's essay, for example, illustrates how digital methods can help bring textual and visual material in conversation with each other. The Life of the Buddha project they discuss annotates large Tibetan murals with both scholarly commentary as well as passages from the texts that inspired the work of art. Similarly, Erin Walcek Averett and Derek B. Counts' contribution highlight how a suite of digital tools have brought fresh insights to archaeological remains from Cyprus. Beyond rich, 3D modeling of artifacts that allow for levels of scrutiny beyond the human eye, digital methods also allow archaeologists to link this data to provenance, geographical information, and scholarly commentary. James Bielo and Claire Vaughn's contribution, meanwhile, demonstrates how the web's ability to keep multiple kinds of data from myriad locations in a single site allows for the creation of specialized research archives. The Materializing the Bible project they discuss has, to date, collected material related to nearly five hundred attractions that made Bible stories real and manifest in some way. Finally, Emily Floyd and Sally Promey's piece on the Material and Visual Cultures of Religion project demonstrates the kind of scholarly outcomes that can come from this kind of critical curation. As previously mentioned, the MAVCOR project includes both a digital archive as well as a peer-reviewed journal that draws upon that archive as a source of reflection and analysis.

In contrast to Parts I and II, Part III on "Places" focuses less on a particular kind of source and more on a mode of thought encouraged by the digital turn. Focusing on questions of space, or using geographic information systems as a means of analyzing material has become such a prominent part of DH work that some have argued that it constitutes its own field.[5] Abhishek S. Amar's essay on the 3D modeling of historic Hindu and Buddhist sites of pilgrimage, for instance, explicitly grapples with the kind of transformative work scholars must do to their data in order to make it machine readable. But the final renderings of these sites, Amar claims, would allow for the study of these sites across time, combining textual, archaeological, and art-historical analysis. Spatial thinking also allowed J. E. E. Pettit, Fenggang Yang, and Yuqian Huang to shed light on territories devoid of other kinds of source material. In creating a database of 72,000 temples, mosques, and churches in China, they were able to approximate the relative size of these religious communities in a country that refuses to release such data. This ability of space to make real that which may seem absent also defines Louis Kaplan and Melissa Shiff's essay on Mapping Ararat. They use augmented and virtual reality software to project failed attempts at establishing a Jewish homeland onto the landscape. In this way, they call forth the religious imaginaries that often animate communities or individuals, instantiating the visions that have motivated people. Finally, where Kaplan and Shiff build speculative topographies, Caleb Elfenbein, Farah Omer, and Julia

---

5 On the importance of "spatial humanities" see Richard White, "What is Spatial History?" *Stanford Spatial History Lab: Working Paper*, February, 2010, http://www.stanford.edu/group/spatialhistory/cgi-bin/site/pub.php?id=29; David J. Bodenhamer, John Corrigan, and Trevor M. Harris, *The Spatial Humanities: GIS and the Future of Humanities Scholarship* (Bloomington: Indiana UNiversity Press, 2010).

Shafer discuss how they also make invisible landscapes visible. By geo-referencing hate crimes against Muslims and other instances of anti-Muslim hostility, Elfenbein, Omer, Shafer, and others demonstrate both the pervasiveness and the regional specificities of American Islamophobia.

Where the collection's first three sections focus on particular types of data or sources, the final section, Part IV, focuses on how scholars are using DH tools and methods to address issues that impact the study of religion more broadly. Christopher R. Cotter and David G. Robertson's essay documents their ambitious plan to create a social media strategy for the academic study of religion. The Religious Studies Project they discuss hosts rigorous debates about theory, method, and interpretation. But it presents this material in genres accessible to wider audience such as blogs and podcasts. S. Brent Plate, meanwhile, reflects upon his experience teaching a Massive Open Online Course, or MOOC. While debates in DH might have moved beyond MOOCs to discuss more practical concerns related to online education, Plate's essay offers a great deal of insight related to how scholars can, and cannot, reach broader audiences online. Wendi Bellar and Heidia A. Campbell's essay, meanwhile, explores how digital platforms can reach new audiences within the academy. Like Floyd and Promey's essay on MAVCOR, Bellar and Campbell's discussion of the Network for New Media, Religion, and Digital Culture Studies site documents how the web has become a place for new subfields to form. In the absence of support from or cooperation between different professional organizations, Bellar and Campbell have created a space online where interlocutors can converse and collaborate. Finally, the collection concludes with a vital, but surprisingly overlooked discussion: the impact digital tools or platforms are having on religious studies departments. An essay by Russell T. McCutcheon discusses one of the oldest attempts to harness new technology to promote the study of religion at the university level. Through blogging, podcasting, and other new media outlets, the University of Alabama's Department of Religious Studies has made the work of generating majors a site of vibrant disciplinary discourse in the field as well.

While the above outline structures the volume, it is, as we mentioned, not the only way to approach the text. We encourage you to engage with the collection in ways that will inform and inspire your own work, diving into those essays that are relevant to you. Those who are interested in the use of digital tools for what might be called "pure" academic research, we recommend you look at the essays by Bingenheimer; Mullen; Elwert; Krawiec and Schroederl Averett and Counts; Amarl; and Pettit, Yange, and Huang. For those who are interested in how digital platforms can help academic research reach broader audiences can consult Mullen; Floyd and Promey; Averett and Counts; Kaplan and Shiff; Elfenbein, Omer, and Shafer; Cotter and Robertson; Plate; and Bellar and Campbell. Those interested in DH pedagogy, meanwhile, should see Bielo and Vaugh; Amar; Elfenbein, Omer, and Shafer; and McCutcheon.

But we hope you will engage with all of the essays, and draw from them the inspiration necessary to develop your own digital work. Because as each of the essays collected here demonstrates, the digital work happening now in the study of religion has built upon the work done in other disciplines across the academy. If the study of religion is to continue to grow, it will need to take up this kind of work and make it its own.

Selected References

Battershill, Claire and Shawna Ross. *Using Digital Humanities in the Classroom: A Practical Introduction for Teachers, Lecturers, and Students*. New York: Bloomsbury Academic, 2017.
Berry, D., editor. *Understanding Digital Humanities*. New York: Palgrave Macmillan, 2012.
Bodenhamer, David J., John Corrigan, and Trevor M. Harris. *The Spatial Humanities: GIS and the Future of Humanities Scholarship*. Bloomington: Indiana University Press, 2010.
Burdick, Anne, et al. *Digital_Humanities*. Cambridge, MA: MIT Press, 2016.
Cantwell, Christopher D. and Hussein Rashid. *Religion, Media, and the Digital Turn*. New York: Social Science Research Council, 2015.
Carnegie Mellon University. *The Digital Humanities Literacy Guidebook*. https://cmu-lib.github.io/dhlg/.
Cohen, Daniel J. and Roy Rosenzweig. *Doing Digital History: A Guide to Gathering, Preserving, and Presenting the Past on the Web*. Philadelphia: University of Pennsylvania Press, 2005.
Gardiner, Elieen. *The Digital Humanities*. New York: Cambridge University Press, 2015.
Gold, Matthew K. "The Digital Humanities Moment." In *Debates in Digital Humanities*, edited by Matthew K. Gold, http://dhdebates.gc.cuny.edu/debates. Minneapolis: University of Minnesota Press, 2012.
Hockey, Susan. "The History of Humanities Computing." In *A Companion to Digital Humanities*, edited by Susan Schreibman, Ray Siemens, John Unsworth, http://digitalhumanities.org:3030/companion/view?docId=blackwell/9781405103213/9781405103213.xml&chunk.id=ss1-2-1&toc.depth=1&toc.id=ss1-2-1&brand=9781405103213_brand. Oxford: Blackwell, 2004.
Kirschenbaum, Matthew. "What Is Digital Humanities and What's It Doing in English Departments?" ADE Bulletin 150 (2010): 2–3.
Rockwell, Geoffrey, and Stefan Sinclair. *Hermeneutica: Computer-Assisted Interpretation in the Humanities*. Cambridge, MA: MIT Press, 2016.
Roy Rosenzweig Center for History and New Media, George Mason University. *Digital Humanities Now*. http://digitalhumanitiesnow.org/.
Sayers, Jentery. Making *Things and Drawing Boundaries: Experiments in the Digital Humanities*. Minneapolis: University of Minnesota Press, 2018.
Social Science Research Council. *Doing Digital Scholarship*. https://labs.ssrc.org/dds/.

White, Richard. *"What is Spatial History?" Stanford Spatial History Lab: Working Paper*, http://www.stanford.edu/group/spatialhistory/cgi-bin/site/pub.php?id=29.

Part I: **Texts**

Marcus Bingenheimer
Digital Tools for Buddhist Studies

Introduction

Buddhists have rarely hesitated to embrace new forms of communication. The earliest Indian epigraphy (3rd century BCE), and the earliest manuscript fragments in Indian languages (1st century BCE/CE) are connected to Buddhism. The earliest extant printed book, dated 868 CE, is a Chinese translation of the Diamond Sutra.[1] Throughout its history Buddhism has used whatever means available to encode, disseminate, and maintain its growing corpus. Buddhist texts were first composed and transmitted in India in a cultural environment that valued the mnemonic techniques of oral transmission. Later Buddhists became eager "early adopters" of two other emerging information technologies—writing and printing. Below I address the shift of Buddhist heritage information into the digital under three main headings: the digitization of Buddhist texts and images, the digitization of scholarly tools (dictionaries, bibliographies etc.), and the application of computational methods on those data.

Data: Digitization of Primary Sources

Digital Editions of Canonical Texts

Through the centuries Buddhists have managed their shifting corpora via the changing media of oral, handwritten, printed, and now digital text. Part of the conceptual apparatus for these endeavors is the notion of canonicity, a central

---

Note: I am grateful for responses by: Venerable Ānandajoti, Rupert Gethin, Paul Hackett, Susan Huntington, Bryan Levman, Charles Muller, Michael Radich, Miroj Shakya, Sam Van Schaik, Venerable Sujato, Venerable Upatissa, Jeff Wallman, and Christian Wittern. Any mistakes in the information about the resources and all opinions expressed about them are my own. The discussion below is meant to be comprehensive, but it is not complete, and I apologize to all deserving projects that I fail to mention. I have chosen to assess only projects of which at least some data is made freely available.

[1] For the connection of Buddhism and the invention of printing in China see T. H. Barrett, *The Woman Who Discovered Printing* (New Haven: Yale University Press, 2008).

and early concern for Buddhists.² Although there is no single, stable Buddhist canon that is used in all traditions, the concept of canonicity, both fluid and robust, has played an important role in shaping how Buddhists perceive of their textual heritage. It is thus not surprising that first digitization efforts were aimed at producing digital editions of the "canon." Most of the digital canonical editions were created independently from each other, and as a result we have several overlapping versions of the Pāli, Chinese, and Tibetan canon. These are often modeled on twentieth century print editions.

Pāli

The Pāli Canon exists in three major independent digital versions, most of which were created in the late 1980s and 1990s. These have been copied across the net, often with minor changes along the way.³ As a result, digital Pāli texts are easy to find online, but their provenance and editorial standards are often undefined. This makes them difficult to cite and to rely on for philological research.

Perhaps the most influential digital edition of Pāli Buddhist texts is the final CD version (Ver. 3) of the *Chaṭṭha Saṅgāyana Edition* that was published by the Vipassana Research Institute (VRI) in late 1999.⁴ As the name Chaṭṭha Saṅgāyana implies, the VRI corpus is a digitization of the printed canon as redacted by the sixth council that was held in Yangon from 1954 to 1956. The strengths of the VRI corpus are that the texts have been proofread, and that it alone among digital editions of the Pāli canon includes the commentaries (*aṭṭhakathā*) and sub-commentaries (*ṭika*). Markup links connect the commentaries to the *mula* text, making it possible to build interfaces that present the *mula* together with two layers of commentaries.

---

2 For a discussion of Buddhist notions of "canonicity" across traditions see Jonathan A. Silk, "Canonicity" In *Brill's Encyclopedia of Buddhism*, ed., Jonathan A Silk, Vol. 1, Languages and Literatures. (Leiden: Brill, 2015), 5–37.
3 E.g. the World Tipiṭaka Edition, produced in Thailand, that aims to improve Burmese Chaṭṭha Saṅgāyana edition. Another large, on-going edition project to watch is the Dhammachai Tipiṭaka Project (2010–) in Thailand. This also aims at a digital edition (Alexander Wynne, "A Preliminary Report on the Critical Edition of the Pāli Canon being prepared at Wat Phra Dhammakāya," *Thai International Journal of Buddhist Studies* 4 (2013): 135–170, and Bryan Levman, "Towards a Critical Edition of the Tipiṭaka," *Journal of Ñāṇasaṃvara Centre for Buddhist Studies (JNCBS)* 87 (2018): 87; but as of today, digital text does not yet seem available.
4 The Pali Tipitaka, https://tipitaka.org. All URLs given here and below were accessed Jun 2019 where not otherwise indicated. The VRI edition was also published in a Devanagari print edition.

It is unclear whether or in how far the online texts currently available on the VRI website, called Chaṭṭha Saṅgāyana Tripitaka Ver. 4.0, were edited beyond the last Chaṭṭha Saṅgāyana CD (Ver. 3) version. As with all digital editions of the Pāli canon there is lack of technical documentation, or indeed any documentation or meaningful metadata. Digital editions need, like their print counterparts, information as to who created the resource, when and where, and what editorial decisions were made (and why) in converting the printed into a digital text. Development on the VRI corpus seems to have stopped some years ago, though a search engine for the corpus (Windows only) and an iPhone app has been made available. These days the best way to use the VRI corpus is via the *Digital Pāli Reader* that is developed and maintained by Yuttadhammo Bhikkhu.[5]

A second digital edition of the Pāli canon is the Sri Lankan *Buddha Jayanti Tripitaka Project*, which has digitized the Pāli Canon from the government sponsored Sinhalese Buddha Jayanti edition (1956–1990), an edition that was created partly in response to the Burmese Chaṭṭha Saṅgāyana. The digital version of the Buddha Jayanti corpus, seems less well proofread than the VRI corpus, but it too has been available since the 1990s, and can be found on various websites. Next to the core texts of the Pāli Tripiṭaka, the Buddha Jayanti corpus comprises a small number of paracanonical and commentarial works, as well as texts on history, grammar and rhetoric. One stable way of accessing the Buddha Jayanti corpus is via the Göttingen Register of Electronic Texts in Indian Languages (GRETIL) (see below).

The third digital Pāli corpus, still hardly noticed by the scholarly community, is the release of the *Pāli Text Society* edition online under a CC License via GRETIL. The digitization is the result of a collaboration between the PTS and the Dhammakaya Foundation in Thailand between 1989 and 1996. The original aim was, as so often in the 1990s, to produce a CD. After two CD versions, this line of distribution was discontinued, and in 2014 the texts were released on GRETIL. The digital PTS corpus so far consists only of the Pāli Vinaya, Sutta and Abhidharma, none of the commentarial and paracanonical works from the PTS print series seem currently available digitally.

To date (June 2019), the files on GRETIL contain the copyright notice: "This file is (C) Copyright the Pali Text Society and the Dhammakaya Foundation, 2015. This work is licensed under a Creative Commons Attribution-ShareAlike 4.0 In-

---

5 "Digital Pali Reader," *Pali.Sirimangalo.Org*, https://pali.sirimangalo.org and Digital Pali Reader repository "Yuttadhammo/Digitalpalireader," GitHub, https://github.com/yuttadhammo/digitalpalireader (the repository also contains XML versions of the Thai and Burmese editions of the Pāli Canon). https://www.digitalpalireader.online/.

ternational License."[6] Moreover, the PTS files in GRETIL contain the following disclaimer: "These files are provided by courtesy of the Pali Text Society for scholarly purposes only. In principle they represent a digital edition (without revision or correction) of the printed editions of the complete set of Pali canonical texts published by the PTS. While they have been subject to a process of checking, it should not be assumed that there is no divergence from the printed editions and it is strongly recommended that they are checked against the printed editions before quoting."

Working with digital Pāli text, at this stage the recommendation is to search the VRI Pāli canon via Yuttodhamma's reader in order to have full access to the commentarial strata, then use the PTS editions in print or pdf to corroborate difficult or doubtful passages.

For simple searches with convenient access to translations and parallels it is best to use the SuttaCentral website (see Sec. 2.2), which hosts emended versions of the VRI corpus, and might at one point add commentarial literature.

Chinese

The history of Buddhist canonical collections in Chinese has long been studied, especially in Japan.[7] The production of digital Chinese Buddhist texts was and is slightly more challenging than for texts in Indian languages or Tibetan, because Chinese cannot be presented meaningfully in alphabetic transcription. For a long time the rendering of Chinese character variants was an endemic problem for Chinese digital text. Only the advent of Unicode in 1993, and especially the addition of Extension B in 2001, put an end to the confusion of encoding systems and normalization strategies that had hindered the digitization of Chinese.

Like with the Pāli and Tibetan corpora, different organizations have produced independent digital editions of Chinese Buddhist texts, often framed around a specific canonical edition. The two main collections to date are the *Chinese Buddhist Electronic Text Association* (CBETA) corpus, and the *SAT Daizōkyō Text Database* (SAT for Saṃgaṇkīkṛtaṃ Taiśotripiṭakaṃ "Society for the Creation of the Taishō Tripiṭaka"). Both started out collaboratively as projects to digitize the Taishō Canon,[8] the canonical edition of the Chinese Buddhist Canon that,

---

[6] Many of the earlier editions are already in the public domain.
[7] Nozawa lists more than 1200 works related to the topic "canon" (*Daizōkyō* 大藏經) published in Japan between 1879 and 2003.
[8] *Taishō shinshū daizōkyō* 大正新脩大藏經. Tokyo: Taishō issaikyō kankōkai 大正一切経刊行会, 1924–1934.

published between 1924 and 1934, became authoritative for East Asian Buddhist Studies.

CBETA was founded in Taipei in 1998 in emulation of the Pāli Text Society, with the aim of providing reliable digital versions of Buddhist texts to a user community that comprised Buddhist believers as well as researchers in Buddhist Studies. CBETA always considered its main task to provide accurate Chinese Buddhist texts in many different formats. The texts can be read online at www.cbeta.org, but are also available for download in multiple formats (epub, mobi, pdf etc.) for different devices and applications. Before the advent of the mobile versions most users availed themselves to the (Windows only) CD version (current: Version 2018) with the dedicated CB Reader search engine. A sophisticated online interface for researchers, with some analytic functions, has been released in 2014.[9]

CBETA published its first CD version containing Vols. 1–55 and 85 of the Taishō Canon in 1999. Since then it has successively added texts from numerous other canonical editions, most importantly the *Manji shinsan zoku zōkyō* 卍新纂續藏經 (Tokyo, 1905–1912), which contains 1230 Chinese Buddhist texts that are not part of the Taishō canon. In addition, the CBETA corpus comprises 285 texts from the Jiaxing Canon 嘉興大藏經, some 250 Buddhist temple gazetteers, and other texts not included elsewhere from various sources and editions. The 2018 version of the corpus contained 4,621 texts.[10]

Crucial for researchers in the digital humanities (DH) is that CBETA provides its corpus in XML/TEI format on github.[11] This is currently the most comprehensive and usable collection of open access Chinese Buddhist texts for the application of corpus linguistics and related forms of analysis.

The SAT corpus, maintained and hosted by a team at Tokyo University, is currently accessible via its websites, the most recent version of which is dated 2018.[12] The interface allows for searches of all or selected sections of the Taishō canon (3,283 texts) as well as the *Jōdoshū zensho* 浄土宗全書, the "Collected Works of the Pure Land School."[13] Especially helpful for translators is

---

9 At: CBETA Online Reader, cbetaonline.dila.edu.tw.
10 The number of files in the archives available at: CBETA Chinese Electronic Buddhist Association, http://cbeta.org/download/ebook.php.
11 "Cbeta-Git/Xml-p5a," Github, https://github.com/cbeta-git/xml-p5a. For textual analysis of the Taishō texts only this version is preferable: "Cltk/chinese_text_cbeta_taf_xml," GitHub, https://github.com/cltk/chinese_text_cbeta_taf_xml.
12 The SAT Daizōkyō Text Database, http://21dzk.l.u-tokyo.ac.jp/SAT/index_en.html.
13 This collection contains 493 texts (Main Collection 306 texts + Supplement 187 texts) from the Japanese Pure Land school. The online texts are hosted at: "浄土宗全書,"テキストデータ

the linking of highlighted texts with the *Digital Dictionary of Buddhism* (see Section 2.1.), and to the corpus of English translations in the *Bukkyō dendō kyōkai* (BDK) corpus of translations. The interface also allows to view scans of the print edition, which is often helpful, especially when dealing with Siddham script, illustrations or variant characters.

Another recent contribution is a search interface for the heavily illustrated volumes 86–96 of the Taishō.[14] The images can be searched by keywords, magnified, and tagged, and are published according to the IIIF standard. Considering the dearth of open data regarding Buddhist Art, this is a welcome addition.

CBETA and SAT are often seen as giving access to the same data, because both started out (and collaborated) on digitizing the Taishō. However, their text base has diverged over the last fifteen years and the actual overlap of searchable text consists only of about 2,270 texts (Taishō Vols. 1–55 & Vol. 85), which is most of the Indian scriptures translated into Chinese, and the works composed in Chinese until c. the 8th century by Chinese, Korean and Japanese Buddhists. Those texts are nearly identical in both corpora, as both rely on the Taishō, however, the CBETA corpus offers revised punctuation for many texts, and has expanded the apparatus. It also provides (transparently marked) emendations where the CBETA editors judge the Taishō text to be erroneous and the mistake does not become apparent in an apparatus entry.[15]

In addition to the 2,270 texts shared between the SAT and the CBETA corpus, the SAT interface searches another c. 1,220 Buddhist texts from the Taishō canon (Vols. 56–84 and Vols. 86–97), which are not contained in the CBETA corpus. These are mostly texts by Japanese authors written after the 8th century, the majority composed in Buddhist Chinese. In contrast, the CBETA corpus contains another c. 2,351 texts from various sources, which are not accessible through the SAT website. These are mostly texts by Chinese authors and written after c. the 8th century.

As a rule of thumb, whoever studies Japanese Buddhism (or the Japanese commentarial tradition on Indian and Chinese texts) ought to work with the

---

ベース, http://jodoshuzensho.jp/jozensearch, where the collection is maintained. Parts of this collection overlaps with the Taishō.

14 At: SAT Taishōzō Image DB, https://dzkimgs.l.u-tokyo.ac.jp/SATi/images.php.

15 The latter practice carries the danger that users may be unaware that they are looking at an emendation. This is mainly an interface issue, as the emendations are transparently marked in the digital master text (XML/TEI) and the original wording is always preserved.

SAT website. For research on Chinese Buddhism one should make use of the latest version of the CBETA corpus.[16]

To date, the main difference between the two projects from a DH perspective is that CBETA aims to produce a wide ranging corpus of Chinese Buddhist texts, and distributes these texts in various formats under a CC license, whereas SAT aims to provide an online research platform for the Taishō edition of the Chinese canon.

A third digital project regarding the Chinese canon is the *The Tripitaka Koreana Knowledgebase Project* by the Research Institute of the Tripitaka Koreana (Seoul). Like SAT it is an attempt to carefully model one particular edition, in this case the first and second printing of the Korean edition of the Chinese canon. The project, however, seems to have been dormant for some years now. The website has been offline at times in the past, and used to be in dire need of maintenance and internationalization.[17] The Tripitaka Koreana Knowledgebase at one point offered scans of the surviving portions of the first printing of the canon, and could be an important resource for research into the printing history of the Buddhist canon, but in its current stage is difficult to use, at least via the English interface.

Tibetan

Various groups have worked on electronic editions of Tibetan canonical collections and the "Collected Works" (*gsung 'bum*) by later authors. The two most widely used digital collections are the *Asian Classics Input Project* (ACIP), and the *Buddhist Digital Resource Center* (BDRC) (formerly Tibetan Buddhist Resource Center).

Since 1987 the ACIP has produced distributables in plain-text format of the Kangyur (bka' 'gyur), the Tangyur (bstan 'gyur) and various Sungbum collections.[18] The text entry of ACIP was accomplished by involving Tibetan communities in India, Tibet and Mongolia. The produced texts lack metadata about the editions from which they were created, which limits their usability for research on the level of individual texts. As a whole, however, the plain-text corpus might be used for corpus linguistic analysis and related forms of research.

---

[16] For searches of the CBETA corpus the recommendation today is: CBETA Online Reader, http://cbetaonline.dila.edu.tw.
[17] I was not able to search the database with current versions of Firefox, Opera, or Chrome.
[18] Downloadable at: Asian Classics Input Project, http://www.asianclassics.org.

BDRC was founded as the Tibetan Buddhist Resource Center in 1999 by E. Gene Smith (1936–2010), who dedicated his life to the preservation and dissemination of Tibetan texts. BDRC has digitized, cataloged, and archived a large number of culturally significant works, securing the once critically endangered Tibetan literary corpus and making it widely accessible. Smith's effort counts among the great success stories in cultural heritage preservation. Most works are distributed under a Creative Commons license. A few texts are restricted based on cultural concerns by stakeholders (e. g. regarding esoteric texts).

Some texts contained in the BDRC corpus are distributed as scans (in PDF) from original documents, others are available as full text. Both come with metadata, which helps to trace provenance, and makes them usable for research on the level of individual texts.[19] Currently, the website allows download only of single texts and even for this a user account is needed. However, the BDRC team is ready to consider requests in case full-text access is needed for DH related analysis.

In 2015, the Board of Directors voted to broaden the Center's preservation mandate to include texts in languages beyond Tibetan, including, among others, Sanskrit, Chinese, and Pāli. To reflect its expanded mission, the Center's name was changed to Buddhist Digital Resource Center.

Many researchers in Tibetan Buddhism make use of both the ACIP and the BDRC collection in one form or another. They often search for material via Paul Hackett's *Buddhist Canons Research Database* (see Sec. 2.2).

Still another Tibetan canon project is less well known, but deserves more attention. The "Resources for Kanjur & Tanjur Studies" (rKTs) provides both transcriptions and scanned images of a range of printed and handwritten editions via a minimalist website.[20] It is part of the Tibetan Manuscripts Project Vienna (TMPV) directed by Helmut Tauscher and a good place for online research into the edition history of the Tibetan canon.

Sanskrit

Though one or more canonical collections in Sanskrit might have existed in India at one point, none have survived *in toto*, and no Sanskrit "canon" as such has been compiled or translated as a whole. Nevertheless, a great wealth of Buddhist

---

19 At: Buddhist Digital Resource Center, https://www.tbrc.org.
20 Resources for Kanjur & Tanjur Studies, https://www.istb.univie.ac.at/kanjur/rktsneu/sub/index.php.

Sanskrit texts have survived, often complete in the monasteries of Nepal and Tibet, sometimes fragmentary in the sands of South and Central Asia.

GRETIL (the "Göttingen Register of Electronic Texts in Indian Languages and related Indological materials from Central and Southeast Asia") is among the oldest and best managed repositories for digital Indian texts.[21] Safely hosted by the Niedersächsische Staats- und Universitätsbibliothek Göttingen, the formidable team has early on avoided the principal mistake of digital resources (overreliance on a particular interface) and distributed well-curated text files with basic metadata. Conceived as a repository for Indology in general, GRETIL has become a valuable resource for digital textual studies of Indian and Pāli Buddhism. The main file format is HTML with some metadata information at the beginning of each file. Longer texts are often split in several files. To date it contains some 250 Buddhist Sanskrit texts that can be downloaded in one single zip-archive. In another major contribution GRETIL recently (2017) has added digitized versions of fourteen Sanskrit dictionaries in CSX format, allowing for aggregated search in platforms such as GoldenDict.

The other Buddhist Sanskrit project of note is the *Digital Sanskrit Buddhist Canon*, since 2003 maintained by the University of the West.[22] The digital texts are produced from available print editions at the Nāgārjuna Institute of Buddhist Studies in Kathmandu. Over the years, the Nāgārjuna Institute has assembled an important collection of texts. The texts have been proofread, have basic metadata associated, and are made available through the project website. The interface, however, is lacking in faceted search and, as so often, does not offer the data archived for download. Fortunately, many of the texts are shared with permission via GRETIL, from which users can assemble their own collections.

Though there is a considerable overlap between GRETIL and the Digital Sanskrit Buddhist Canon there are texts which are unique to either of the repositories. At this stage, the recommendation is to get the latest version of the Buddhist Sanskrit texts from GRETIL and search the files via a text editor or a grep-like command line tool. In addition, one should search the Digital Sanskrit Buddhist Canon.

---

21 GRETIL, http://gretil.sub.uni-goettingen.de/.
22 Digital Sanskrit Buddhist Canon Project, www.dsbcproject.org.

Thematic Collections

Next to collections of "canonical" texts—however defined—there are a number of important digital collections that do not work with the canon as a primary category for organizing text, but which are centered in various ways on geographic regions, single material collections, topics or genres.

The *Huntington Photographic Archive of Buddhist and Asian Art* is the largest independent archive of Buddhist Art. It represents the field documentation efforts by John and Susan Huntington from 1969 to the present.[23] The archive was first established at Ohio State University in 1986. The collection is currently being accessioned by the University of Chicago Libraries, where it will be permanently housed and maintained.

The material Huntington Archive contains more than 250,000 original slides and photographs documenting the artistic traditions of Asia from ancient to modern times. The collection emphasizes Buddhist material, but also includes significant holdings on Asian art in general. Currently, some 60,000 photographs are available online,[24] with the goal to have all remaining images online by 2020. In addition to the image database, the Archive's website includes online exhibitions and educational materials on Asia and Asian art, including useful maps. The database is currently transitioning its metadata to the VRA Core (Ver. 4) standard that will help to unify the Archive's terminology and classification system.

The data is so far not published under an open license, but made available to researchers online without charge. To reproduce the images in publications, researchers still need permission from the archive and fees might apply.

The *Tibetan and Himalayan Library* (THL) was begun in 2000 under the leadership of David Germano and was one of the first large online collections of scholarly information about Tibet.[25] Today the THL is designed, according to its website, as "a publisher of websites, information services, and networking facilities relating to the Tibetan plateau and southern Himalayan regions." Its interface provides access to a large collection of c. 70,000 photographs, audio and visual material, a map collection, and Tibetan language tools. Among these the "THL Tibetan to English Translation Tool" is especially noteworthy.[26]

---

23 The Huntington Archive of Buddhist and Asian Art, http://www.huntingtonarchive.org/.
24 The Huntington Archive of Buddhist and Asian Art, http://huntingtonarchive.org/database.php.
25 The Tibetan & Himalayan Library, http://www.thlib.org/.
26 "THL Tibetan to English Translation Tool," The Tibetan & Himalayan Library, http://www.thlib.org/reference/dictionaries/tibetan-dictionary/translate.php.

Tied to the canonical catalogs are helpful bibliographies of secondary literature for many texts in the canon. The THL is designed as online library and offers no distributable data. The emphasis is on linking historical and textual information to be accessible online in the THL interface.

Another project that provides image data is the *International Dunhuang Project* (IDP) that aims at making the richness of Central Asian manuscripts available. These manuscripts are indispensable for the study of late Indian, medieval Chinese, and early Tibetan Buddhism. IDP was established in 1994 to coordinate international teams of conservators, catalogers, researchers and digitization professionals to ensure the preservation of the Eastern Silk Road collections and to make them freely available online. Hosted by the British Library, IDP has brought together collections and stakeholders from the UK, France, Germany, Russia, China and Japan. Although not all collections have been fully digitized so far (about 30% of the Stein collection remains unscanned), and not all that is digitized is released, much has been made available and is distributed via the IDP website.[27] IDP currently offers access to over half a million images of over 100,000 manuscripts, paintings, artifacts, and photographs.

Also focused on manuscripts is the *Digital Library of Lao Manuscripts*, which aims to preserve the rich heritage of Laotian Buddhist manuscripts (15th to 20th century).[28] It contains images of c. 12,000 texts, which are findable by title, ancillary term, language, script, category, material, location, and date via an exemplary faceted search function. The data is the happy result of the Preservation of Lao Manuscripts Programme of the Lao Ministry of Information & Culture, which was supported by the German Ministry of Foreign Affairs from 1992 until 2004. According to the website the criteria for the selection for microfilming were "historico-cultural importance, cultural diversity or regional representation, age (all manuscripts over 150 years old) and quality of the manuscript. Within these general guidelines, priority for microfilming was given to extra-canonical literature, all manuscripts which were thought to represent indigenous literary traditions, and all texts of a non-religious nature."[29] The texts were originally preserved in microfilm format, but have been digitized and are now made available both

---

27 International Dunhuang Project, http://idp.bl.uk. Single collections can sometimes be accessed through other portals as well. E.g. the Pelliot collection has been made available through Gallica, gallica.org. The Berlin Turfan Collection can be found at: Digitales Turfan Archiv, http://turfan.bbaw.de/dta/.
28 Digital Library of Lao Manuscripts, http://www.laomanuscripts.net.
29 "About DLLM: Background," Digital Library of Lao Manuscripts, http://www.laomanuscripts.net/en/about/background (Sept. 2020).

via an online interface and packaged with professional metadata for download in pdf format.

In Taiwan, there are the projects conducted at the *Library and Information Center of the Dharma Drum Institute of Liberal Arts.* Over the last fifteen years a steady series of some twenty projects have produced open data on various aspects of Buddhist culture.[30] To mention only five:

- Among the larger projects at the Library and Information Center was a visualization platform for *Gaoseng zhuan* collections, i.e. Buddhist biographical literature, that allowed users to explore the information in different views, e.g. on a map or as social network. The social network information that was produced during this project is the largest of its kind in Buddhist Studies.[31]
- The *Digital Archive of Buddhist Temple Gazetteers* resulted in the full-text digitization of some 250 local histories of Buddhist temples, which are important sources for the study of Buddhism in late imperial China.[32] Data & metadata is published in the form of METS archives, the full text is encoded in XML-TEI.
- The *Catalog Database of Republican Era Buddhist Journals* is an detailed online catalog of two large print collections of Buddhist periodicals published between 1912 and 1950. It allows for searches by topic and genre.
- *Buddhist Temples in Taiwan* is a geo-referenced dataset of c. 5,500 temple in Taiwan, which includes historical and religious information about most sites. Data & metadata is published in XML. An online interface visualizes the distribution of temples on a time line. The data is joined with the largest image database of temples on Taiwan.
- The *Buddhist Authority Databases* were designed to provide authority data for various projects at Dharma Drum that needed to disambiguate the names of persons and places, text titles, and East Asian calendar dates.[33]

---

30 Digital Archive Project, http://lic.dila.edu.tw/digital_archives_projects.
31 A follow-up project added figures from the Song, Yuan, Ming and Qing dynasties. The current size of the network is at c. 17,500 actors (data available at: Socnet Resources, http://mbingenheimer.net/tools/socnet/.)
32 Digital Archive of Chinese Buddhist Temple Gazetteers, http://buddhistinformatics.dila.edu.tw/fosizhi/. Potential DH uses of the corpus are outlined in Marcus Bingenheimer, "The Digital Archive of Buddhist Temple Gazetteers and Named Entity Recognition (NER) in Classical Chinese," *Lingua Sinica* 1, no. 8 (November 2015): 1–19.
33 Buddhist Studies Authority Database Project, http://authority.dila.edu.tw/. The calendar authority is described in Marcus Bingenheimer et al., "Modeling East Asian Calendars in an Open Source Authority Database," *International Journal of Humanities and Arts Computing* 10, no. 2 (September 2016): 127–144.

The data is made available in XML in packaged archives which are updated on a monthly basis.³⁴ There is also an open API for outside projects that wish to work with linked data. As of 2020, the Dharma Drum Buddhist Person Authority, which is continually developed and expanded, is the largest digital onomasticon for Buddhist Studies.

Next to long term projects that enjoy strong institutional support, sometimes outstanding data collections are provide by individual researchers. Special mention should be made of the well-curated *Ancient Buddhist Texts* of Venerable Ānandajoti, who has prepared a number of annotated digital editions of Pāli and Sanskrit texts.³⁵ All the material, which includes rare works from the grammatical tradition, is available via his website and downloadable in pdf, epub and mobi format. Value is added to many of the better known sūtra texts by the annotation. The material is well packaged, and, one would hope, will one day find its way into a long-term archive such as Zenodo.

Whether individual or institutional projects, long term sustainability is always an issue for digital resources; as is the creation of an environment for collaborative research.³⁶ Thankfully, data producers are increasingly aware of this and Dharma Drum, *Ancient Buddhist Texts, Suttacentral* and many others make archival packages of their data available. More and more open Buddhist data can be found archived on Github and other version controlled platforms. Such repositories open the possibility of new forms of collaborative research, e.g. the joint development of digital editions.

---

34 The Place Name Authority contains a large number of entries originally created by the GIS team at Academia Sinica. This part of the data is not included in the downloadable archives.
35 At: Ancient Buddhist Texts, https://ancient-buddhist-texts.net. Ānandajoti has also assembled a rich collection of over 13,000 photos of Buddhist art and sites at: Photo Dharma, https://www.photodharma.net.
36 In the field of Buddhist Studies, we were e.g. lucky that the translations from the Pāli Canon collected on the *Access to Insight* webspace have been incorporated into the Suttacentral corpus (s.b.). Access to Insight, https://accesstoinsight.org. was started 1993 by John Bullitt, but, although still online, the website is not longer maintained and since 2013 users are advised to download the Legacy Edition: "Download the Whole Website ATI Offline Edition," Access to Insight, https://accesstoinsight.org/tech/download/bulk.html. Suttacentral has also incorporated the comprehensive site of German translations of Pāli texts, which was created by Wolfgang Greger: Tipitaka, der Palikanon die Lehre des Theravada, http://palikanon.de and first went online in 1998.

One repository for premodern Chinese texts that is designed along those lines is *Kanripo*, developed since 2013 by Christian Wittern.[37] The repository combines texts produced in-house with texts collected from other projects and sources on the internet. It combines the large *Siku quanshu* 四庫全書 collections, which contain the output of 2,500 years of Confucian literati culture, the Daoist canon and the CBETA corpus, thus allowing to research terminology across genres and traditions.[38] The total number of texts in Kanripo is currently close to 10,000 individual items. Texts that are available multiple times in these collections are consolidated into one entry and, where available, digital facsimiles of the text are juxtaposed with the full text. The texts are released under a Creative Commons (BY-SA) license. The collection is working with users who want to add texts to the repository and works towards a "sinological common" that can provide reliable sources for research.[39]

Most of the resources mentioned so far concentrate on collecting and presenting texts in only one Buddhist language. However, in Buddhist Studies almost every text is a cluster of texts and research often involves comparing different translations. To prepare aligned text that assists with such comparisons is time consuming, and only a few multi-lingual digital editions have been attempted so far. The largest project is the Thesaurus Literaturae Buddhicae (TLB), which was developed by Jens Braarvig as part of the Bibliotheca Polyglotta.[40] The website presents dozens of multilingual Buddhist text clusters in Sanskrit, Tibetan, Chinese and English. The texts are chunked in (loosely defined) sentence or paragraph units. So far the data is limited to online use, one would hope that the dataset of linked text could one day be made available for download. The alignment of sentence-size chunks would be very helpful for e. g. the computational analysis of translation vocabulary.

For a Chinese Āgama text (T.100) a detailed, aligned TEI edition with all Chinese, Pāli, Sanskrit and Tibetan parallels has been prepared for a project at

---

[37] Kanripo is short for "Kanseki 漢籍 Repository". For information about the provenance of the digital corpora aggregated in Kanripo.

[38] The project aims to collect all texts from the *Siku quanshu* 四庫全書, *Sibu congkan* 四部叢刊, *Zhengtong Daozang* 正統道藏, *Daozang jiyao* 道藏輯要 and the CBETA corpus 電子佛典集成. A few texts are still missing as of Sep. 2017.

[39] "Kanseki Repository 漢籍リポジトリ," GitHub, https://www.kanripo.org. https://github.com/kanripo. The texts can be queried and read at: Tipitaka, der Palikanon, die Lehre des Theravada, https://www.kanripo.org.

[40] Bibliotheca Polyglotta, https://www2.hf.uio.no/polyglotta/index.php?page=library&bid=2.

Dharma Drum.[41] Similarly Chinese, Sanskrit and Tibetan versions of the Yogācārabhūmi are available in a dedicated interface.[42]

Digital Tools for Scholarship

After the digitization of primary sources, the other obvious target of digitization were research tools such as dictionaries, catalogs, and bibliographies, which could be modeled relatively easily as databases.

Lexicography

The multilingual character of the Buddhist tradition has resulted in a large number of dictionaries, glossaries and encyclopedias. Lexicography is not merely an concern for modern Buddhist Studies, but has been part of Buddhist scholasticism for centuries.[43]

While, on the one hand, dictionaries originally designed for print have been digitized,[44] the most widely used online dictionary of Eastern Buddhism is the "digital-native" *Digital Dictionary of Buddhism* (DDB), an original and innovative creation by Charles Muller.[45] Started as a private dictionary lookup tool in the late 1980s, Muller took the DDB online in the 1990s making it one of the earliest surviving online tools for Buddhist Studies. Conceived of as a collaborative project, contributions of various sizes and types have since been made by over 300 scholars. For each Chinese lemma, the DDB provides an array of definitions, which are individually credited to their contributors. It offers pronunciations in Mandarin, Korean, Japanese and Vietnamese, as well as pointers to print dic-

---

41 T.100 別譯雜阿含 project, http://buddhistinformatics.dila.edu.tw/BZA/.
42 Yogācārabhūmi Database, http://ybh.chibs.edu.tw/.
43 Norman dates the earliest extant work of Pāli lexicography, the *Abhidhānappadīpika*, to the late 12th century. For Buddhist Chinese, where the prolific translation and transliteration of Indian terms made glossaries indispensable, the earliest glossary is the *Fanfanyu* 翻梵語 (T.2130), at least parts of which can be dated to the 6th century. For Tibetan—and later Mongolian and Manchu—Buddhist texts the most influential glossary that unified translation practice was the *Mahāvyutpatti*, compiled in the 8th to 9th century.
44 Notable for its wide availability on different platforms is e. g. the *Foguang dacidian* 佛光大辭典 (first printed 1988).
45 Digital Dictionary of Buddhism, http://www.buddhism-dict.net/ddb.

tionaries that contain the lemma.⁴⁶ The DDB has incorporated the Soothill-Hodous *Dictionary of Chinese Buddhist Terms* and Lewis Lancaster's *The Korean Buddhist Canon: A Descriptive Catalogue* and, as of 2019, contains c. 72,000 entries. The DDB is a subscription service, but visitors can query up to ten terms per day by using a generic log-in ("guest") without password.

Another dictionary that was created digitally by and for Buddhist scholars is the *Dictionary of Gāndhārī* by Stefan Baums and Andrew Glass. As of 2016, this deeply erudite work had c. 6,700 entries. It is based on the growing corpus of manuscripts written in the Gāndhārī prakrit that was used in northwestern India, Pakistan and Central Asia between c. 300 BCE and 400 CE. The entries provide a gateway to editions of some of the earliest surviving Buddhist texts. The data is currently not available outside of the interface.

For those who prefer to use offline lookup tools, currently the best solution is to use a dictionary platform such as GoldenDict or Babylon and add Buddhist dictionaries and glossaries in formats such as StarDict, Babl, or CSX, many of which are available on the web. The DILA collection of *Glossaries for Buddhist Studies* alone provides fourteen free Buddhist dictionaries and glossaries, which can be searched simultaneously, e. g. in GoldenDict.⁴⁷ Together with the newly digitized Sanskrit dictionaries available through GRETIL, concurrent search across a large number of digitized dictionaries is a convenient way to get an overview of the semantics of a term.

However, whether online or offline, digital lexicography has drawbacks for learners. Dictionary platforms reduce a dictionary to its entries, i.e. users can rarely consult the editorial principles laid out in the introductions, or even access

---

46 The latter is based on the important "all_index.xml" data that is a thorough index of many East Asian dictionaries of Buddhism. Started by Urs App and Christian Wittern, it is now maintained by Charles Muller who makes the 2010 version available here: Composite Index of East Asian Buddhist Lexicographical Sources, http://www.buddhism-dict.net/ddb/allindex-intro.html. The all_index is an important tool for computational linguistics on East Asian Buddhist texts. It contains the largest number of indexed terms (c. 290,000) of all tools. There is also a stardict version for it.

47 Glossaries for Buddhist Studies, http://glossaries.dila.edu.tw/. Among the available glossaries are: A Chinese Translation of A.P. Buddhadatta's *Concise Pali-English Dictionary*, Jeffrey Hopkins' *Tibetan-Sanskrit-English Dictionary*, the *Mahāvyutpatti* in Sanskrit, Tibetan and Chinese, a *Pentaglot Dictionary of Buddhist Terms* (only Sanskrit, Manchu, Mongolian and Chinese), Digital Index of Noun & Verb Ending from Rod Bucknell's *Sanskrit Manual*, Soothill-Hodous' *A Dictionary of Chinese Buddhist Terms*. Produced painstakingly from the original files of the author, Dharma Drum has collaborated with Seishi Karashima to produce distributable versions of his fine glossaries. Karashima created glossaries for Dharmarakṣa's and Kumārajīva's translations of the Lotus Sūtra, Lokakṣema's Translation of the Aṣṭasāhasrikā Prajñāpāramitā, and others.

its list of abbreviations. Working with a dictionary one needs to understand its editorial intent and organizing principles, which are obscured when dozens of entries from different works appear in the same interface. Nevertheless, the aggregation of dictionary data from dozens of reference works into concurrent look-up tools saves much time and shelf space.

Catalogs

In Chinese and Tibetan Buddhism canonical editions comprised more than a thousand, sometimes more than two or three thousand, different texts. Without a catalog as finding aid such editions would have been very difficult to use. In the 20th century, researchers have created meta-catalogs which establish the connections between the Pāli, Sanskrit, Chinese, and Tibetan corpora. In the digital realm the two largest of these, which have incorporated many of the printed catalogs, are the *Buddhist Canon Research Database*, which specializes in Tibetan and Sanskrit material, and *SuttaCentral*, which specializes in early Buddhist texts (i.e. roughly the content of the Pāli canon and its parallels).

The Buddhist Canon Research Database was started by Paul Hackett in the mid-1990s as a digital catalog to the Tibetan Buddhist canon with full-text searching and linked dictionary look-up. In 2010, the collection catalog, together with a bibliography of secondary literature, and accompanied by hypertext links to online resources, was made publicly available on servers at Columbia University.[48] The following year, localization of the interface for nine different languages was implemented, together with the addition of bibliographic data for select sets of indigenous Tibetan commentaries. In 2013, an updated interface with full-text search capability accessing the full text of the complete Tibetan canon was added.

At present, the online resource contains approximately 10,000 bibliographic records for primary texts, with another 12,000 bibliographic records for the associated secondary literature, while the full-text interface offers search features for a large corpus of Tibetan canonical texts. The database is maintained and continues to be developed, but is so far not shared beyond the interface.

Under the leadership of Venerable Sujato the team at *SuttaCentral* has assembled a comprehensive database of early Buddhist texts, linking texts in Pāli, Chinese, Sanskrit and Tibetan with modern translations in a delightful rich-

---

[48] The Buddhist Canons Research Database, http://databases.aibs.columbia.edu.

ness of languages that includes not only the usual suspects, but also Malay/Indonesian, Vietnamese, Korean, Czech, Hungarian and many more.[49]

SuttaCentral was founded in 2005 by Bhikkhu Sujato, Rod Bucknell, and John Kelly as a web service for the sutta parallel tables developed by Rod Bucknell and Venerable Anālayo. As of 2017, SuttaCentral lists nearly 50,000 parallels and hosts over 60,000 texts in 39 languages. It also provides several dictionaries, including an ongoing revision of Buddhadatta's concise dictionary (now as the New Concise Pāli-English Dictionary).

Most of the translations have originally been adapted from pre-existing, open access work, but new translations are now being prepared specifically for the site. A new translation of the four Pāli *nikāyas* has been completed by Bhikkhu Sujato and is scheduled to be published in 2018. In addition, a new translation of the Pāli Vinaya by Bhikkhu Brahmali is underway and is being progressively added to the site. These new translations use a segmented approach, in which text and translation are matched segment by segment, resulting in addressable passages. This design can support a new generation of semi-automated and/or crowd sourced translations and, if widely adopted, could lead to a new canonical reference system.

Material on SuttaCentral is covered by a variety of licensing conditions. The original texts are in the public domain. Legacy translations are available under a variety of licenses, mostly permitting non-commercial use. New translations are dedicated to the public domain via Creative Commons Zero. All data and software is freely available on Github. SuttaCentral encourages reuse and copying of its data.

Both the Buddhist Canon Research Database and SuttaCentral are aggregate sites that have grown out of comparative catalogs. They have incorporated large amounts of digitally available material, improved accessibility, and continue to curate the data by adding links and corrections. A long term trend seems to be that projects that started out as digitization of canonical editions like CBETA and BDRC tend to add on catalogs and multilingual linking as they develop, while projects that start out as cataloging databases such as SuttaCentral tend to add on full text.

A catalog resource focusing on the Chinese canon was created by the late Aming Tu. The *Digital Database of Buddhist Tripitaka Catalogs* is still the best way to check online which of the many Chinese canonical editions contain a given sutra.[50]

---

[49] SuttaCentral, https://suttacentral.net. "SuttaCentral," GitHub, https://github.com/suttacentral.
[50] Digital Library of Buddhist Scripture Catalog, http://jinglu.cbeta.org/.

Michael Radich and Jamie Norrish have developed the "Chinese Buddhist Canonical Attributions Database" that synthesizes past research on the vexing problems that surround the translatorship attributions in the Chinese Buddhist catalog tradition.[51]

Bibliography

Bibliographic control of secondary scholarship is a fundamental part of research. In order to make a contribution we must know whether our questions have been addressed before. The struggle to avoid repetition is after all what separates modern academic from traditional scholastic practice. Thus we consult bibliographies, which in the case of Buddhist Studies are of irritating range. Buddhist Studies began in the 19th century and many early works are still useful, citable, and now, with the Internet Archive, more easily available than ever. Already the earliest monographic bibliography of Buddhist secondary literature (Held 1916) contained c. 2,500 items. Academic communication in Buddhist Studies is still conducted internationally in Japanese, Chinese, English, and French. German and Italian on the other hand used to be read internationally, but today—like Korean, Dutch, Russian, Nepali, Vietnamese, Thai, Sinhalese and many others—are hardly ever used beyond their national boundaries. Nevertheless important secondary literature exists in all these languages and should ideally been taken into account.

In the 20th century quite a few Buddhist bibliographies appeared in print *Bibliography of Buddhist Studies Bibliographies* lists some 148 items.[52] In the digital, two important long-term initiatives, the NTU *Digital Library and Museum of Buddhist Studies* and the *Indian and Buddhist Studies Treatise Database* (IN-BUDS), provide an entry point into the vast amount of secondary literature in Chinese and Japanese. The more recent *H-Buddhism Bibliography Project* has the potential to become a strong, community-maintained bibliographic database.

The *Digital Library and Museum of Buddhist Studies*, hosted at National Taiwan University (NTU), is neither a library nor a museum, but rather the most comprehensive online bibliography of Buddhist Studies.[53] It started as a cooperation between Dharma Drum and the NTU Philosophy Department and is now

---

51 Chinese Buddhist Canonical Attributions database, http://dazangthings.nz/cbc.
52 Bibliographies on Buddhism in Western Languages, http://mbingenheimer.net/tools/bibls/biblBibl.html.
53 Digital Library & Museum of Buddhist Studies, http://buddhism.lib.ntu.edu.tw/DLMBS/.

maintained and developed by the NTU Library. In its early stages an agreement with INBUDS allowed for the inclusion of much of the INBUDS dataset. As of June 2019, according to the project website, the database contains c. 405,000 entries and 60,000 full-text articles.[54] The Digital Library and Museum of Buddhist Studies collects data in any language from all fields of Buddhist Studies, but its coverage of Chinese secondary literature is especially advanced. The interface is adequate, but a download of the dataset is not possible. One can, however, obtain one's query results via email.

The *Indian and Buddhist Studies Treatise Database* (INBUDS), was a visionary endeavor when it was conceived in the late 1980s, before the days of the World Wide Web. Going online in 1998, it still maintains the largest bibliography of Japanese secondary literature on Buddhist Studies. Western and Chinese literature on the other hand are not well represented. It seems that in recent years INBUDS has become integrated into the national Japanese CiNii database service, the meta-catalog for Japanese university libraries and academic journals published in Japan. Very laudably and hardly noticed, INBUDS has made a snapshot of its dataset (as of 2015.03) available for download that contains c. 69,000 entries.[55] A lot of interesting things could be done with that (building author-topic networks, identifying research trends over time, incorporating it into the H-Buddhism Bibliography Project etc.).

The *H-Buddhism Bibliography Project* was started in 2012 by Charles Muller, the creator of the Digital Dictionary of Buddhism (see Sec 2.1), who is also responsible for more than 90% of the entries. It is an Zotero based attempt that aims to pool the many private bibliographies that users of the H-Buddhism mailing list have developed, but contribution from the community has been relatively low, as few researchers keep their bibliographies in structured formats (BibTeX, EndNote, RIS etc.). Currently, there are c. 9,300 items on file. This is less than either the NTU Digital Library and Museum or INBUDS, but for DH purposes the difference is that the records are available in an easily computable format and via the Zotero platform can be easily converted to other formats e.g. for use in LaTex, or research in trend analysis.

Apart from these general bibliographies for Buddhist Studies a few specialized digital bibliographies created by individual researchers are openly available. Part of the "Epistemology and Argumentation in South Asia and Tibet" (EAST) project led by Birgit Kellner is a meticulous, multi-lingual bibliography

---

[54] These figures are somewhat inflated. Even cursory use reveals numerous redundant records, i.e. different entries that reference the same item. Also, not all of the full-text material is accessible.

[55] NBUDS Download, http://www.inbuds.net/jpn/down.html.

on scholarship about late Buddhist Indian texts.⁵⁶ There is also Dan Martin's Tib-Skrit, these days available on Dropbox.⁵⁷ My own *Bibliography of Translations from the Chinese Buddhist Canon* first went online in 2001 and collects c. 1,200 translations of c. 550 texts into "Western" languages.⁵⁸

Going Forward: New Paths for Research

Detractors at times accuse the DH of failing to make good on their promise to deliver new results,⁵⁹ and indeed it seemed at times as if the growing amount of data was inverse to the amount of new approaches or research questions that are asked of it. This criticism, however, overlooks two things.

First, we simply are still very much at the beginning—lasting developments in academic methodology often span generations, and the juggernaut of digitization and related technologies of the last 20 years has been nothing like the relatively stable phase in academic communication that preceded it—the age of print. It is quite possible that we will not see a defined and stable set of research methods for some time to come. Like the editions of the Buddhist canon, such as the PTS or the Taishō, that became authoritative in the 20ᵗʰ century, have dissolved into a moving array of digital corpora, our methods to tackle those texts too might stay very much in flux for the foreseeable future.

Second, just like the technologies of digitization, methodological developments in the DH have originated elsewhere. The conceptual tools to deal with digital cultural heritage data were almost never invented by Humanists, but adopted and adapted from existing methods in data science. There is no reason why Humanists should not make occasional use of Principal Component Analysis or a chi-squared test, considering such methods are quite common elsewhere in academia. Still, there is an extra effort involved in the adoption of these methods. In many neighboring fields (e. g. geography, archeology, sociology, computational linguistics), as well as in more distant departments (e. g. the life sciences), researchers have long been adapting statistical and computational methods

---

56 EAST, http://east.uni-hd.de/abstract/.
57 Dan Martin, "Tibskrit 2014.doc," Dropbox, https://www.dropbox.com/s/zwsf1bv6upp376d/Tibskrit%202014.doc. Other versions can be found on the web, including a stardict glossary from the 2008 version.
58 Marcus Bingenheimer, Bibliography of Translations, http://mbingenheimer.net/tools/bibls/transbibl.html.
59 See for example the penetrating critical review of computational literary studies ("what is robust is obvious (in the empirical sense) and what is not obvious is not robust") by Da (2019).

to meet their own needs—albeit with less resistance from their peers. One problem is that training in even basic statistical and computational methods is almost never part of graduate programs in the Humanities, and students have to rely on their own initiative to acquire the necessary skills. Thus, although digitization has certainly changed the way Humanists query their data, and has made a vast amount of hitherto neglected works available, the application of computational methods to analyze the data has been lagging. Larger fields such as literature, medieval studies, and classics, which are a few years ahead of Buddhist Studies in the creation of their corpora, have in recent years seen a steady stream of DH informed analysis.[60]

However, even in the field of Buddhist Studies a few initiatives have begun to use computational methods. In the field of East Asian Buddhism, for instance, one of the more pressing research questions is the translatorship of pre-Tang dynasty texts in the canon, many of which are unattributed or were wrongly attributed by traditional catalogs. We are still in the early stages of exploring how best to assess and compare Buddhist sutras computationally, but first results are quite encouraging.

Michael Radich and Jamie Norrish have developed a set of programs called TACL ("Textual Analaysis for Corpus Linguistics"), for the large-scale analysis of digitized texts in the Chinese Buddhist canon and similar corpora.[61] The core functionality is simple—TACL compares two or more user-defined texts or corpora of any size, to find either (a) strings common to all sides of the comparison (intersect), or (b) strings unique to one side (difference). At present, TACL operates only on literal contiguous strings, i.e. it cannot handle fuzzy matching or patterns that bracket intervening text. Further functionality includes filtering and sorting of the results; concatenation of multiple tests (feeding of results from one test into a new test) etc.

TACL has significant power to help scholars locate evidence bearing on problems of ascription, dating, textual history, earlier sources, later reception and impact, textual circulation, and other aspects of intertextual relations. In

---

60 See for example the "pamphlets" series at: "Pamphlets," Stanford Literary Lab, https://litlab.stanford.edu/pamphlets. or the recent supplement "The Digital Middle Ages" to *Speculum* Vol. 92 (2017) (S1) edited by D. Birnbaum, S. Bonde, and M. Kestemont. For recent applications of distant reading methods to Chinese religious texts see Edward Slingerland et al., "The Distant Reading of Religious Texts: A 'Big Data' Approach to Mind-Body Concepts in Early China," *Journal of the American Academy of Religion* 85, no. 4 (2017): 985–1016.
61 Freely available at: "ajenhl/tacl," GitHub, documentation at http://pythonhosted.org/tacl. TACL is written in Python.https://github.com/ajenhl/tacl; documentation at: TACL's documentation, http://pythonhosted.org/tacl. TACL is written in Python.

a series of publications, Radich has piloted application of the tool to a range of typical research problems.⁶²

Jen-jou Hung and myself have been interested in similar issues and are working towards improving algorithms that can show textual anomalies and distinctive differences between translation styles. After developing a variable n-gram algorithm for Classical Chinese Texts, we have applied clustering algorithms to identify translatorship and translation date.⁶³ The Digital Archive of Buddhist Gazetteers can be used as a benchmark corpus for name and entity recognition as well as allowing for diachronic analysis of how certain places were perceived or the association of individuals with places.⁶⁴ Parallel to this we have built various datasets based on markup that contain facets which can be utilized in two other prominent fields of DH application: geographic and social network analysis (SNA).

Both geographic and social network analysis used to be the preserve of specialists, who had to invest significant time and effort to master the software and methodology. Moreover, neither GIS nor SNA data relevant to the study of Buddhism was readily available in digital form. This has changed in recent years. Datasets have been created and made openly available, while open source tools such as QGIS and GEPHI have lowered the threshold for their use.⁶⁵ Applied to the study of Buddhism geographic perspectives can e.g. elucidate regional pat-

---

62 See e.g. Michael Radich, "Tibetan Evidence for the Sources of Chapters of the Synoptic *Suvarṇaprabhāsottama-sūtra* T664 Ascribed to Paramārtha," *Buddhist Studies Review* 32, no. 2 (2015): 245–270. Michael Radich, "On the *Ekottarikāgama* 增壹阿含經 T 125 as a Work of Zhu Fonian 竺佛念," *Journal of Chinese Buddhist Studies* 30 (July 2017): 1–31. Michael Radich and Anālayo Bhikkhu, "Were the *Ekottarika-āgama* 增壹阿含經 T 125 and the *Madhyama-āgama* 中阿含經 T 26 Translated by the Same Person? An Assessment on the Basis of Translation Style," In *Research on the Madhyama-āgama*, ed. Dhammadinnā (Taipei: Dharma Drum Publishing Corporation, 2017), 209–237.
63 Jen-jou Hung, Marcus Bingenheimer, and Simon Wiles, "Quantitative Evidence for a Hypothesis regarding the Attribution of early Buddhist Translations," *Literary and Linguistic Computing* 25, no. 1 (April 2010): 119–134; and Marcus Bingenheimer, Jen-Jou Hung, and Cheng-en Hsieh, "Stylometric Analysis of Chinese Buddhist Texts—Do Different Chinese Translations of the Gaṇḍavyūha Reflect Stylistic Features That Are Typical for Their Age?," *Journal of the Japanese Association for Digital Humanities* 2, no. 1 (2017): 1–30.
64 Marcus Bingenheimer, "The Digital Archive of Buddhist Temple Gazetteers and Named Entity Recognition (NER) in Classical Chinese," *Lingua Sinica* 1, no. 8 (November 2015): 1–19. The archive itself makes c. 250 temple gazetteers available http://buddhistinformatics.dila.edu.tw/fosizhi.
65 Datasets for the application of historical GIS to the study of Buddhism can be found at: http://mbingenheimer.net/tools/socnet/. http://mbingenheimer.net/tools/histgis/. SNA data is available here: http://mbingenheimer.net/tools/socnet/.

terns in the growth or decline of Buddhist institutions, or visualize patterns of pilgrimage travel.[66] Historical network analysis can help us to identify key-players or cliques in past and present Buddhist networks, and answer questions regarding patronage or the flow of information over time through the networks of monastic and lay-practitioners.[67]

Conclusion

Looking back over what has been accomplished in the creation of digital resources for Buddhist Studies, the successes with regard to the digitization of texts are spectacular. Canonical collections of most Buddhist traditions (we are still missing a few outliers such as Tangut or Khotanese) are now available in digital form. Crucially, these digital repositories now surpass any single canonical print tradition in terms of volume, acquisition cost, searchability, and portability. The latter is rarely mentioned, but deeply effecting many of us, as digitization has freed researchers to work wherever they like. 20 years ago only a few large research libraries held paper copies of all canonical editions. Today we can search and compare sutra passages in Chinese, Sanskrit, and Tibetan in a café overlooking the ocean, in a monastery deep in the Himalayas, or on an airplane in transit between the two. Reliability of transcription, which in the beginning seemed a serious concern, is now hardly ever mentioned, as researchers can complement their full text searches with large facsimile collections in pdf or djvu.

Digitization, beyond the confines of academic practice, also stands to affect the tradition itself. As primary texts have been made widely available, they have found more readers than ever. It is for future generations of researchers to under-

---

[66] For the application of a GIS informed perspective on Buddhist history see e.g. Jen-Jou Hung, 洪振洲, Marcus Bingenheimer 馬德偉, and Zhi-Wei Xu 許智偉, "漢文佛典的語意標記與應用：《高僧傳》文獻的時空資訊視覺化和語意搜尋—Semantic markup for Chinese Buddhist texts and its Application- A Platform for Querying and Spatio-temporal Visualization of the Biographies of Eminent Monks," National Chengchi University Journal of Librarianship and Information Studies 國立政治大學圖書與資訊學刊 2, no.3 (Aug 2010): 1–24. Marcus Bingenheimer, "Knowing the Paths of Pilgrimage'—The Network of Pilgrimage Routes in 19th Century China According to the Canxue Zhijin 參學知津," Review of Religion and Chinese Society 3, no. 2 (2016): 189–222; and other articles in the special issue of geo-spatial studies of the Review of Religion and Chinese Society Vol. 3-2 (2016) (edited by J. Pettit & J. Protass), or the contribution of Pettit, Yang and Huang to this volume.

[67] Marcus Bingenheimer, "Who Was 'Central' for Chinese Buddhist History?—A Social Network Approach," International Journal of Buddhist Thought and Culture 28, no. 2 (December 2018): 45–67.

stand how this opening of the archive will influence Buddhism itself, where traditionally the canon as a whole has remained out of reach for most believers (mostly for practical reasons, but in esoteric Buddhism also on doctrinal grounds). Writing this on the 500th anniversary of the Reformation, for which vernacularization played such a crucial role, one cannot help but wonder what will happen to the teachings of the Buddha when all ancient and modern versions of all texts become instantly accessible.

One change that is already discernible is that the authoritative canonical editions which for many decades guided scholars into studying certain texts, have been dissolved into corpora and now live on merely as conventions of citation. The canon is dead, long live the corpus.

While the digitization of primary texts has been overwhelmingly successful, the digitization of images, objects, and spaces has just begun. Apart from the Huntington Archive and Ānandajoti's *Photodharma* so far no archive of Buddhist art has succeeded. Many museums today make images of their holdings available, but an archive with faceted search across institutions and geared to Buddhist iconography still needs to be built. Technologies for 3D scanning and printing of objects are perhaps too new and still too much in flux to warrant a major initiative, but they have strong potential both for teaching and for research. The digitization of Buddhist sacred spaces is also still in its early stages, but as the essay of Quintman and Schaeffer in this volume illustrate, can provide researchers and students with plenty of new data.[68] Digitized text might save a trip to the library, but well documented digital spaces can save a trip to India or China.

Now that much data has moved from print to digital, Humanists can complement their methods with statistical and computational approaches. As our traditional methods continue to work just fine, we can take our time with this (and we do). Indeed only time will tell whether the application of analytic methods that depend on digital data can substantially advance Buddhist Studies. Judging from the success of past digitization initiatives that have profoundly changed the scope of primary data that we now can query and access, there is reason for optimism.

---

68 First steps towards digital models of the Dunhuang Caves look very promising. At: Digital Dunhuang https://www.e-dunhuang.com.

Selected References

Barrett, T. H. *The Woman Who Discovered Printing*. New Haven: Yale University Press, 2008.

Bingenheimer, Marcus. "The Digital Archive of Buddhist Temple Gazetteers and Named Entity Recognition (NER) in Classical Chinese." *Lingua Sinica* 1, no. 8 (November 17, 2015): 1–19.

Bingenheimer, Marcus. "Knowing the Paths of Pilgrimage'—The Network of Pilgrimage Routes in 19th Century China According to the Canxue Zhijin 參學知津." *Review of Religion and Chinese Society* 3, no. 2 (2016): 189–222.

Bingenheimer, Marcus. "Who Was 'Central' for Chinese Buddhist History?—A Social Network Approach." *International Journal of Buddhist Thought and Culture* 28, no. 2 (December 2018): 45–67.

Bingenheimer, Marcus, Jen-jou Hung, Simon Wiles, and Bo-yong Zhang. "Modeling East Asian Calendars in an Open Source Authority Database." *International Journal of Humanities and Arts Computing* 10, no. 2 (September 2016): 127–44.

Bingenheimer, Marcus, Jen-jou Hung, and Cheng-en Hsieh. "Stylometric Analysis of Chinese Buddhist Texts—Do Different Chinese Translations of the Gaṇḍavyūha Reflect Stylistic Features That Are Typical for Their Age?" *Journal of the Japanese Association for Digital Humanities* 2, no. 1 (2017): 1–30.

Chen, Chin-chih. "*Fan fan-yü: ein Sanskrit-chinesisches Wörterbuch aus dem Taishō-Tripiṭaka*." Unpublished PhD diss., Rheinische Friedrich-Wilhelms-Universität Bonn, 2004.

Da, Nan Z. "The Computational Case against Computational Literary Studies." *Critical Inquiry* 45, no. 3 (2019): 601–39.

Held, Hans Ludwig. *Deutsche Bibliographie Des Buddhismus. Eine Übersicht Über Deutschsprachliche Buddhistische Und Buddhologische Buchwerke, Abhandlungen, Vorträge, Aufsätze, Erwähnungen, Hinweise Und Rezensionen Mit Ausschliesslicher Berücksichtigung Des Buddhismus Als Religionswissenschaft*. München, Leipzig, 1916.

Held, Hans Ludwig. 1916. *Deutsche Bibliographie des Buddhismus*. München, Leipzig: Hans Sachs, 1916.

Hung, Jen-jou, Marcus Bingenheimer, and Simon Wiles. "Quantitative Evidence for a Hypothesis regarding the Attribution of early Buddhist Translations." *Literary and Linguistic Computing* 25, no. 1 (April 2010): 119–134.

Hung, Jen-Jou 洪振洲, Marcus Bingenheimer 馬德偉, and Zhi-Wei Xu 許智偉. "漢文佛典的語意標記與應用：《高僧傳》文獻的時空資訊視覺化和語意搜尋—Semantic markup for Chinese Buddhist texts and its Application- A Platform for Querying and Spatio-temporal Visualization of the Biographies of Eminent Monks." *National Chengchi University Journal of Librarianship and Information Studies* 國立政治大學圖書與資訊學刊 2, no.3 (Aug 2010): 1–24.

Hung, Jen-Jou 洪振洲, Marcus Bingenheimer 馬德偉, Zhi-Wei Xu 許智偉. 2010. "漢文佛典的語意標記與應用：《高僧傳》文獻的時空資訊視覺化和語意搜尋—Semantic markup for Chinese Buddhist texts and its Application- A Platform for Querying and Spatio-temporal Visualization of the Biographies of Eminent Monks", *National Chengchi University Journal of Librarianship and Information Studies* 國立政治大學圖書與資訊學刊. Vol.2, No.3 (No.74) (Aug 2010): 1–24.

Levman, Bryan. "Towards a Critical Edition of the Tipiṭaka." *Journal of Ñāṇasaṃvara Centre for Buddhist Studies (JNCBS)* 87 (2018): 87.

Levman, Bryan (Forthcoming). "Towards a Critical Edition of the Tipiṭaka." *Thai International Journal of Buddhist Studies.*

Muller, Charles A. "The Digital Dictionary of Buddhism and CJKV-English Dictionary: A Brief History." In *Introductions to Digital Humanities and Buddhism,* edited by Daniel Veidlinger, 143–156. Berlin, Boston: De Gruyter, 2019.

Muller, Charles A. forthcoming 2018. "The Digital Dictionary of Buddhism and CJKV-English Dictionary: A Brief History." In D. Veidlinger (ed.) *Introductions to Digital Humanities: Buddhism* New York & Berlin: DeGruyter.

Norman, Kenneth Roy. *Pāli Literature: Including the Canonical Literature in Prakrit and Sanskrit of All the Hīnayāna Schools of Buddhism.* Wiesbaden: Harrassowitz, 1983.

Nozawa, Harumi 野澤晴美 (ed.). *Daizōkyō kankei bunken mokuroku* 大藏經關係文獻目錄. *Risshō Daigaku Shigaku* 立正大学史學. 2003.

Radich, Michael. "Tibetan Evidence for the Sources of Chapters of the Synoptic *Suvarṇaprabhāsottama-sūtra* T664 Ascribed to Paramārtha." *Buddhist Studies Review* 32, no. 2 (2015): 245–270.

Radich, Michael. "On the *Ekottarikāgama* 增壹阿含經 T 125 as a Work of Zhu Fonian 竺佛念." *Journal of Chinese Buddhist Studies* 30 (July 2017): 1–31.

Radich, Michael and Anālayo Bhikkhu. "Were the *Ekottarika-āgama* 增壹阿含經 T 125 and the *Madhyama-āgama* 中阿含經 T 26 Translated by the Same Person? An Assessment on the Basis of Translation Style." In *Research on the Madhyama-āgama,* edited by Dhammadinnā, 209–237. Taipei: Dharma Drum Publishing Corporation, 2017.

Silk, Jonathan A. "Canonicity" In *Brill's Encyclopedia of Buddhism,* edited by Jonathan A Silk, Vol. 1, Languages and Literatures, 5–37. Leiden: Brill, 2015.

Slingerland, Edward, Ryan Nichols, Kristoffer Neilbo, and Carson Logan. "The Distant Reading of Religious Texts: A 'Big Data' Approach to Mind-Body Concepts in Early China." *Journal of the American Academy of Religion* 85, no. 4 (2017): 985–1016.

Slingerland, Edward, Ryan Nichols, Kristoffer Neilbo, and Carson Logan. 2017. "The Distant Reading of Religious Texts: A "Big Data" Approach to Mind-Body Concepts in Early China." *Journal of the American Academy of Religion* 2017: 1–32.

Winter, Thomas N. "Roberto Busa, S.J., and the Invention of the Machine-Generated Concordance." *The Classical Bulletin* 75, no. 1 (1999): 3–20.

Wittern, Christian. 2016. センター研究年報*2015 CIEAS Research Report 2015*—特集漢籍リポジトリ Special Issue: Kanseki Repository. Kyoto: Center for Informatics in East Asian Studies, Institute for Research in the Humanities, Kyoto University.

Wynne, Alexander. "A Preliminary Report on the Critical Edition of the Pāli Canon being prepared at Wat Phra Dhammakāya." *Thai International Journal of Buddhist Studies* 4 (2013): 135–170.

Lincoln A. Mullen
The Making of *America's Public Bible*: Computational Text Analysis for Religious History

America's Public Bible: A Commentary is a website that charts biblical quotations in U.S. newspapers over the nineteenth and early twentieth centuries.[1] The prototype version uses the *Chronicling America* corpus of over 12 million newspaper pages as a source base. It finds and identifies quotations from the King James Version of the Bible. The prototype was created for the *Chronicling America* Data Challenge hosted by the National Endowment for the Humanities.[2] The initial version, which will remain available until the revised version is published, is available at the project website: https://americaspublicbible.org (figure 1). A much expanded version—featuring an additional newspaper corpus, more versions of the Bible, and expanded interpretation in the form of visualizations and prose—is in progress.[3]

America's Public Bible as a digital history or digital religious studies project could be discussed in a variety of ways. The project uses the techniques of machine learning and text analysis to find the quotations, and it uses data analysis and visualization to make sense of them. It is therefore a kind of computational humanities project, like Frederik Elwert's use of network analysis of ancient Egyptian and Indian religion or Marcus Bingenheimer's digital analysis of Buddhist texts, both described in this volume. Since such projects are rather unusual for humanities scholars (though decidedly less so in recent years), the specific computational techniques and methods that they use are understandably an object of curiosity: what did the data look like? what methods were used to analyze it?

But it is also possible to talk about the methods behind these projects in a more humanistic way, to understand how computational methods drawn from other disciplines can be used within humanities disciplines. In other words, how does one use computational methods to make a historical interpretation?

---

[1] Lincoln A. Mullen, *America's Public Bible: A Commentary* (Stanford University Press, forthcoming): (http://americaspublicbible.org).
[2] "NEH Announces the Winners of the Chronicling America Data Challenge," National Endowment for the Humanities, 27 July 2016, https://www.neh.gov/news/press-release/2016-07-25.
[3] The expanded version will be published by Stanford University Press as a part of their digital publishing program: https://www.sup.org/digital/.

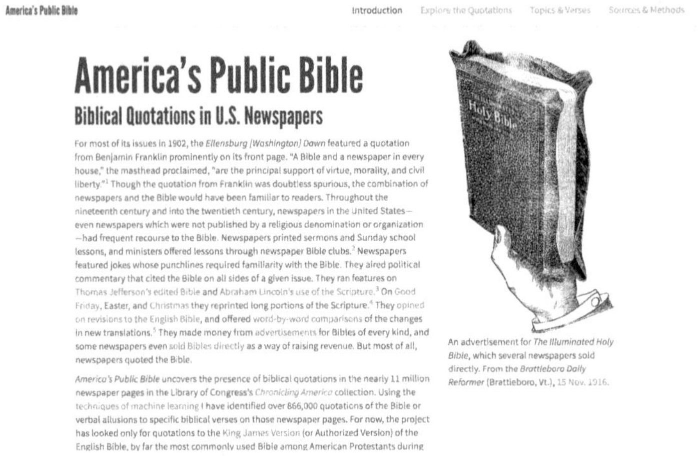
America's Public Bible

Introduction Explore the Quotations Topics & Verses Sources & Methods

America's Public Bible

Biblical Quotations in U.S. Newspapers

For most of its issues in 1902, the *Ellensburg [Washington] Dawn* featured a quotation from Benjamin Franklin prominently on its front page. "A Bible and a newspaper in every house," the masthead proclaimed, "are the principal support of virtue, morality, and civil liberty."[1] Though the quotation from Franklin was doubtless spurious, the combination of newspapers and the Bible would have been familiar to readers. Throughout the nineteenth century and into the twentieth century, newspapers in the United States—even newspapers which were not published by a religious denomination or organization—had frequent recourse to the Bible. Newspapers printed sermons and Sunday school lessons, and ministers offered lessons through newspaper Bible clubs.[2] Newspapers featured jokes whose punchlines required familiarity with the Bible. They aired political commentary that cited the Bible on all sides of a given issue. They ran features on Thomas Jefferson's edited Bible and Abraham Lincoln's use of the Scripture.[3] On Good Friday, Easter, and Christmas they reprinted long portions of the Scripture.[4] They opined on revisions to the English Bible, and offered word-by-word comparisons of the changes in new translations.[5] They made money from advertisements for Bibles of every kind, and some newspapers even sold Bibles directly as a way of raising revenue. But most of all, newspapers quoted the Bible.

America's Public Bible uncovers the presence of biblical quotations in the nearly 11 million newspaper pages in the Library of Congress's *Chronicling America* collection. Using the techniques of machine learning I have identified over 866,000 quotations of the Bible or verbal allusions to specific biblical verses on those newspaper pages. For now, the project has looked only for quotations to the King James Version (or Authorized Version) of the English Bible, by far the most commonly used Bible among American Protestants during

An advertisement for *The Illuminated Holy Bible*, which several newspapers sold directly. From the *Brattleboro Daily Reformer* (Brattleboro, Vt.), 15 Nov. 1916.

Figure 1: The front page of the prototype version of America's Public Bible, created for the Chronicling America Data Challenge.

And then it is also possible to discuss such projects in terms of the contributions they make to a specific field of study. What does this project tell us about nineteenth-century American religious history?

In this essay, I invert the typical way of discussing such projects. Instead of discussing methods, then modes of interpretations, then results, I begin with a case study of an interpretative result, describe a general pattern of interpretation that is useful to humanities scholars working with digital methods, show how *America's Public Bible* offers an interface that enables such interpretations, and finally describe the computational methods for finding quotations. Scholars in history and religious studies who are contemplating a digital project must acquaint themselves with the various computational methods available to them, to be sure. But the more pressing problem is finding a mode of argumentation which usefully applies those methods to make meaningful interpretations in one's given field.

In explaining what my digital project is and how it functions, I hope to show two things. First, the project creates serendipitous findings through computational history by surfacing sources that would otherwise go unnoticed. And second, the project disciplines those searches by setting the results in a much broader chronological context in which the typical and the exceptional can be identified. This disciplined serendipity constitutes a method of approaching the past that is relevant to religious studies.[4]

To see how disciplined serendipity can be used as an interpretative method, let's take up the case of the McKinley assassination.

Interpretation: The McKinley Assassination and 2 Chronicles 7:14

William McKinley, the twenty-fifth president of the United States, was assassinated on September 6, 1901. *The Cameron County Press* reported on the sermons preached in the churches of Emporium, Pennsylvania, to commemorate the slain president. The Methodist, Catholic, Episcopal, and Presbyterian churches all paid tribute to McKinley. The Rev. Robert McCaslin called his Presbyterian congregation to prayer and repentance, juxtaposing prosperity and repentance. "In our national prosperity we were forgetting God and we were becoming self-reliant," the minister claimed. He called his hearers to repent using 2 Chronicles 7:14: "If my people, which are called by my name, shall humble themselves, and pray, and seek my face, and turn from their wicked ways; then will I hear from heaven, and will forgive their sin, and will heal their land."

There could be no mistaking who McCaslin and his hearers thought the "people" in that biblical verse were. "The people" were not ancient Israelites, or even the Presbyterians sitting in the pews of Emporium, but the citizens of the United States. McCaslin drove home this point by encouraging his listeners to heed "the loud call of God to nations to rise and stamp out the curse of anarchy," regretting that it was "deeply humiliating that such a deed could take place in this christian land."[5]

This brief scene should be familiar to any historian of American religion. Christian nationalism has long been a subject of concern to religious historians, and this episode in Emporium is a classic example of the jeremiad.[6] A public event in the wake of a national tragedy linked Christianity and the state, and

---

[4] This idea of disciplined serendipity comes out of conversations with my colleague, Mike O'Malley, who shared an unpublished essay on the topic.

[5] *Cameron County Press* (Emporium, PA), 19 Sept. 1901, p. 1. All citations to newspapers in this chapter come from *Chronicling America: Historic American Newspapers*, Library of Congress, http://chroniclingamerica.loc.gov/.

[6] Sacvan Bercovitch, *The American Jeremiad* (Madison: University of Wisconsin Press, 1978); Andrew R. Murphy, *Prodigal Nation: Moral Decline and Divine Punishment from New England to 9/11* (New York: Oxford University Press, 2009).

the sacred scriptures were used to undergird the power of the state and condemn its enemies. Like a Puritan fast day, the day was set aside for prayer, preaching, and repentance. And the scriptural verse that was used would later become popular in the late twentieth-century rise of the religious right.

The question, though, is whether this use of the Bible was typical or unusual. When the citizens of Emporium opened their Bibles to 2 Chronicles, would the verse calling them to humility have naturally seemed to mean what their ministers said it meant that day? Or might it have held other meanings that they saw as more obvious? We understand that the cultural significance of that text in the light of our recent history, but can we uncover the assumptions of the somewhat more distant past?[7]

In the case of 2 Chronicles 7:14, an answer is readily apparent. In the decade on either side of the McKinley assassination, the connection between that text and Christian nationalism was fairly infrequent. Instead the verse had two other uses which were far more common.

Far more frequently, the verse was used as a call to humility for Christians, occasionally in the context of a revival. The *Watchman and Southron* in Sumter, SC, ran an unsigned column on "Humility," assembling the verse from 2 Chronicles and other texts, encouraging its readers to "never mind the showy array and costly equipage of the worldly, but to be clothed with humility."[8] In St. Louis, the pastor of the Presbyterian church preached on the text, reminding his congregants that, as the newspaper headline put it, "to give up habits of sin … is harder than pulling teeth."[9] In Sacramento, a Congregationalist pastor surprisingly used the text to recommend the practice of Lent, and less surprisingly to encourage preparation for the famous preacher D. L. Moody's upcoming revival in that city. The pastor did regard the Lenten observance as tied to the responsibility of Christians in the state, observing that "as a nation and people we need many things," among them "higher ideals of business integrity" and "higher standards of political morality," besides the "deeper consciousness of God."[10]

The verse was also popular in revivals. The *Pacific Commercial Advertiser* in Honolulu tried to gather "the Christian people of our city" in a "call to prayer," quoting that text as a means of "putting away of sin."[11] A cooperative revival in

---

[7] On the varieties of ways in which Christian terminology has been associated with very different political and cultural assumptions in American history, see Matthew Bowman, *Christian: The Politics of a Word* (Cambridge, MA: Harvard University Press, 2018).
[8] *Watchman and Southron* (Sumter, SC), 19 Dec. 1882, p. 5.
[9] *St. Louis Republic* (St. Louis, MO), 17 Feb. 1902, p. 6.
[10] *Record-Union* (Sacramento, CA), 6 Mar. 1899, p. 4.
[11] *Pacific Commercial Advertiser* (Honolulu, HI), 5 Apr. 1905, p. 2.

Kentucky of "God's own people," meaning clearly the members of churches and not the citizens of the United States, was announced by the *Mt. Sterling Advocate*.[12] The *Monroe City Democrat* in Missouri specifically addressed the verse to Christians, quoting it as the solution "when a church wants a real revival."[13]

The second common use of 2 Chronicles 7:14 was as a prayer for rain in response to drought. After all, verse 13 framed the call to repentance as being in a time when "there is no rain." A Baptist pastor in Clinton, Missouri, promised that "God will answer prayer with rain."[14] In Utah the *Deseret Evening News* quoted the verse while worrying about a drought's effect on the corn crop.[15] In 1901, the governor of Nebraska proclaimed a day of prayer "for relief from destructive winds and drouth." Ten years later the *Omaha Daily Bee* ran a commemoration, claiming that by 1pm on the day of prayer it began to rain and that in a few days "the whole state was wet down." The verse from 2 Chronicles was used to bolster that meteorological claim.[16]

The point of undertaking this brief history of 2 Chronicles 7:14 is not to suggest that Christian nationalism was not a significant factor in American life, but to put it in a larger context. That verse did not provoke automatic and easy assumptions of the role of God's favor on the United States or his promised defense against anarchists and other enemies. For farmers in Utah or Missouri, the danger against which God might defend them was drought. Their link to ancient Israel was not a theological identity as the people of God but their common agricultural occupation. A pattern of quotation that became a favorite of Jerry Falwell and the Moral Majority in the 1980s had antecedents at the turn of the twentieth century, but those antecedents were unusual rather than typical.

Pattern of Argumentation: The Typical and the Exceptional

This vignette of a civil religious event in the McKinley assassination combined with the more complicated history of a biblical text highlights a fundamental tension in how scholars approach the study of religion and history. The problem lies in knowing whether some phenomenon that we are interested in studying is

---

12 *Mt. Sterling Advocate* (Mt. Sterling, KY), 22 March 1905, p. 7.
13 *Monroe City Democrat* (Monroe, MO), 22 Mar. 1900, p. 6.
14 *St. Louis Republic* (St. Louis, MO), 15 July 1901, p. 7.
15 *Deseret Evening News* (Salt Lake City, UT), 15 July 1901, p. 4.
16 *Omaha Daily Bee* (Omaha, NE), 23 July 1911, p. 3.

typical or exceptional. Arguments in the humanities tend to be structured around common patterns. One common pattern explicates some unusual text or event. An equally important strand of research aims at explicating the typical, the everyday, and the ordinary.

But how can we know whether the object of our study is one or the other? Scholars often assert but seldom prove their claims about whether something is exceptional or typical. Such claims typically rest on the basis of the scholar's expertise. This is not without justification. After all, a career spent immersed in the sources does give scholars some ability to distinguish between the ordinary and the extraordinary.

Our intuitive understanding, though, is inadequate in the face of the scarcity and abundance of our sources—a problem faced by all humanistic disciplines, but perhaps especially by historians. Historical sources are scarce in that we always have the problem of an incomplete set of sources, and must therefore be attentive to the silences of the archives.[17] The art of being a historian is taking sources that were created for one purpose and reading against their grain to answer the questions which interest us.

Yet historians also face the problem of an abundance of sources. However partial they may be, the archival and the printed record is vast. Despite all the labors of librarians and archivists, our sources are inadequately cataloged and indexed. For even the most narrowly targeted scholarly question, the sources available outstrip the historian's time and ability to read. There are no two ways about it: the way we go about finding our sources are inextricably tied up with chance and happenstance.

This problem of scarcity and abundance is rendered all the more acute by the rise of digitized sources. Digitized sources, as everyone knows, hardly represent the whole of the human record available in libraries and archives, nor do the collections in libraries and archives represent everything that was created in the past. The easy availability of digitized sources leads historians to use those sources rather than other sources which are available in the archive but remain undigitized.[18] Digital sources thus exacerbate the problem of scarcity, in the sense that they are an incomplete and partial record of the past which ab-

---

[17] Roy Rosenzweig, "Scarcity or Abundance? Preserving the Past in a Digital Era," *American Historical Review* 108, no.3 (2003): 735–762, https://doi.org/10.1086/ahr/108.3.735.

[18] For a useful example, see Ian Milligan, "Illusionary Order: Online Databases, Optical Character Recognition, and Canadian History, 1997–2010," *Canadian Historical* Review 94, no. 4 (2013): 540–569.

sorbs scholarly attention.[19] But digitized sources also exacerbate the problem of abundance. Scholars can access a much larger source base than ever before thanks to collections of primary source materials such as newspapers, photographs, government documents, as well as large book collections including the HathiTrust and Google Books.

We should not have a rosy view of this digitization of sources, not least because of two problems that limit their usability. One is the enclosure of our cultural heritage by the large, for-profit publishers often doing the digitizing. The importance of large scale, publicly funded projects like the *Chronicling America* collection of over 15.3 million newspaper pages cannot be understated. These projects are free for scholars both in terms of cost and free in terms of the way that scholars can use them for any purpose, including computational research. (They are both *gratis* and *libre*, in the parlance of open-source software.) But the norm for digital collections is the subscription database from a for-profit company. Though paid for at great cost by university libraries, such databases are not usually available for text mining.[20]

The second limitation of digitized sources is the way that increasing the scale of a source base tends to decrease the diversity of sources used. A thousand, a million, or ten million newspaper pages are all still just newspaper pages. Historians and most humanities scholars tend to rely on combining many different kinds of sources in order to make useful interpretations, but using large-scale text corpora, for instance, paradoxically narrows the kinds of source scholars use.[21]

The abundance of digitized sources has already transformed historical research.[22] As Lara Putnam has pointed out in an article on "The Transnational and the Text Searchable," searching digitized collections is now a basic scholarly practice. Putnam argues that the ability to search for sources without the constraints of the national archive has allowed new angles of vision on transnational history, because "transnational approaches among historians did not become

---

[19] Michael O'Malley, "Evidence and Scarcity," blog post, 2 October 2010, http://theaporetic.com/?p=176; and Sean Takats, "Evidence and Abundance," blog post, 18 October 2010, http://quintessenceofham.org/2010/10/18/evidence-and-abundance/.

[20] Thomas Padilla, "Text and Data Mining: Seeking Traction," LIS Scholarship Archive, 7 March 2018, https://osf.io/preprints/lissa/qxs9j.

[21] For a discussion of this problem, see the Arguing with Digital History working group, "Digital History and Argument," white paper, Roy Rosenzweig Center for History and New Media (13 November 2017): https://rrchnm.org/argument-white-paper/.

[22] Jennifer Rutner and Roger Schonfeld, "Supporting the Changing Research Practices of Historians" Ithaka S+R, 11 August 11, 2015, https://doi.org/10.18665/sr.22532, p. 9, 14–15.

commonplace until technology radically reduced the cost of discovering information about people, places, and processes outside the borders of one's prior knowledge." Yet as Putnam points out, "Digital search makes possible radically more decontextualized research" and makes it possible to find examples of what we are looking for without a sense of its significance. To deal with this problem of context, Putnam observes that "computational tools can discipline our term-searching if we ask them to. By measuring proximity and comparing frequencies, topic modeling [or other text analysis methods, we might add] can balance easy hits with evidence of other topics more prevalent in those sources."[23]

In addition to its contribution to the history of the Bible in the United States, *America's Public Bible* is a work of scholarship whose whose form as a digital project is intended to implement Putnam's idea of text analysis that facilitates the discovery of new sources at the same time that it disciplines searching. To understand how it accomplishes that end, let me explain how it the site's interface works, and how it was put together.

Interface: Disciplined Serendipity through Interactive Visualization

The prototype version of *America's Public Bible* has as its centerpiece a visualization that lets users interactively explore a time series visualization of the trend in quotations for over one thousand of the most quoted verses in the Bible (figure 2).

Most important, users can disaggregate the time series and find each quotation in the context of the newspaper page at *Chronicling America*. A table below the visualization shows a row for each instance of a quotation. Users can follow links to that specific newspaper page in *Chronicling America* (figure 3).

On arriving at the newspaper page in *Chronicling America*, the key words in the quotation are highlighted on the page. This allows the user to readily identify the quotation on the page, which would otherwise be quite difficult.

---

[23] Lara Putnam, "The Transnational and the Text-Searchable: Digitized Sources and the Shadows They Cast," *American Historical Review* 121, no. 2 (2016): 377–402, https://doi.org/10.1093/ahr/121.2.377. Quotations at p. 383, 392. See also Tim Hitchcock, "Confronting the Digital: or How Academic History Writing Lost the Plot," *Cultural and Social History* 10, no. 1 (2013): 9–23; Ted Underwood, "Theorizing Research Practices We Forgot to Theorize Twenty Years Ago," *Representations* 127, no. 1 (2014): 64–72.

The Making of America's Public Bible — 39

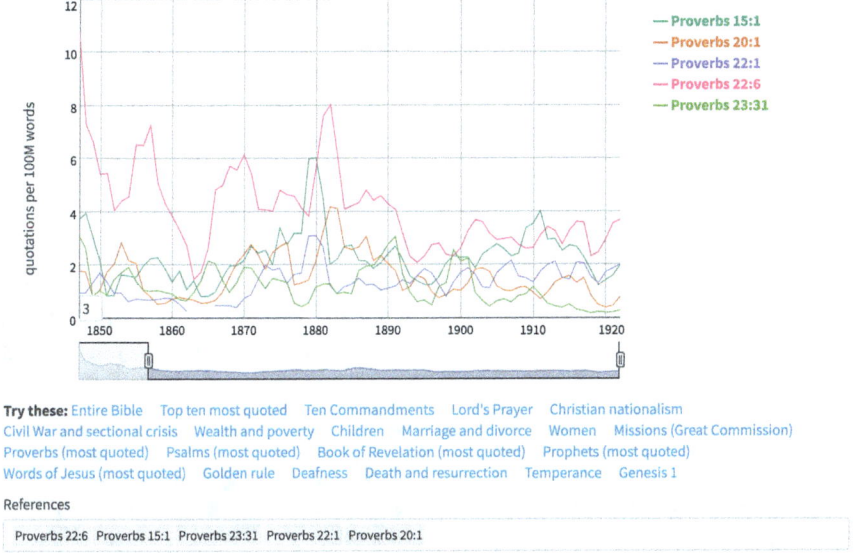

Figure 2: An example of the interactive verse browser, showing time series of the rate of quotations for the five most quoted verses from Proverbs. Users can enter verse references on their own, or they can choose from pre-selected collections of verses, such as the Lord's Prayer or verses on marriage and divorce.

Figure 3: After identifying an interesting verse and see the trend in quotations, users can find every quotation which was used to compute that trend line. The links to the right take users to that specific newspaper page in *Chronicling America*.

The ability to move between the trend line of the verse's quotation and its location in the actual primary source is the way that the site enables disciplined serendipity. The serendipity lies in how the site surfaces hundreds of thousands of instances of quotations which the user can readily browse. This may be a subjective judgement, to be sure, but this has been the most fun project that I have ever worked on because I am constantly surprised by the quotation finder. (And not just that it works at all!) For example, how could I have known to look for the time when a Democratic newspaper thought Samuel Tilden had been elected in the disputed presidential race of 1876, and plastered the banner "The Lord called Samuel" across the paper (figure 4). Biblical jokes are another frequent category that I did not expect.[24] I take this as a sign that the method truly is serendipitous.

Figure 4: In the disputed 1876 election for the president of the United States between Rutherford B. Hayes and Samuel Tilden, several Democratic newspapers announced Tilden's supposed victory using the verse 1 Samuel 3:8 ("The Lord called Samuel"). *Stark County Democrat* (Canton, OH), 7 December 1876.

The discipline in this approach lies in how the site contextualizes those quotations. Each text is set within two different contexts. The first context is the trend in the verse's quotation. This context allows the user to see both how often the verse was quoted in an absolute sense, and to compare that trend to other verses to get a sense of its relative popularity.

The second context is the place of the text on the newspaper page itself (figure 5). This context allows the scholar to understand how the Bible verse was used. The Bible was a common yet contested text, and the fact that a verse

---

[24] "Our own inability to get the joke is an indication of the distance that separates us from the workers of preindustrial Europe. ... When you realize that you are not getting something—a joke, a proverb, a ceremony—that is particularly meaningful to the natives, you can see where to grasp a foreign system of meaning in order to unravel it." Robert Darnton, *The Great Cat Massacre: And Other Episodes in French Cultural History* (Basic Books, 1984), 77–78.

was quoted only raises the questions of why it was quoted and what meaning was imparted to the text. To add to the example of how 2 Chronicles 7:14 was used in the wake of the McKinley assassination, take the trend for John 15:13 ("Greater love hath no man than this, that a man lay down his life for his friends"). This verse exploded in popularity around World War I, and looking at how the verse was used in specific newspapers confirms that it was popular because of obituaries. Investigating earlier uses of the verse shows that it was not associated with the military in any significant way until the Great War. It was more likely to be used to memorialize medical personnel who died taking care of people infected with cholera or yellow fever.

Figure 5: On arriving at the *Chronicling America* newspaper page containing a biblical quotation, the key words of the quotation are highlighted on the page. On this page from the *Cameron County Press* (cited above), the quotation from 2 Chronicles 7:14 is highlighted beneath the image of McKinley and the program for his memorial service in Emporium, Pennsylvania. The other highlights on the page are uses of words in the verse such as "humble" and "people."

This prototype version of the site has implemented an interface that enables disciplined serendipity, but the site needs further work to more fully advance historical interpretations about the role of the Bible in public life.

This expanded version still in progress will extend the prototype in several ways. First, it will broaden the source base and improve the reliability of the method used to find the quotations. This will include extending the source base to Gale's corpus of *19th Century U.S Newspapers*.[25] This corpus is available only by license from Gale, which in my case was made possible because my university library purchased the text mining rights. Restrictions on how the corpus can be used, in particular a requirement that only the briefest of snippets be reproduced from the text of the newspapers, means that displaying the full text of the quotation and allowing back-and-forth interaction between contexts is simply not possible using the Gale dataset. The Gale corpus is also an order of magnitude smaller than *Chronicling America*. But it has cleaner OCR than *Chronicling America*, and it is segmented into articles rather than pages. It is therefore useful for validating the trends identified in *Chronicling America*. Article- rather than document-level segmentation may allow me to use computational analysis to understand the topics of articles in which the verses were quoted.

Second, the new version will extend the finding of quotations to other versions of the Bible. The prototype version used the King James Version since it was a standard text for English-speaking Protestants in the United States. The revised version will use other versions of the Bible as well. It is an extremely difficult problem both identify quotations *and* distinguish between versions of the Bible which have only small (if significant) verbal differences. Nevertheless this is an important addition to the project, since the rise of different translations of the Bible was a crucial moment for American Protestants, and since Catholics and Jews (and non-English speakers, for that matter) have all used versions other than the KJV.[26]

Third, the site will take the exploratory analysis of the prototype and add interpretative work on the history of the Bible. In digital history there is a widespread sense that digital work has failed to make historical interpretations.[27] This digital monograph will feature a series of chapters that interpret various aspects of the history of the Bible, using prose and visualizations. This interpreta-

---

25 Gale, *19th Century U.S. Newspapers*, https://www.gale.com/c/19th-century-us-newspapers

26 Peter J. Thuesen, *In Discordance with the Scriptures: American Protestant Battles Over Translating the Bible* (Oxford University Press, 1999).

27 Cameron Blevins, "Digital History's Perpetual Future Tense," in *Debates in the Digital Humanities 2016*, eds. Matt Gold and Lauren Klein (University of Minnesota Press, 2016), http://dhdebates.gc.cuny.edu/debates/text/77; "Digital History and Argument" white paper.

tive process will necessarily involve more data analysis of the quotations, such as clustering the time series of quotation trends in order to find verses that had similar trajectories over time. But at its root will be much close reading of the context of the verses. The project will thus rely on the serendipity of the quotations that are turned up but also on the discipline of its empirical approach.

Method: Text Analysis and Machine Learning for Quotation Identification

Having described the public facing component of the project, and considered the disciplined serendipity it provides, I want to go behind the scenes of the project to show how it was created. In particular, I want to show how sources, methods, and questions came together to shape the project. The availability of large digital corpora and knowledge of digital text analysis methods are not sufficient to create a meaningful intellectual project. Nor does having a worthwhile intellectual question in one's discipline imply that that one can apply the right methods to the right sources to answer that question in a meaningful way. Sources, methods, questions—these three must come together to make a worthwhile interpretative project. To the extent that *America's Public Bible* is useful as a model—let the reader be the judge—I want to show how that project combined those three considerations.

The initial problem was formulating a worthwhile historical question that was amenable to some kind of computational research. When the National Endowment for the Humanities opened a data challenge for the best project using *Chronicling America*, I had an incentive to compete but not a question worth asking. The sources were predetermined, but they were not obviously "religious" and so there was no clear connection to my primary research area. I was familiar with an array of methods that might be applied to the dataset. (Topic modeling! the newly initiated to DH is likely to suggest.) But without a question to answer there was no purchase for the methods.

My question came after hearing a paper that Mark Noll gave at the 2016 annual meeting of the American Society of Church History. The paper compared how the Bible was used in sermons after the deaths of Presidents Washington, Lincoln, and Garfield.[28] Noll had counted scriptural references in those sermons,

---

[28] Mark A. Noll, "Presidential Death and the Bible: 1799, 1865, 1881" (American Society of Church History, Atlanta, 2016). The latter two presidents died from an assassin's bullet, so con-

then analyzed the way the quotations were used. That paper, along with the work he had done in his books *In the Beginning Was the Word* and *The Civil War as a Theological Crisis*, established that the Bible could be studied as a text that had been used in the public sphere.[29] Other historians have of course studied the Bible from a number of angles, not least its reception by Christians and its use as a material object.[30] The intellectual frame that Noll was using, however, was one that both justified the source base and pointed to the methods that should be used. Why was a large corpus of newspapers worth investigating for a project on the history of the Bible in America? Because it could provide evidence of how the Bible was used in the public sphere. What kind of method would be worth using? Finding quotations was a task that could be turned over to the computer.

America's Public Bible is an example of a methodological approach that I would label computational history. The project uses the techniques of machine learning to find examples of biblical quotations, then uses data analysis combined with close reading to understand the meaning of those quotations.

Finding the quotations computationally required a series of steps.[31] The first step was to download the plain text versions of the roughly 12 million newspaper

---

sider the opening of this essay on the McKinley assassination an homage to the conference paper which gave rise to this project.

29 Mark A. Noll, *In the Beginning Was the Word: The Bible in American Public Life, 1492–1783* (Oxford University Press, 2016); Mark A. Noll, *The Civil War as a Theological Crisis* (University of North Carolina Press, 2006), ch. 3.

30 James P. Byrd, *Sacred Scripture, Sacred War: The Bible and the American Revolution* (Oxford University Press, 2013); Valerie C. Cooper, *Word, Like Fire: Maria Stewart, the Bible, and the Rights of African Americans* (University of Virginia Press, 2011); Philip Goff, Arthur E Farnsley II, and Peter J. Thuesen, eds., *The Bible in American Life* (Oxford University Press, 2017); Paul C. Gutjahr, *An American Bible: A History of the Good Book in the United States, 1777–1880* (Stanford University Press, 1999); Nathan O. Hatch and Mark A. Noll, eds., *The Bible in America: Essays in Cultural History* (Oxford University Press, 1982); Colleen McDannell, *Material Christianity: Religion and Popular Culture in America* (Yale University Press, 1995); Seth Perry, "Scripture, Time, and Authority among Early Disciples of Christ," *Church History* 85, no. 4 (December 2016): 762–83, https://doi.org/10.1017/S0009640716000780; Jonathan D. Sarna and Nahum M. Sarna, "Jewish Bible Scholarship and Translations in the United States," in *The Bible and Bibles in America*, ed. Ernest S. Frerichs (Scholars Press, 1988), 83–116; Stephen J. Stein, "America's Bibles: Canon, Commentary, and Community," *Church History* 64, no. 2 (June 1, 1995): 169–84, https://doi.org/10.2307/3167903; Peter J. Wosh, *Spreading the Word: The Bible Business in Nineteenth-Century America* (Cornell University Press, 1994); John Fea, *The Bible Cause: A History of the American Bible Society* (Oxford University Press, 2016); Paul Gutjahr, ed., *The Oxford Handbook of the Bible in America* (Oxford University Press, 2017).

31 Users who are interested in how the project was created are welcome to investigate the code for themselves: "America's Public Bible," GitHub, https://github.com/lmullen/americas-public-pages available at the time. While for many online resources this would be an enormous chore or impossible due to licensing restrictions, the data for *Chronicling America* is easily downloaded thanks to the Library of Congress's bulk data downloads. *Chronicling America* also offers JSON and RDF application programming interfaces (APIs) for downloading the machine-readable metadata for each newspaper, such as its title, dates of publication, and location.[32] I parsed this API into a database of newspaper pages and publication information. From there I also computed other basic metadata such as the word count for each page.

Once the data was in hand (or, more precisely, on a RAID array) it was time to apply a series of digital methods. The second step was to identify potential quotations on the page. Detecting quotations can be thought of as a problem of detecting text reuse.[33] While this is a well-known area of research, for my particular problem it was necessary to invent a new kind of method.

The idea behind my method is straightforward. A common process in text analysis is to tokenize the text, that is, to turn it into single words or phrases (called n-grams).[34] The corpus can then be represented in a matrix, often called a document-term matrix, where the rows are documents (in this case a newspaper page), the columns are tokens, and the counts in the cells are the number of times that token appears in that document. I created a document-term matrix for both newspapers and for the Bible. As long as one limits the tokens in the matrix to only those words and phrases which appear in the Bible, it is possible to multiply one matrix by the transpose of the other matrix. The result is a document-document matrix with Bible verses down the rows, newspaper pages across the columns, and the count of the number of biblical words and phrases in the cells. If all that talk of tokenizing and matrix multiplication sounds confusing, just remember that at the root it is all just counting: the new matrix has the counts of how many words and phrases from each Bible verse appeared on each newspaper page. That is a common-sense understanding of what a quotation looks like.

---

bible. The prototype project has a methodological appendix: "Sources and Methods," America's Public Library, http://americaspublicbible.org/methods.html.

32 *Chronicling America*'s bulk data and API: "Chronicling America: Library of Congress," News about Chronicling America RSS, https://chroniclingamerica.loc.gov/about/api/

33 For a useful overview of the problem, see David A. Smith, Ryan Cordell, and Abby Mullen, "Computational Methods for Uncovering Reprinted Texts in Antebellum Newspapers," *American Literary History* 27, no. 3 (2015): 1–15, https://doi.org/10.1093/alh/ajv029.

34 For an overview of the text analysis process generally, see Kasper Welbers, Wouter Van Atteveldt, and Kenneth Benoit, "Text Analysis in R," *Communication Methods and Measures* 11, no. 4 (October 2, 2017): 245–65, https://doi.org/10.1080/19312458.2017.1387238.

In practice it is not always possible to distinguish between actual quotations and false positives using only the counts of tokens. So I created other features that could indicate a quotation. I weighted the matrices by their TF/IDF score.[35] This technique gives more weight to phrases which are unusual and characterize a document. For example, the four-word phrase "went into the city" appears several times in the Bible and many times in newspapers, but the mere fact that phrase appears on a page is unlikely to indicate that it is a quotation from the Bible. But the four-word phrase "through a glass darkly" cannot be anything other than a biblical allusion to 1 Corinthians 13:13. Another useful feature is the percentage of the Bible verse that is quoted. If every word or phrase from a verse appears on a page, then it is more likely to be a genuine quotation than if a single phrase appears. Then too, a bunch of random phrases from the Bible scattered around the newspaper page are unlikely to be a quotation, but if those phrases are concentrated in a single location, then they are far likelier to be a quotation. The spread of phrases around the page can be computed with a statistical test of randomness called a runs test.

Applying this algorithm to the newspapers using my university's high-performance computing cluster resulted in millions of possible quotations from the Bible. But many of these are false positives, as could be verified by looking up the newspaper page and seeing if the biblical text actually appeared. What was necessary was a way of weeding through all of these potential matches and identifying the correct quotations and discarding the incorrect ones. The computational tool for the job is called machine learning, or more precisely, supervised classification.

The aim of supervised classification is to take inputs (potential quotations) which have certain features (the count of tokens, how unusual those tokens were, etc.) and assign them a label ("quotation" or "not-a-quotation"). Typically when writing a program one takes input data and creates a series of rules about what to do with it. To make up some rules, for instance, one might instruct the computer that matches with more than five tokens are genuine quotations and the rest are noise. However such rules are extremely difficult to write. Ten meaningless tokens might not be a quotation, but one unusual token could be a quotation.

Machine learning takes a different approach. Instead of writing rules and getting answers, you give the computer a set of answers and get back the rules, called a model. In my case, I went through a number of potential matches

---

[35] Jure Leskovec, Anand Rajaraman, and Jeff Ullman, *Mining of Massive Datasets*, 2nd ed. (Cambridge University Press, 2014), section 1.3.1, http://www.mmds.org/.

and labeled them as quotations or not (see figure 6). The idea behind supervised classification is that by showing a machine-learning model what a bunch of real quotations and a bunch of false quotations look like, the model can learn to tell the difference between them.[36] The accuracy of this model can be estimated to test its precision (what proportion of its results are true matches) and recall (the proportion of all matches that it has found).

	A	B	D	G	H	I
1	match	reference	url	token_count	proportion	tfidf
2	FALSE	Luke 10:31 (KJV)	http://chroniclingamerica.loc.gov/lccn/sn83040198/1897-12-23/ed-1/seq-3/#	1	0.02631578947	0.1992378928
3	TRUE	Romans 3:23 (KJV)	http://chroniclingamerica.loc.gov/lccn/sn85034235/1883-01-25/ed-1/seq-1/#	5	0.2941176471	2.763034851
4	FALSE	Judges 19:1 (KJV)	http://chroniclingamerica.loc.gov/lccn/sn88076523/1911-08-18/ed-1/seq-2/#	2	0.01612903226	0.2649320585
5	FALSE	Proverbs 10:28 (KJV)	http://chroniclingamerica.loc.gov/lccn/sn84026853/1886-04-28/ed-1/seq-1/#	3	0.125	1.170029201
6	FALSE	Galatians 3:15 (KJV)	http://chroniclingamerica.loc.gov/lccn/sn83025121/1897-03-02/ed-1/seq-2/#	1	0.02222222222	0.1671054807
7	FALSE	1 Maccabees 13:7 (KJV)	http://chroniclingamerica.loc.gov/lccn/sn82014780/1852-03-19/ed-1/seq-1/#	2	0.05882352941	0.9353564175
8	TRUE	John 7:17 (KJV)	http://chroniclingamerica.loc.gov/lccn/sn84026688/1903-02-05/ed-1/seq-1/#	9	0.2195121951	2.129326177
9	TRUE	Ecclesiastes 1:2 (KJV)	http://chroniclingamerica.loc.gov/lccn/sn86083274/1889-12-13/ed-1/seq-2/#	1	0.05882352941	0.5777842192
10	TRUE	Matthew 6:13 (KJV)	http://chroniclingamerica.loc.gov/lccn/sn86072143/1876-12-01/ed-1/seq-1/#	30	0.6511627907	6.657510873
11	FALSE	Acts 10:28 (KJV)	http://chroniclingamerica.loc.gov/lccn/sn82015387/1895-12-17/ed-1/seq-4/#	1	0.01234567901	0.1162576126
12	FALSE	Isaiah 28:15 (KJV)	http://chroniclingamerica.loc.gov/lccn/sn84020104/1853-12-16/ed-1/seq-4/#	1	0.002409638554	0.1183413461
13	FALSE	Ecclesiasticus 17:5 (KJV)	http://chroniclingamerica.loc.gov/lccn/sn86072143/1872-03-15/ed-1/seq-3/#	1	0.008172839506	0.1615503599
14	FALSE	Tobit 7:15 (KJV)	http://chroniclingamerica.loc.gov/lccn/sn82016419/1877-07-27/ed-1/seq-1/#	1	0.3333333333	3.274110575
15	FALSE	1 Kings 6:30 (KJV)	http://chroniclingamerica.loc.gov/lccn/sn97071026/1888-06-08/ed-1/seq-1/#	1	0.05263157895	0.4804833971
16	FALSE	Baruch 2:26 (KJV)	http://chroniclingamerica.loc.gov/lccn/sn88078726/1854-06-15/ed-1/seq-1/#	1	0.008928571429	0.1681583325
17	FALSE	Luke 1:63 (KJV)	http://chroniclingamerica.loc.gov/lccn/sn83016942/1838-07-21/ed-1/seq-3/#	3	0.05172413793	1.016103282
18	FALSE	1 Samuel 17:3 (KJV)	http://chroniclingamerica.loc.gov/lccn/sn86063943/1910-06-24/ed-1/seq-5/#	3	0.06382978723	0.4916472619
19	FALSE	Ecclesiasticus 19:15 (KJV)	http://chroniclingamerica.loc.gov/lccn/sn85025620/1888-03-01/ed-1/seq-2/#	1	0.04347826087	0.4270579011
20	FALSE	Galatians 4:9 (KJV)	http://chroniclingamerica.loc.gov/lccn/sn84024716/1908-06-03/ed-1/seq-3/#	1	0.01754385965	0.1723216092
21	FALSE	1 Maccabees 2:21 (KJV)	http://chroniclingamerica.loc.gov/lccn/sn83016810/1905-08-05/ed-1/seq-2/#	3	0.2	1.825838909

Figure 6: Some sample data used to train the classification model. The combination of the "reference" and "url" columns indicate a single potential quotation. I then filled out the "match" column to indicate whether that row was a genuine quotation ("TRUE") or just noise ("FALSE").

The result was a table, where each row was an instance of a Bible verse quoted on a specific newspaper page, along with the probability that it was a quotation, according to the model.[37] This table was then able to be used for data analysis. The most obvious approach, and the one which is at the center of the prototype site, was to chart time series of the popularity of the verses. Another approach is to look for verses which often appear next to one another, omitting verses which appear in the same passage. And ultimately the research involves lots of reading of the context of the quotations to see how the Bible was actually used.

---

36 This description of machine learning is indebted to Francois Chollet and J.J. Allaire, *Deep Learning with R* (Manning, 2018), section 1.1.2, especially figure 1.2.

37 Technically any quotation with a probability above 50% was predicted to be a quotation by the model. But I chose to use only predicted quotations above 90% for the initial studies, since I judged the problem of false matches to be worse than the problem of missing genuine quotations. This cutoff will change for the final version of the project.

To return to the initial contention of this chapter, this method of using machine learning to identify biblical quotations is both like and unlike the kinds of keyword searching and, more broadly, the lack of context and evidence of typicality that I was critiquing.

This method is unlike that kind of keyword searching in that it provides a rigorous contextualization of the quotations I've found. Unlike a keyword search, which in a corpus of 60+ billion words is guaranteed to turn up at least something, this shows me the relative frequency of occurrence, so that I can see the context of a quotation in terms of the change in quotations over time and in the context of the Bible as a whole.

But this method is still a kind of search. In particular, the way that machine learning works requires you to feed it a bunch of examples, and the model then finds other things that look like the examples. It is impossible for my model to find things other than biblical quotations. Fundamentally, the model cannot surprise because it only ever returns biblical quotations. It is thus subject to the limitations of the question that I have framed.

Broader Applications

The method of quotation detection that I have described is only one example of a kind of computational text analysis applicable to the humanities. Other forms of identifying the reuse of text can show the migration of ideas.[38] Another category of textual analysis relies on understanding discursive structures. Topic modeling shows how categories of topics change over time at the document or corpus level.[39] Word-embedded models can show the structure of the relationships between words, allowing a kind of multidimensional map of concepts which can also be compared over time.[40] In short, there are a range of text-analytical methods which can be applied to questions in religious studies.

---

38 Ryan Cordell, "Reprinting, Circulation, and the Network Author in Antebellum Newspapers," *American Literary History* 27, no. 3 (2015): 417–45, https://doi.org/10.1093/alh/ajv028; Kellen Funk and Lincoln A. Mullen, "The Spine of American Law: Digital Text Analysis and U.S. Legal Practice," *American Historical Review* 123, no. 1 (2018): 132–64, https://doi.org/10.1093/ahr/123.1.132.
39 Andrew Goldstone, Susana Galán, C. Laura Lovin, Andrew Mazzaschi, and Lindsey Whitmore, *An Interactive Topic Model of Signs*, part of *Signs at 40*. http://signsat40.signsjournal.org/topic-model.
40 Ben Schmidt, "Word Embeddings for the Digital Humanities," blog post, 25 October 2015, http://bookworm.benschmidt.org/posts/2015-10-25-Word-Embeddings.html; Ben Schmidt, "Rejecting the Gender Binary: a Vector-Space Operation," 30 October 2015, http://bookworm.benschmidt.org/posts/2015-10-30-rejecting-the-gender-binary.html.

What does it take to learn these kinds of methods? At a minimum it requires learning some kind of programming language suitable for data analysis such as R or Python. There are plenty of digital humanities (DH) approaches where one can work without ever seeing, let alone writing, a single line of code. But any serious text analysis or data analysis requires the ability to write scripts in a programming language. There are now abundant resources targeted at the humanist who wishes to learn these skills.[41] And in fact learning to program is the least difficult part of what it takes to do computational work in the humanities. More important is gaining a conceptual familiarity with methods such as data analysis or machine learning. And most important of all is the problem how to frame worthwhile disciplinary questions in the humanities and religious studies.[42]

And since everyone asks, what does this project cost? Both the prototype version and the eventual digital monograph have been created with a minimal budget, apart from a couple hundred dollars I spent out of pocket on web hosting for the first year or so of the project. The lack of apparent costs, however, obscures the very real costs behind the project. The *Chronicling America* dataset was provided for free by the NEH and the Library of Congress, while the George Mason University Libraries entered into an agreement with Gale to allow text mining of their newspaper collection. I have access to the high-performance computing cluster at George Mason, along with web servers at the Roy Rosenzweig Center for History and New Media to host the site. And it is not to be taken for granted that my department affords me the time to undertake research and has the intellectual flexibility to regard such experimental digital projects as scholarship.

But in this chapter I have set these details about writing code to do text analysis and within the large conceptual problems in computational research. I have described the process of creating a large-scale text analysis project. This project is a kind of a search, and therefore subject to the limitation that it depends on the question asked. But on the other hand the project is generative of new knowledge in that it offers a disciplined serendipity in searching, rigorously contextualizing its results but continually turning up surprises. Whatever methods reli-

---

41 E. g., Taylor Arnold and Lauren Tilton, *Humanities Data in R* (Springer, 2015), http://link.-springer.com/10.1007/978-3-319-20702-5; Matthew L. Jockers, *Text Analysis with R for Students of Literature* (Springer, 2014), http://link.springer.com/10.1007/978-3-319-03164-4.
42 For that reason my own contribution to the literature on learning computational methods, which is still very much at the beginning stages, focuses primarily on the patterns of argumentation in computational history. Lincoln A. Mullen, *Computational Historical Thinking: With Applications in R* (2018–): http://dh-r.lincolnmullen.com.

gious studies scholars employ, to the extent that the try to cope with large data sets and text analysis, they may find the problem of distinguishing between the typical and the exceptional a useful starting question, and the kind of disciplined serendipity that I have described a useful method.

Selected References

Arnold, Taylor and Lauren Tilton. *Humanities Data in R.* Springer, 2015. http://link.springer.com/10.1007/978-3-319-20702-5.

Bercovitch, Sacvan. *The American Jeremiad.* Madison: University of Wisconsin Press, 1978.

Blevins, Cameron. "Digital History's Perpetual Future Tense." In *Debates in the Digital Humanities 2016*, edited by Matt Gold and Lauren Klein. University of Minnesota Press, 2016, http://dhdebates.gc.cuny.edu/debates/text/77.

Bowman, Matthew. *Christian with The Politics of a Word.* In America, Cambridge, MA: Harvard University Press, 2018.

Byrd, James P. *Sacred Scripture, Sacred War: The Bible and the American Revolution.* Oxford University Press, 2013.

Chollet, Francois and J.J. Allaire. *Deep Learning with R.* Manning, 2018.

Cooper, Valerie C. *Word, Like Fire: Maria Stewart, the Bible, and the Rights of African Americans.* University of Virginia Press, 2011.

Cordell, Ryan. "Reprinting, Circulation, and the Network Author in Antebellum Newspapers." *American Literary History* 27, no. 3 (2015): 417–45. https://doi.org/10.1093/alh/ajv028.

Darnton, Robert. *The Great Cat Massacre: And Other Episodes in French Cultural History.* Basic Books, 1984.

Fea, John. *The Bible Cause: A History of the American Bible Society.* Oxford University Press, 2016.

Funk, Kellen and Lincoln A. Mullen. "The Spine of American Law: Digital Text Analysis and U.S. Legal Practice." *American Historical Review* 123, no. 1 (2018): 132–64. https://doi.org/10.1093/ahr/123.1.132.

Goff, Philip, Arthur E Farnsley II, and Peter J. Thuesen, eds. *The Bible in American Life.* Oxford University Press, 2017.

Goldstone, Andrew, Susana Galán, C. Laura Lovin, Andrew Mazzaschi, and Lindsey Whitmore. *An Interactive Topic Model of Signs*, http://signsat40.signsjournal.org/topic-model.

Gutjahr, Paul C., ed. *The Oxford Handbook of the Bible in America.* Oxford University Press, 2017.

Gutjahr, Paul C. *An American Bible: A History of the Good Book in the United States, 1777–1880.* Stanford University Press, 1999.

Hatch, Nathan O. and Mark A. Noll. eds. *The Bible in America: Essays in Cultural History.* Oxford University Press, 1982.

Hitchcock, Tim. "Confronting the Digital: or How Academic History Writing Lost the Plot." *Cultural and Social History* 10, no. 1 (2013): 9–23.

Jockers, Matthew L. *Text Analysis with R for Students of Literature.* Springer, 2014. http://link.springer.com/10.1007/978-3-319-03164-4.

Leskovec, Jure, Anand Rajaraman, and Jeff Ullman, *Mining of Massive Datasets*. 2nd ed. Cambridge University Press, 2014. http://www.mmds.org/.

McDannell, Colleen. *Material Christianity: Religion and Popular Culture in America*. Yale University Press, 1995.

Murphy, Andrew R. *Prodigal Nation: Moral Decline and Divine Punishment from New England to 9/11*. New York: Oxford University Press, 2009.

Noll, Mark A. *In the Beginning Was the Word: The Bible in American Public Life, 1492–1783*. Oxford University Press, 2016.

Noll, Mark A. *The Civil War as a Theological Crisis*. University of North Carolina Press, 2006.

Perry, Seth. "Scripture, Time, and Authority among Early Disciples of Christ." *Church History* 85, no. 4 (December 2016): 762–83. https://doi.org/10.1017/S0009640716000780.

Putnam, Lara. "The Transnational and the Text-Searchable: Digitized Sources and the Shadows They Cast." *American Historical Review* 121, no. 2 (2016): 377–402. https://doi.org/10.1093/ahr/121.2.377.

Rosenzweig, Roy. "Scarcity or Abundance? Preserving the Past in a Digital Era." *American Historical Review* 108, no.3 (2003): 735–762. https://doi.org/10.1086/ahr/108.3.735.

Sarna, Jonathan D. and Nahum M. Sarna. "Jewish Bible Scholarship and Translations in the United States." In *The Bible and Bibles in America*, edited by Ernest S. Frerichs, 83–116. Scholars Press, 1988.

Smith, David A., Ryan Cordell, and Abby Mullen. "Computational Methods for Uncovering Reprinted Texts in Antebellum Newspapers." *American Literary History* 27, no. 3 (2015): 1–15. https://doi.org/10.1093/alh/ajv029.

Stein, Stephen J. "America's Bibles: Canon, Commentary, and Community." *Church History* 64, no. 2 (June 1, 1995): 169–84. https://doi.org/10.2307/3167903.

Thuesen, Peter J. *In Discordance with the Scriptures: American Protestant Battles Over Translating the Bible*. Oxford University Press, 1999.

Underwood, Ted. "Theorizing Research Practices We Forgot to Theorize Twenty Years Ago." *Representations* 127, no. 1 (2014): 64–72.

Welbers, Kasper, Wouter Van Atteveldt, and Kenneth Benoit. "Text Analysis in R." *Communication Methods and Measures* 11, no. 4 (October 2, 2017): 245–65. https://doi.org/10.1080/19312458.2017.1387238.

Wosh, Peter J. *Spreading the Word: The Bible Business in Nineteenth-Century America*. Cornell University Press, 1994.

Frederik Elwert
Network Analysis of Religious Texts. Case Studies on Ancient Egyptian and Indian Religion

Introduction: Dipping into the Ocean of Digital Humanities

In this essay, I recapitulate the genesis and the development of the SeNeReKo project[1]. While I had an interest in technology and programming before, SeNeReKo was the first real project that would intersect my research in religion and my interest in technology. It was initiated by a grant scheme by the German ministry for education and research that targeted as the then-called "eHumanities."[2] It invited applications for projects where researchers from the humanities and the qualitative branch of social sciences would partner with researchers from computer science in order to advance their studies. The call was interesting for two reasons: Firstly, in contrast to earlier funding schemes that focused on digitization, this one looked for projects that would use previously digitized resources to answer actual research questions. Secondly, these research questions should be rooted in the humanities and social sciences, with computer science and other technical disciplines providing tools that allow for new methodological approaches—thus "enhanced humanities."

Digital humanities (DH) were already a thing in 2011, with the first ADHO conference being held as early as 1989,[3] but for me this was my first contact with the field. It is also probably fair to say that in the years since then the contours of DH as a discipline became much clearer, especially in the German con-

---

[1] The SeNeReKo project was funded by the German Federal Ministry of Education and Research under the project number 01UG1242 A. The author of this paper is responsible for its content.
[2] Bundesministerium für Bildung und Forschung, "Bekanntmachung des Bundesministeriums für Bildung und Forschung von Richtlinien zur Förderung von Forschungs- und Entwicklungsvorhaben aus dem Bereich der eHumanities," May 10, 2011, https://web.archive.org/web/20180214170128/https://www.bmbf.de/foerderungen/bekanntmachung-643.html.
[3] Alliance of Digital Humanities Organizations, "Conference," accessed July 13, 2019, https://web.archive.org/web/20190713044353/http://adho.org/conference.

text.[4] But at its inception, our project did not position itself as part of the DH per se. Our starting point was the application of network analysis methods to the study of religions. Network analysis is a popular part of the methodological canon of DH.[5] At the same time, there are many different lines of research in how network analysis is applied in the humanities. A comprehensive overview is beyond the scope of this article, but I will briefly sketch out some of the dominant strands what were relevant for developing our approach.

The study of religions in its current form is methodologically hybrid. It has a strong philological tradition, but it is also interested in historical developments beyond the text, and increasingly engages with contemporary phenomena, using methods from the social sciences. Similarly, very different traditions of network analysis can potentially be applied to study religions. A strong tradition of network analysis exists in social sciences, where it is called Social Network Analysis (SNA).[6] Increasingly, SNA has also been applied to historical phenomena in the form of Historical Network Research (HNR). A prominent use case are early modern correspondence networks, e. g. the "republic of letters."[7] Like in SNA, in HNR still individual persons (and sometimes organizations) and their interactions are the main constituents of the network. The further back in history you go, the sparser the evidence about individuals. Still, archeology has been actively adopting network analysis, though more often based on the connections that can be reconstructed or assumed between historical places.[8] A second field increasingly embracing network analysis are literary studies.[9] An influential early essay in

---

[4] The German association DHd was founded 2013, its first conference took place in 2014. See digital humanities im deutschsprachigen raum, "Über DHd," accessed April 21, 2019, https://web.archive.org/web/20190421064131/http://dig-hum.de/ueber-dhd.

[5] See e.g. Scott B. Weingart, "Demystifying Networks, Parts I & II," *Journal of Digital Humanities* 1, no. 1 (2011), http://journalofdigitalhumanities.org/1-1/demystifying-networks-by-scott-weingart/.

[6] See e.g. John Scott, *Social Network Analysis: A Handbook* (1–1), 2nd ed. (Los Angeles [u.a.]: Sage, 2009); David Knoke and Song Yang, *Social Network Analysis*, 2nd ed., Quantitative Applications in the Social Sciences 154 (Los Angeles [u.a.]: SAGE Publ, 2008).

[7] Caroline Winterer, "Where Is America in the Republic of Letters?" *Modern Intellectual History* 9, no. 3 (2012): 597–623, doi:10.1017/S1479244312000212.

[8] For an application with regard to religion, see e.g. Anna Collar, "Re-Thinking Jewish Ethnicity Through Social Network Analysis," in *Network Analysis in Archaeology: New Approaches to Regional Interaction*, ed. Carl Knappett (Oxford University Press, 2013), 223–45. For a broader overview, see Anna Collar et al., "Networks in Archaeology: Phenomena, Abstraction, Representation," *Journal of Archaeological Method and Theory* 22, no. 1 (2015): 1–32, doi:10.1007/s10816-014-9235-6.

[9] Peer Trilcke, "Social Network Analysis (SNA) als Methode einer textempirischen Literaturwissenschaft," in *Empirie in der Literaturwissenschaft*, ed. Philip Ajouri, Katja Mellmann, and Christoph Rauen, Poetogenesis 8 (Münster: mentis, 2013), 201–47.

this regard was "Network Theory, Plot Analysis" by Franco Moretti.[10] Especially stage plays have been studied using network analysis approaches, but also novels and other literary works can be analyzed using SNA. A very different application of network analysis can be found in linguistics, especially corpus linguistics and computational linguistics. Here, not (actual) people or (fictional) characters are the elements of the network, but words. The network does not model social systems, but language. This can be used to find out more about the historical meaning of a word and its change by looking at the context words it appears together with.[11] All these different approaches can be fruitful for the study of religions, as we study contemporary and historical persons, accounts of mythical figures and transcendent beings in religious texts, but also discourse and the historical meaning of certain concepts, including religion itself.

I myself had been introduced to network analysis methods a while ago by Alexander Nagel for whom I had briefly worked in a project on EU policy networks.[12] For his work, he had developed a semiotic method of structural connotation.[13] This method could be used to identify actors in texts and the different kinds of relations between the actors that the texts talked about. Using that information, one could extract network data and perform network analysis. His method relied on manual content analysis, so in the end a researcher would go through collections of texts and identify actors and their relations. I was interested to see if this could be automated through the use of computational methods.

The Center for Religious Studies (CERES) at the Ruhr University Bochum was a great place to develop a project like this. The Center's director and my PhD supervisor, Volkhard Krech, supported the project from the beginning and provided valuable input with regard to the theoretical framework. But we also needed scholars who worked with large text collections and were willing to try out

---

10 Franco Moretti, "Network Theory, Plot Analysis," *New Left Review*, no. 68 (2011): 80–102, https://newleftreview.org/II/68/franco-moretti-network-theory-plot-analysis.
11 See e.g. Alexander Mehler et al., "Inducing Linguistic Networks from Historical Corpora: Towards a New Method in Historical Semantics," in *New Methods in Historical Corpora*, ed. Paul Bennett et al., Korpuslinguistik Und Interdispiplinäre Perspektiven Auf Sprache 3 (Tübingen: Narr, 2013), 257–74.
12 See Alexander-Kenneth Nagel, *Politiknetzwerke und politische Steuerung: institutioneller Wandel am Beispiel des Bologna-Prozesses*, Staatlichkeit im Wandel 12 (Frankfurt am Main: Campus, 2009).
13 Alexander-Kenneth Nagel, "Analysing Change in International Politics: A Semiotic Method of Structural Connotation," TranState Working Papers (University of Bremen, Collaborative Research Center 597: Transformations of the State, 2008), http://hdl.handle.net/10419/24983.

new methods of approaching them. We found them in Beate Hofmann, an Egyptologist who had worked on structural genre analysis, and Sven Wortmann, who was working on a PhD project on interreligious contact during the early Buddhist period. Later in the project, Egyptologist Simone Gerhards and Indologist Sven Sellmer joined the team. For the part of Egyptology, a database of various hieroglyphic texts existed at the Berlin-Brandenburg Academy of Sciences and Humanities, the Thesaurus Linguae Aegyptiae (TLA) (http://aaew.bbaw.de/tla/). For the project on Buddhism, the Pāli Canon existed in digital form, like the one provided by the Vipassana Research Institute.[14]

Structure and Semantics, or: Structure of Semantics

One of CERES' main areas of research is interreligious contact in Eurasian history of religions. We are interested in the conditions, modes and consequences of interreligious encounter. So it was a natural starting point to look for descriptions of interreligious encounter and analyze how they were described in the texts. The Pāli Canon seemed to be a good example for this, as it contains a series of encounters between the Buddha and members of other religious groups. But also Ancient Egyptian sources contained descriptions of other peoples.

An analytical distinction between structure and semantics served as a theoretical framework.[15] On the one hand we were interested in seeing the structural relations between various groups and actors, something network analysis seemed well suited for. But structural formations alone probably will not explain the dynamics of religious history. A layer of semantics in the sense of "ideas"

---

[14] *Pāḷi Tipiṭaka*, http://tipitaka.org/. For a more detailed review of the digital resources availble for Buddhist studies, see Bingenheimer's chapter in this volume.

[15] See Volkhard Krech, "Dynamics in the History of Religions. Preliminary Considerations on Aspects of a Research Programme," in *Dynamics in the History of Religions Between Asia and Europe in Past and Present Times*, ed. Volkhard Krech and Marion Steinicke (Leiden: Brill, 2012), 15– 70, p. 27. The terms "structure" and "semantics" are borrowed from systems theory, see e.g. , Rudolf Stichweh, "Semantik und Sozialstruktur: Zur Logik einer Systemtheoretischen Unterscheidung," *Soziale Systeme. Zeitschrift für Soziologische Theorie* 6, no. 2 (2000). Similar distinctions are common in the sociology of religions. See e.g. Max Weber's distinction between "ideas" and "interests" in Max Weber, "Die Wirtschaftsethik der Weltreligionen. Vergleichende religionssoziologische Versuche. Einleitung," in *Max Weber Gesamtausgabe*, vol. 19 (1920; repr., Tübingen: Mohr Siebeck, 1989), 101. See also Georg Simmel's distinction between "content" and "form" in Georg Simmel, *Soziologie. Untersuchungen über die Formen der Vergesellschaftung* (Leipzig, Duncker & Humblot, 1908), 5–6.

also has to be taken into account: Of what sort are the relations between actors? How do they talk about each other? What are the abstract concepts they refer to? Nagel's method accounted for a basic level of semantic annotation by differentiating between different types of relations, e.g. legitimization, cooperation or financial transaction. But this requires to define a set of possible relation types in advance. In contrast, we wanted to study the semantics used in the religious sources themselves, and thus try a more inductive approach that would *discover* emic categories instead of *defining* etic ones. So in a sense, we did not only want to study structure and semantics of religious encounter, we also wanted to study the structure *of* semantics in historical religious sources.

As a technical partner, we chose the Trier Center for Digital Humanities. They had a lot of experience with preparing digital editions from historical sources, so we were confident they could help us answer new questions based on the sources that were accessible to us.

The Basics I: Network Analysis

Our goal was to analyze relations—between actors, but also between actors and ideas. The method we wanted to apply for this purpose was network analysis. Without going too much into the methodological details here, I want to give a brief overview of what network analysis is and how it matched our aims.

Network analysis is a method for studying relational data. Relational data allow us to answer different kinds of questions. In traditional statistical datasets, we usually have distinct elements, e.g. persons, and their attributes, e.g. age, gender and income. This allows us to answer questions like "do you earn more the older you get?" or "do men earn more than women?" In contrast, relational data contain information about the position of an element in relation to all the other elements. This allows us to answer questions like "who is the most central actor in a network?" or "which sub-groups can I observe in the larger system?"

Humanities questions can often be phrased in relational term: We usually think of cultural systems as woven nets and entangled processes.[16] However, this perspective often remains metaphorical: We use a regulative idea of webs or networks, but we don't use formal methods that operationalize the metaphor.

---

16 For a history of the metaphor, see Sebastian Giessmann, *Netze und Netzwerke: Archäologie einer Kulturtechnik, 1740–1840* (Bielefeld: Transcript, 2006).

Network analysis in its strict sense[17] uses a formal mathematical model of what a network is: In its most basic sense, a network consists of distinct elements, the "nodes," and relations between them, the "edges." This simple model already enables a series of calculations that can answer questions about the network as a whole or about the position of individual nodes. Extensions of this most basic network model cover varying strength of relations ("edge weight"), directions of relations ("A likes B, but B doesn't like A", "directed network"), different types of relations like "personal" or "professional" ("multiplex networks"), or different types of nodes, e.g. "persons" and "ideas" ("multi-modal networks").

In social networks, the model can be applied in a straight-forward manner: People are nodes, and relations between them form the edges. This still requires a series of decisions during research design, e.g., how do I observe relations? Do I ask people about their friends, and if so, what do they understand as friendship? What are the limits of a network, an organization, a country, or potentially the whole world? But in the case of cultural networks that include abstract ideas in general, and in religious networks in particular, these questions become even more difficult: Are non-physical entities (e.g., gods or ancestors) parts of the network? What is the place of an "idea" in a network, is it a quality of the relations (as different edge types) or are they distinct nodes? If the latter, what is an "idea?"

Additionally, we wanted to generate these networks from textual sources in an automated manner. The kind of hermeneutic work that a researcher could bring in, as in Nagel's approach, had to be translated into a simple set of rules—or at least into something a computer could infer from the data.[18] But in order for a computer to understand what is going on in a text, we needed the computer to understand more about language: What are the meaning bearing words in a sentence (e.g., nouns, adjectives and verbs in contrast to particles)? How are relations between those elements expressed?

---

[17] There is a discussion about "qualitative network analysis" as a method that does not use formal network analysis methods, but still uses a relational perspective as its methodological basis. See e.g. Betina Hollstein, "Qualitative Approaches," in *The Sage Handbook of Social Network Analysis*, ed. Peter J. Carrington and John Scott (Los Angeles: Sage, 2011), 404–16.

[18] When we started, we knew little about the possibilities of recent machine learning algorithms. They change quite drastically how we think of the computer's work, because they allow us to teach a machine by examples, as we would instruct human encoders, instead of formulating deterministic rules.

The Basics II: Linguistic Groundwork and Data Preparation

Working with text on a linguistic level—instead of, e. g., a simple full-text search—requires thorough data preparation.[19] A project requirement was that the text collections we used had to be digitized already. So at least we did not have to scan or photograph texts, apply OCR or handwriting recognition, or things like these which often are required when working with sources beyond the large digitized text collections. But the level of additional linguistic information was very different between the collections.

The TLA contained manual annotations for the Egyptian texts. The edited texts were split into sentences and words—not a trivial task for a language with no punctuation or spaces. Each word was linked to the corresponding *lemma* (base form) in a dictionary which already accounted for different senses of a single word. Additionally, each word carried detailed information about its morphology, e.g. part of speech, gender, or number. However, the information was encoded in an arcane system of numerical codes that was developed some twenty years ago when the work on the TLA started. So in order to make use of the information, we first had to decipher these codes with the help of a lengthy manual—and the patient support of the colleagues from the Berlin-Brandenburg Academy. We completely underestimated the effort that was required to simply use the information that was already there. On the other hand, we had a series of fruitful conversations with colleagues from multiple Egyptological research projects about a more accessible, shared encoding schema that would help projects like ours in the future.[20]

The Pāli Canon, on the other hand, was available with only minimal markup. Visible elements like section headings or verse were identifiable, but no linguistic information was given. In order to get linguistic information about the text, we decided to re-use existing tools from computational linguistics to automatically enrich the textual markup. But since—to our knowledge—no one previously developed a part-of-speech tagger for Pāli, we had to train one ourselves.[21] This

---

19 See Krawiec's and Schroeder's chapter on Coptic Scriptorium in this volume for an example of the required steps for building a digital corpus.
20 Laurent Coulon et al., "Towards a TEI Compliant Interchange Format for Ancient Egyptian-Coptic Textual Resources" (Annual Meeting of the TEI Consortium: Connect, Animate, Innovate, Lyon, 2015), http://orbi.ulg.ac.be/handle/2268/187518.
21 Modern part-of-speech taggers work for multiple languages. In order to add support for a new language, the basic program itself does not have to be modified. Instead, the program is

required substantial work by the Indologist on our Team, Sven Wortmann, and his student, Manuel Pachurka. We had a partially completed version at the end of the project. We learned a lot about the current state of linguistic tools,[22] but we finished that part of the work too late to actually analyze the canon with regard to our original research questions.

Luckily, we received a generous "data donation" from Indologist and computational linguist Oliver Hellwig. He had developed a lemmatizer and part-of-speech tagger for Sanskrit[23] and provided us with an automatically annotated version of the Indian epic Mahābhārata. This allowed us to analyze an important text from Indian religious history, albeit a different one than originally planned.

In retrospect, we substantially underestimated the amount of work that went into data preparation and technical infrastructure. The computer scientist on the team, Jürgen Knaut, spent a considerable amount of work on data transformation and programming tools that allowed us to prepare the texts for automatic processing. On the one hand, this cost us time we would have rather spent on actually working on our research questions. On the other hand, we all learned a lot in terms of text annotation and interoperability. This knowledge has already proven itself helpful in different occasions, and generally seems to be increasingly important for digitization projects—a development that hopefully will make it easier for projects that come after ours to make use of the ever-growing treasure of digitized sources.

Methods: Network Generation and Analysis

Our goal was to automatically extract networks from the text corpora. Those networks could then be analyzed using standard network analysis tools. We experimented a lot with different approaches.

One of the major questions was to determine what we wanted to see as "nodes" in our networks: Do we aim at social network analysis and regard

---

provided with a manually encoded subset of the text from which it "learns" the rules it then applies to the rest of the corpus.
22 See Frederik Elwert et al., "Toiling with the Pāli Canon," in *Proceedings of the Workshop on Corpus-Based Research in the Humanities*, ed. Francesco Mambrini, Marco Passarotti, and Caroline Sporleder (Warsow, 2015), 39–48, http://crh4.ipipan.waw.pl/index.php/download_file/view/13/152/.
23 Oliver Hellwig, "Performance of a Lexical and POS Tagger for Sanskrit," in *Sanskrit Computational Linguistics*, ed. Girish Jha (Berlin / Heidelberg: Springer, 2010), 162-72, doi:10.1007/978-3-642-17528-2_12.

only persons as nodes? In that case, do we include gods and other transcendent beings? Or do we choose a semantic network approach where we also include more abstract concepts as nodes, allowing to track the position of ideas in the network? Answering this question depended not only on methodological and technical considerations, but also on the characteristics of the texts: In order to get a meaningful person network, we would want a fixed set of persons whose repeated interactions are described in the text. In the case of the Egyptian texts, we often had tomb inscriptions describing the relation of the decedent to the gods. Here, the chance of finding mentions of the same person again were relatively slim. More promising were narratives like "the contendings of Horus and Seth"[24]. These stories, however, were often relatively short, making them less suitable for computational analysis. In contrast, the Pāli canon frequently mentions interactions between the Buddha and more or less random people he meets during his journey. Those people then were also only mentioned once in the description of that particular encounter, while the Buddha appeared over and over again. As a network, this would result in a star-like figure: The Buddha in the center with relations to various other persons who themselves have little or no relations to each other.

A Semantic Social Network of the Mahābhārata

On the other hand, the Mahābhārata turned out to be fit for this type of personal network analysis: It is mainly a continuous and coherent narrative with limited *dramatis personae* and repeated interactions between the parties. This allowed us to construct a social interaction network from the Mahābhārata. We used a very simple heuristic in order to determine if two persons have some sort of connection and thus should be linked through an "edge" in the network: If any two persons were mentioned within the same verse (usually two lines), we added an edge between them. This already revealed interesting patterns: By applying standard network analysis measures like degree centrality, we could identify the central actors of the narrative. More interestingly, applying a community detection algorithm[25] also revealed the major factions of the narrative: The gods are identified as one coherent group (with the exception of Yama, the god of death, who

---

24 Chester Beatty I, recto (Dublin, Chester Beatty Library).
25 Vincent D. Blondel et al., "Fast Unfolding of Communities in Large Networks," *Journal of Statistical Mechanics: Theory and Experiment* 2008, no. 10 (2008): P10008, doi:10.1088/1742-5468/2008/10/P10008.

is more closely linked to the main protagonists), as are groups of heroes with particularly close relations among themselves.

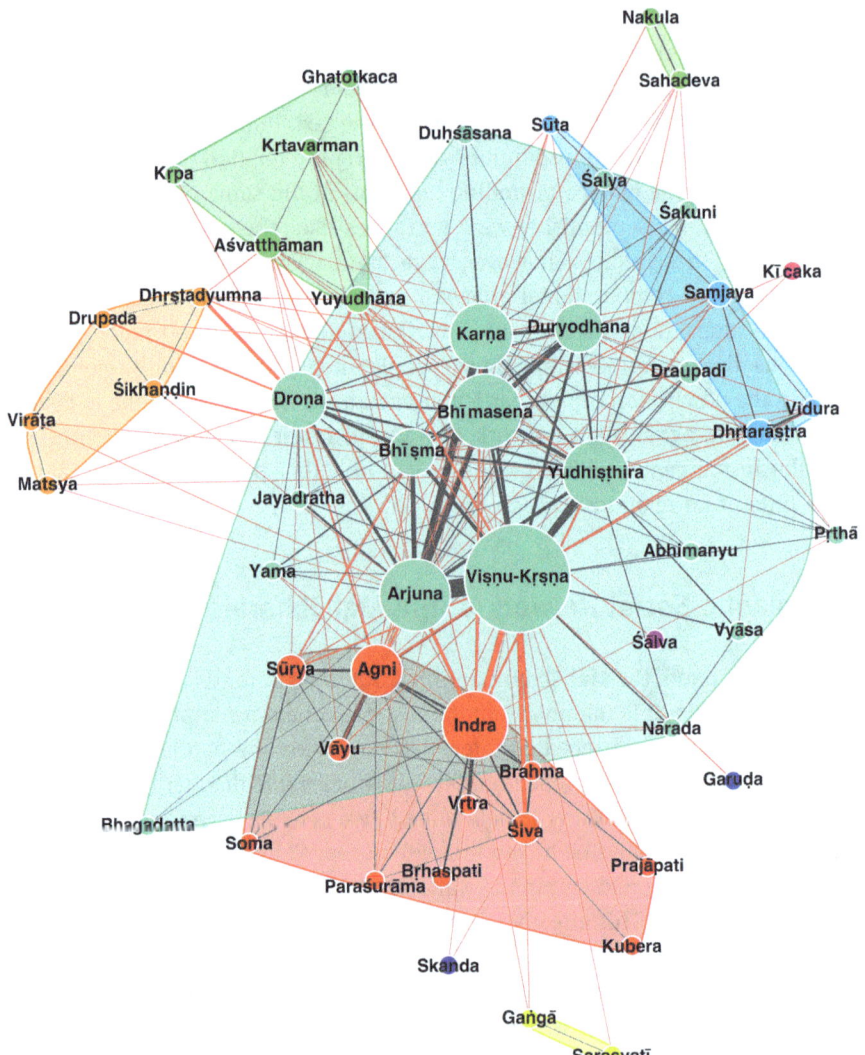

Figure 1: Identifying communities in the Mahābhārata.

This information stemmed from the structural patterns of relations alone. What was missing from this was any information about the kind of relations the actors have. Were they allies or enemies? Were they friends or family?

Which kinds of relationship existed between the gods and the humans? One approach would be to define a limited set of such relation types, as Nagel[26] did in his work. Promising implementations exist that allow us to train a computer to identify these relations from texts, basically training them using a set of examples like one would train a human coder.[27] We chose a different strategy: We did not want to impose our own categories on the material. Instead, we wanted to inductively identify which different connotations personal relations had in the Sanskrit text itself. An algorithm based on topic modeling[28] allowed us to do exactly that: By taking the text that surrounded the mentioning of a pair of persons into account, the computer identified sets of words that are indicative of the relation qualities.[29] Additionally, it identified similar sets of words for the individual actors, allowing to characterize their role in the story. These word lists or "topics" still require interpretation and are less clear-cut than pre-defined categories. But one can argue they are more grounded in the textual data themselves. And they turned out to be quite nuanced: We identified two relation topics that comprised battle-related words. We could not really tell their semantic difference until we added that information to the network and looked at who was connected by those topics. It turned out that one connected primarily the allies who fought side by side in the battles, while the other connected the enemies who fought against each other.

Socio-Semantic Networks of Egyptian Pyramid Texts

For the Ancient Egyptian texts, we chose a different strategy: Here, we included the semantic aspect in the network as nodes, not as edge types. This allowed us to see the structural position of certain concepts in the fabric of the network. We implemented an algorithm for text network analysis that takes a complete text

---

26 "Analysing Change in International Politics."
27 Andre Blessing, Jens Stegmann, and Jonas Kuhn, "SOA Meets Relation Extraction: Less May Be More in Interaction," in *Service-Oriented Architectures (SOAs) for the Humanities: Solutions and Impacts*, 2012, 6–11.
28 For an introduction to topic modeling, see Megan R. Brett, "Topic Modeling: A Basic Introduction," *Journal of Digital Humanities* 2, no. 1 (2012), http://journalofdigitalhumanities.org/2-1/topic-modeling-a-basic-introduction-by-megan-r-brett/.
29 Jonathan Chang, Jordan Boyd-Graber, and David M. Blei, "Connections Between the Lines: Augmenting Social Networks with Text," in *Proceedings of the 15th ACM SIGKDD International Conference on Knowledge Discovery and Data Mining* (ACM, 2009), 169–78, http://dl.acm.org/citation.cfm?id=1557044.

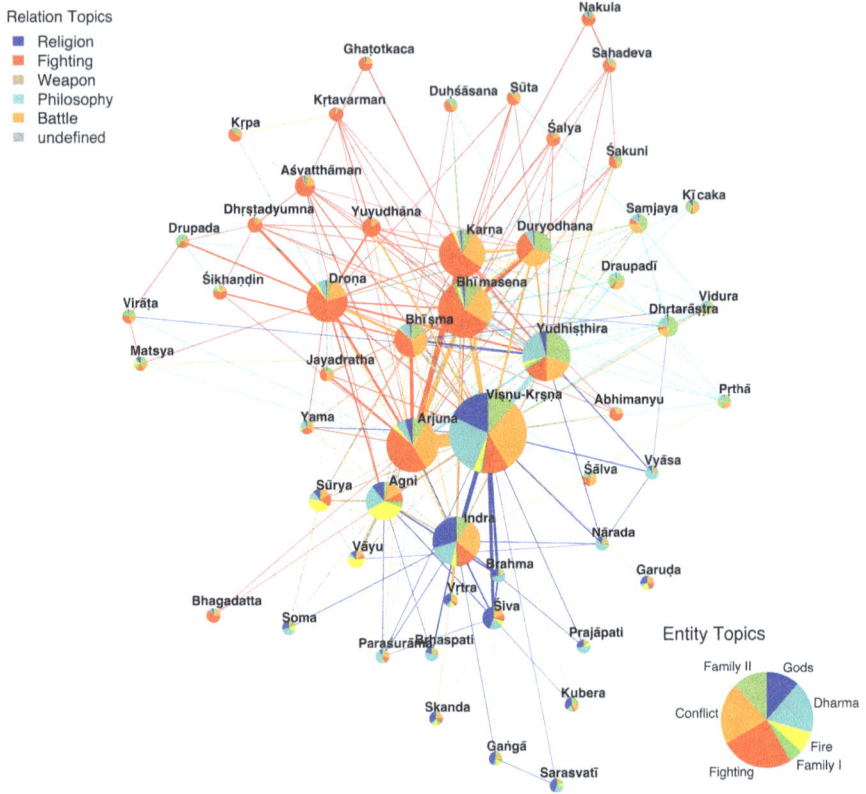

Figure 2: A topic model of network relations in the Mahābhārata.

and represents the co-occurrences of the individual words as a network.[30] This is a completely different kind of network than the one before: It is not a social network, enhanced with semantic information about actors (nodes) and relations (edges), but it is a semantic network that represents the text itself as a web of words[31]. Personal names can also be part of the network, but they are not necessarily privileged over other words.

In our application, we found that this type of network representing a whole text quickly becomes unreadable as the texts (or text corpora) get larger. Thus,

---

30 Dmitry Paranyushkin, "Identifying the Pathways for Meaning Circulation Using Text Network Analysis" (Nodus Labs, 2011), http://noduslabs.com/research/pathways-meaning-circulation-text-network-analysis/.
31 Or as a "textexture," to quote the website that first made the algorithm popular: http://textexture.com/.

we adapted the methodology to focus on the context analysis of individual words. Building upon ideas from computational linguistics, we assumed that the different contexts a word is used in and with that its different facets become visible when we map other words used in its vicinity. In a case study, we studied the use of the name of the god Horus in the pyramid texts. We took all nouns which appear in any sentence that mentions Horus and created a text network based on this sample. This shows how Horus is contextualized in the hieroglyphic inscriptions. Running a community detection algorithm on the network gives a hint about different contextual domains.[32]

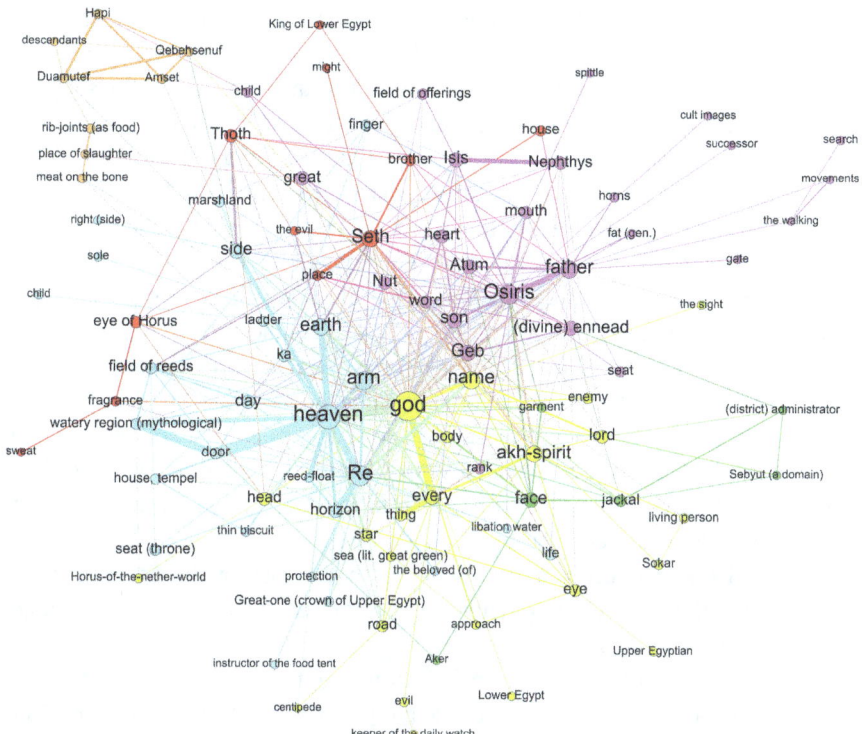

---

[32] In a more complex case study, we concurrently analyzed the usage context of two important terms for Ancient Egyptian cosmology, namely Maat and Heka Beate Hofmann and Frederik Elwert, "Heka und Maat. Netzwerkanalyse als Instrument ägyptologischer Bedeutungsanalyse," in *"Vom Leben umfangen". Ägypten, das Alte Testament und das Gespräch der Religionen. Gedenkschrift für Manfred Görg*, ed. Georg Gafus and Stefan Wimmer, Ägypten und Altes Testament 80 (Münster: Ugarit, 2014), 235–45.

Figure 3: A context network of Horus in the pyramid texts.[33]

Lessons Learned

SeNeReKo started in 2012 and ended in 2015. In hindsight, it was a valuable experience and for many of us the first serious endeavor in DH. Saying that we learned a lot during these three years also hints at what we did not know or expect when we envisioned the project. In the end, the project was a test bed for experimenting with different methodologies in text network analysis. But as the focus shifted to basic methodological research, we also somewhat lost focus on our initial research questions. The search for interreligious contact in the sources became somewhat secondary. This was also affected by the selection of the corpora we worked with: The Pāli canon, from which we hoped to learn a lot about that question, turned out to be most difficult to work with in terms of linguistic annotations. But also its narrative structure turned out to be less suited for network analysis than we initially had expected. The Mahābhārata on the other hand, which was a very valuable "data donation" we received during the course of the project, resulted in some of the most advanced network analyses, but revealed next to nothing about interreligious contact.

Next to the increased methodological focus, we also learned a lot about what one might deem dry technical matters like data preparation, linguistic annotation, and interoperability. Only because the texts were already available in digital form and did not have to be scanned or OCR'd did not mean they were ready to be processed. Developing a part of speech tagger for Pāli would probably have been a reasonable research project on its own. But also in cases where we had extensive and rigorous annotations, as in the case of the Egyptian texts, transforming the TLA database structure into something we could run our algorithms on turned out to be a major share of our work.

But even these difficulties were fruitful in the end. Being part of a working group that designed a TEI XML based exchange format for Egyptological projects did not contribute much to our initial research goals, but it feels like probably one of the lasting and valuable contributions to the broader field. The experiences we made during the project now fundamentally shape how we approach new

---

[33] English translations are used for easier comprehension. The actual analysis was performed on the transliterated text in its original language.

research projects and how we think about open research data and interoperability in general.³⁴

Selected References

Alliance of Digital Humanities Organizations. "Conference." Accessed July 13, 2019. https://web.archive.org/web/20190713044353/http://adho.org/conference.
Blessing, Andre, Jens Stegmann, and Jonas Kuhn. "SOA Meets Relation Extraction: Less May Be More in Interaction." In *Service-Oriented Architectures (SOAs) for the Humanities: Solutions and Impacts*, 6–11, 2012.
Blondel, Vincent D., Jean-Loup Guillaume, Renaud Lambiotte, and Etienne Lefebvre. "Fast Unfolding of Communities in Large Networks." *Journal of Statistical Mechanics: Theory and Experiment* 2008, no. 10 (2008): P10008. doi:10.1088/1742-5468/2008/10/P10008.
Brett, Megan R. "Topic Modeling: A Basic Introduction." *Journal of Digital Humanities* 2, no. 1 (2012). http://journalofdigitalhumanities.org/2-1/topic-modeling-a-basic-introduction-by-megan-r-brett/.
Bundesministerium für Bildung und Forschung. "Bekanntmachung des Bundesministeriums für Bildung und Forschung von Richtlinien zur Förderung von Forschungs- und Entwicklungsvorhaben aus dem Bereich der eHumanities," May 10, 2011. https://web.archive.org/web/20180214170128/https://www.bmbf.de/foerderungen/bekanntmachung-643.html.
Chang, Jonathan, Jordan Boyd-Graber, and David M. Blei. "Connections Between the Lines: Augmenting Social Networks with Text." In *Proceedings of the 15th ACM SIGKDD International Conference on Knowledge Discovery and Data Mining*, 169–78. ACM, 2009. http://dl.acm.org/citation.cfm?id=1557044.
Collar, Anna. "Re-Thinking Jewish Ethnicity Through Social Network Analysis." In *Network Analysis in Archaeology: New Approaches to Regional Interaction*, edited by Carl Knappett, 223–45. Oxford University Press, 2013.
Collar, Anna, Fiona Coward, Tom Brughmans, and Barbara J. Mills. "Networks in Archaeology: Phenomena, Abstraction, Representation." *Journal of Archaeological Method and Theory* 22, no. 1 (2015): 1–32. doi:10.1007/s10816-014-9235-6.
Coulon, Laurent, Frederik Elwert, Emmanuelle Morlock, Stéphane Polis, Vincent Razanajao, Serge Rosmorduc, Simon Schweitzer, and Daniel A. Werning. "Towards a TEI Compliant Interchange Format for Ancient Egyptian-Coptic Textual Resources." Lyon, 2015. http://orbi.ulg.ac.be/handle/2268/187518.
digital humanities im deutschsprachigen raum. "Über DHd." Accessed April 21, 2019. https://web.archive.org/web/20190421064131/http://dig-hum.de/ueber-dhd.
Elwert, Frederik. "Open Data, Open Standards and Open Source: Field Notes from the SeNeReKo Project." *Omega Alpha | Open Access*, May 5, 2015. https://oaopenaccess.

---

34 Frederik Elwert, "Open Data, Open Standards and Open Source: Field Notes from the SeNeReKo Project," *Omega Alpha | Open Access*, May 5, 2015, https://oaopenaccess.wordpress.com/2015/05/05/open-data-open-standards-and-open-source-field-notes-from-the-senereko-project/.

wordpress.com/2015/05/05/open-data-open-standards-and-open-source-field-notes-from-the-senereko-project/.

Elwert, Frederik, Sven Sellmer, Sven Wortmann, Manuel Pachurka, Jürgen Knauth, and David Alfter. "Toiling with the Pāli Canon." In *Proceedings of the Workshop on Corpus-Based Research in the Humanities*, edited by Francesco Mambrini, Marco Passarotti, and Caroline Sporleder, 39–48. Warsow, 2015. http://crh4.ipipan.waw.pl/index.php/download_file/view/13/152/.

Giessmann, Sebastian. *Netze und Netzwerke: Archäologie einer Kulturtechnik, 1740–1840*. Bielefeld: Transcript, 2006.

Hellwig, Oliver. "Performance of a Lexical and POS Tagger for Sanskrit." In *Sanskrit Computational Linguistics*, edited by Girish Jha, 162-72. Berlin / Heidelberg: Springer, 2010. doi:10.1007/978-3-642-17528-2_12.

Hofmann, Beate, and Frederik Elwert. "Heka und Maat. Netzwerkanalyse als Instrument ägyptologischer Bedeutungsanalyse." In *"Vom Leben umfangen". Ägypten, das Alte Testament und das Gespräch der Religionen. Gedenkschrift für Manfred Görg*, edited by Georg Gafus and Stefan Wimmer, 235–45. Ägypten und Altes Testament 80. Münster: Ugarit, 2014.

Hollstein, Betina. "Qualitative Approaches." In *The Sage Handbook of Social Network Analysis*, edited by Peter J. Carrington and John Scott, 404–16. Los Angeles: Sage, 2011.

Knoke, David, and Song Yang. *Social Network Analysis*. 2nd ed. Quantitative Applications in the Social Sciences 154. Los Angeles [u. a.]: SAGE Publ, 2008.

Krech, Volkhard. "Dynamics in the History of Religions. Preliminary Considerations on Aspects of a Research Programme." In *Dynamics in the History of Religions Between Asia and Europe in Past and Present Times*, edited by Volkhard Krech and Marion Steinicke, 15–70. Leiden: Brill, 2012.

Mehler, Alexander, Silke Schwandt, Rüdiger Gleim, and Alexandra Ernst. "Inducing Linguistic Networks from Historical Corpora: Towards a New Method in Historical Semantics." In *New Methods in Historical Corpora*, edited by Paul Bennett, Martin Durrell, Silke Scheible, and Richard J. Whitt, 257–74. Korpuslinguistik Und Interdispiplinäre Perspektiven Auf Sprache 3. Tübingen: Narr, 2013.

Moretti, Franco. "Network Theory, Plot Analysis." *New Left Review*, no. 68 (2011): 80–102. https://newleftreview.org/II/68/franco-moretti-network-theory-plot-analysis.

Nagel, Alexander-Kenneth. "Analysing Change in International Politics: A Semiotic Method of Structural Connotation." TranState Working Papers. University of Bremen, Collaborative Research Center 597: Transformations of the State, 2008. http://hdl.handle.net/10419/24983.

Nagel, Alexander-Kenneth. *Politiknetzwerke und politische Steuerung: institutioneller Wandel am Beispiel des Bologna-Prozesses*. Staatlichkeit im Wandel 12. Frankfurt am Main: Campus, 2009.

Paranyushkin, Dmitry. "Identifying the Pathways for Meaning Circulation Using Text Network Analysis." Nodus Labs, 2011. http://noduslabs.com/research/pathways-meaning-circulation-text-network-analysis/.

Scott, John. *Social Network Analysis: A Handbook*. 2nd ed. Los Angeles [u. a.]: Sage, 2009.

Simmel, Georg. *Soziologie. Untersuchungen über die Formen der Vergesellschaftung*. Leipzig, Duncker & Humblot, 1908.

Stichweh, Rudolf. "Semantik und Sozialstruktur: Zur Logik einer Systemtheoretischen Unterscheidung." *Soziale Systeme. Zeitschrift für Soziologische Theorie* 6, no. 2 (2000).
Trilcke, Peer. "Social Network Analysis (SNA) als Methode einer textempirischen Literaturwissenschaft." In *Empirie in der Literaturwissenschaft*, edited by Philip Ajouri, Katja Mellmann, and Christoph Rauen, 201–47. Poetogenesis 8. Münster: mentis, 2013.
Weber, Max. "Die Wirtschaftsethik der Weltreligionen. Vergleichende religionssoziologische Versuche. Einleitung." In *Max Weber Gesamtausgabe*, 19:83–127. 1920. Reprint, Tübingen: Mohr Siebeck, 1989.
Weingart, Scott B. "Demystifying Networks, Parts I & II." *Journal of Digital Humanities* 1, no. 1 (2011). http://journalofdigitalhumanities.org/1-1/demystifying-networks-by-scott-weingart/.
Winterer, Caroline. "Where Is America in the Republic of Letters?" *Modern Intellectual History* 9, no. 3 (2012): 597–623. doi:10.1017/S1479244312000212.

Rebecca Krawiec and Caroline T. Schroeder
Digital Approaches to Studying Authorial Style and Monastic Subjectivity in Early Christian Egypt

The study of religion in late antiquity sits at the crossroads of multiple disciplines: Classics, History, Papyrology, Linguistics, Literature, Religious Studies, Egyptology, Archaeology, Art History. Across all of these, research into the literature and culture of Egypt plays an important role. During the Roman period of Egyptian history, at the same time Christianity rose to prominence in the region, the Coptic language emerged. Coptic is the last phase of the Egyptian language family, having evolved from Demotic and ultimately from the hieroglyphs. Important, even irreplaceable, sources for the history of religion—Roman, traditional Egyptian, and Christian—survive to us in Coptic. The library of Nag Hammadi, which contains early Christian apocryphal and non-canonical texts, is in Coptic. Our largest corpus of early monastic sources from a single monastery was composed in Coptic. Magical spells, saints' lives, letters, sermons, homilies, prayers all exist in Coptic and provide a rich and understudied source for Religious Studies research. Moreover, Coptic documentary sources in papyri and ostraca, such as wills, letters, and contracts, provide detailed records of daily life and social history in late antiquity. In linguistics, the study of Coptic historically facilitated the eventual translation of ancient Egyptian hieroglyphs and it continues to be an important language for research in historical linguistics, bilingualism, and language change. Indeed, studying Coptic language and literature allows us to study the layers of social and religious dynamics that characterize late antiquity.

In creating the project *Coptic SCRIPTORIUM* (http://copticscriptorium.org/), we sought to produce a resource that would enable research across disciplines, especially in these three: Linguistics, Religious Studies, and History. As we have realized during our research, an interdisciplinary digital and computational environment enables more than conducting research in multiple disciplines; it enables using multiple disciplinary methods simultaneously for particular Religious Studies research questions.

In what follows, we will describe the structure, data models, and principles in the creation of Coptic Scriptorium.[1] We will then provide a case study using

---

[1] Funding for this research has been provided by the National Endowment for the Humanities Division of Preservation and Access, the NEH Office of Digital Humanities, the German Research

Coptic Scriptorium's tools and corpora to research the writing style of a figure often cited as Coptic's most important author, Shenoute of Atripe. As researchers of early Christianity, especially asceticism and monasticism in Egypt, we work with primary sources under-researched when compared to Greek and Latin ascetic authors such as Augustine or Jerome. For these ancient authors, multiple editions and studies exist, which explore their rhetoric, social context, gender ideologies, ascetic theory, and biblical interpretation (to name a few research areas). Minimal similar scholarship exists for Egyptian monastic sources. Much of Coptic literature survives in a dismembered form; pages from the same codex and even the same work reside in different libraries, with extensive lacunae and scattered publication records.[2] We need a platform that will bring together sources previously published in print volumes (books or articles) *and* born-digital editions of primary sources still unpublished, all searchable in one location. Coptic Scriptorium provides a more efficient tool than paging through a cribbed corpus of different editions of published texts and purchased photographs of unpublished manuscripts. Even more, though, it provides context and linguistic analysis that enables researchers—ourselves included—to make connections among various parts of a text we might otherwise miss. The natural language processing tools (born of the field of Linguistics) that tag for lemma and morphologically analyze Coptic enable digital philological research core to scholarship in Religious Studies. For example, Caroline T. Schroeder's research on children in early Egyptian monasteries can be greatly enhanced by access to a large digital corpus where we can find key words associated with the terms for boys or girls or use algorithmic methods to detect how often those references to children appear in quotations or citations of earlier texts (such as biblical passages). Rebecca Krawiec's research on gender and discourse benefits from digital corpus tagged by lemma and parts of speech to understand the rhetorical strategies authors use when talking about gender and sexuality. Thus, collaborative interdisciplinary methods enrich even disciplinary scholarship.

Federation (Deutsche Forschungsgemeinschaft), the German Federal Ministry of Education and Research (Bundesministerium für Bildung und Forschung), the University of the Pacific, Georgetown University, and Canisius College. We also thank the Perseus Digital Library (especially Greg Crane, Bridget Almas, Alison Babeu, and Lisa Cerrato), Amir Zeldes, Anke Lüdeling, Thomas Krause, Tito Orlandi, David Brakke, and Heike Behlmer.

2 See, for example, Stephen Emmel, *Shenoute's Literary Corpus*, CSCO 599–600, Subsidia 111–12 (Leuven: Peeters, 2004); Tito Orlandi, "The Library of the Monastery of Saint Shenute at Atripe," In *Perspectives on Panopolis: An Egyptian Town from Alexander the Great to the Arab Conquest*, eds. A. Egberts, B.P. Muhs, and J. van der Vliet (Leiden: Brill, 2002), 211–31; Paola Buzi, and Stephen Emmel, "Coptic Codicology." In *Comparative Oriental Manuscript Studies: An Introduction*, ed. Alessandro Bausi et al. (Hamburg: Comparative Oriental Manuscript Studies, 2015), 137–53.

Tools and Technology of Coptic SCRIPTORIUM

The premise of Coptic SCRIPTORIUM is to provide a digital environment for digital and computational research into Coptic language and literature using a variety of methods and for multiple disciplinary questions, beginning with the classical dialect of Sahidic.[3] The project originally was conceived and designed by Amir Zeldes and Caroline T. Schroeder, but quickly expanded with collaboration from others. While our goal was to produce a full suite of processing tools (including natural language processing tools) and a substantial digitized corpus annotated with these tools, we faced a significant first hurdle: when we began, few digitized Coptic texts existed in the Unicode. Thanks to projects such as the Perseus Digital Library, researchers in Greek and Latin have access to large scale, open access digitized texts for their reuse. In addition, several databases (such as the *Thesaurus Linguae Graecae*, *Brepols Latin Library*, among others) provide large corpora of ancient texts for search and querying, albeit behind subscription services. For Coptic, a few projects began digitization of texts on the open web. The Papyrological portal, papyri.info, created and maintained by papyrologists, contained a few Coptic papyri. The Marcion site, a hobby project of a non-academic programmer, had begun to put documents online in Unicode. The New Testament was available at the Sahidica site created by Warren Wells, another non-academic interested in Coptic. Other scholars and heritage groups had also digitized Coptic texts, especially biblical texts, on CDs; the St. Shenouda Society sold CDs, and the Packard Humanities Institute CDs contained the New Testament and Nag Hammadi. While these resources enabled more access to Coptic sources, they were either limited, not encoded in Unicode characters, or not formatted consistently for robust search. The Corpus dei Manuscritti Copti Litterari (CMCL), created by the early digital humanist Tito Orlandi, had also digitized a number of Coptic manuscripts, albeit not in Unicode; in addition, to support its work, the CMCL was a subscription site.

Our second challenge was developing tools for processing Coptic text. We were embarking on creating the first open source natural language processing tools for any phase of the Egyptian language family. Such a task was complicated by Coptic being an agglutinative language; to create a searchable database, we would need to break bound groups of text into individual words and morphemes (called tokenization in natural language processing). Linguist Wolf-Peter Funk and staff at Université Laval had developed a lemmatizer for Coptic,

---

[3] Caroline T. Schroeder, and Amir Zeldes, "Raiders of the Lost Corpus," *Digital Humanities Quarterly* 10, no. 2 (2016), http://digitalhumanities.org/dhq/vol/10/2/000247/000247.html.

which separated terms and linked them to their dictionary forms in order to create concordances. This software operated only on an old laptop with a long defunct operating system; it was neither open source nor easily ported.

Our work was significantly advanced by the generosity of colleagues. Tito Orlandi shared with us a digital lexicon, with each word annotated for part of speech, which he had created for the CMCL. This resource shaved a year or more off of our work as we created natural language processing tools dependent on this information (see more below). Coptologist Stephen Emmel contributed digitized transcriptions of some text. Additionally, Rebecca Krawiec joined the project, providing a translation and expertise in a text we chose for a pilot. Janet Timbie also transcribed and annotated some text. In March 2013, we released a pilot, proof-of-concept corpus: it consisted of digitized text of most of the manuscript witnesses to Shenoute's letter known as *Abraham Our Father* and a few apothegms from the Coptic *Sayings of the Desert Fathers*. It was tokenized with annotations for normalization, loan word vocabulary from Greek and other languages, manuscript information, an English translation aligned by sentence or phrase, and rich metadata.

Our pilot corpus and subsequent corpora were annotated with technology known as natural language processing tools. We developed a tokenizer to break Coptic text into words and a normalizer to normalize orthography, spelling and punctuation. We trained an open source, cross-language part-of-speech tagger on a set of Coptic training data[4] and a language of origin tagger to annotate loan words originating from Greek, Latin, and other languages. Each of these tools required substantial manual development of either lexica to inform these automated tools and/or training corpora to "teach" the automated tools the patterns of Coptic. A handful of sayings from the *Sayings of the Desert Fathers* were also manually annotated for dependency syntax and entities. In addition, prior to any of this processing, all texts needed to be saved as text files using the Unicode Coptic character set; thus, we created and applied our own converters to convert texts transcribed in legacy fonts into Unicode. To enable the querying of all this annotated data, we customized an open source search and visualization tool (ANNIS), which Zeldes had helped design for linguistics research; all

---

[4] Amir Zeldes, and Caroline T. Schroeder, "Computational Methods for Coptic: Developing and Using Part-of-Speech Tagging for Digital Scholarship in the Humanities," *Digital Scholarship in the Humanities*, vol. 30, no. suppl 1, (Dec. 2015): i164–76.

the text and annotations (words, language of origin tags, part of speech tags, etc.) can be searched online in a web-based installation of ANNIS.[5]

As an open source project, Coptic Scriptorium's textual data and annotations are freely available on GitHub (http://github.com/CopticScriptorium/corpora) In order to facilitate the sharing of our textual data, we both adapted and created converters to convert our data into various archival formats. Our full dataset—all Coptic text and annotations (manuscript information, normalization of text, part of speech tags, language of origin tags, etc.)—is available for download in PAULA XML, a format of XML (extensible markup language) used especially by linguists. We also archive the text with core annotations in the most popular encoding format for digital editions in digital humanities (DH), TEI- XML. TEI-XML is a type of markup language created by the Text Encoding Initiative; while TEI-XML cannot handle all linguistic annotations (especially for syntax), it is ideal for archiving essential metadata, manuscript information (where applicable), and some linguistic annotations. Finally, we also release our relational database files (the relANNIS format for the ANNIS search and visualization tool); this way, interested users can install the ANNIS tool on their own computers and create their own relational database for search and querying.

As of this writing, we have expanded substantially, again due to collaborations and building on prior work in the field. A significant outcome of the project has been to develop a diverse network of researchers working together, each contributing in different degrees and to different aspects based on interest, availability, and scholarly training (see also Bellar and Campbell in this volume). Over a dozen editors and translators have contributed to the publication of 152 documents with over 84,000 words. The documents have been machine-annotated by our suite of tools and manually edited and corrected by experts in Coptic who have volunteered to contribute to our site, or in a small number of cases have been funded by one of our grants. Metadata, including reference identifiers for citation, was added.[6] Each document was then peer-reviewed by a senior editor on the team. Anyone who has worked on a document is named in the metadata, and all past and present contributors to the project as a whole are named on our "About" page (http://copticscriptorium.org/about). In addition, the full New Testament in the Sahidic dialect and existing digitized books of the Coptic Old Testament in Sahidic have been published with only machine-annotations

---

[5] Thomas Krause, and Amir Zeldes, "ANNIS3: A New Architecture for Generic Corpus Query and Visualization," *Literary and Linguistic Computing* (October 2014): fqu057. llc.oxfordjournals.org, doi:10.1093/llc/fqu057.

[6] Bridget Almas, and Caroline T. Schroeder, "Applying the Canonical Text Services Model to the Coptic SCRIPTORIUM," *Data Science Journal* 15, (2016): 13. http://doi.org/10.5334/dsj-2016-013

(corpora totaling over 400,000 more words). This biblical data is messier, since it has not been manually edited, but we re-publish these two corpora with every significant update to our tools. We have a full natural language processing pipeline for most of our textual annotations, complete with both an API and an online web service.[7] Consequently, *any* researcher can use the NLP pipeline to process, analyze, and annotate their own Coptic text.

Our corpora are now annotated for lemmas, and each lemma links to an online dictionary (accessible at https://corpling.uis.georgetown.edu/coptic-dictionary/). The lexicon file for the Egyptian vocabulary in the dictionary was designed and created primarily by Frank Feder. The lexicon for the Greek loan words was produced from the Database and Dictionary of Greek Loanwords in Coptic, led by Tonio Sebastian Richter with contributions by many other researchers. Thus, scholars and students at multiple institutions "contributed lexical data" and worked on the online implementation.[8]

Additional tools are enabling us to expand collaboration and annotations. An online environment allows for collaborative transcription and annotation of Coptic texts. A corpus of seventeen documents has been annotated for linguistic dependencies following the standards of the cross-language Universal Dependencies project (http://universaldependencies.org/cop/; http://copticscriptorium.org/treebank.html).

Many research possibilities are enabled by our tools and corpora; this essay examines only philologically oriented research enhanced by digital and computational methods. Other applications include use of our tools in So Miyagawa's and Marco Büchler's digital methods for detecting text reuse (quotations, citations, and allusions of other work) in Coptic literature and Paul Dilley's "distant

---

[7] Amir Zeldes, and Caroline T. Schroeder, "An NLP Pipeline for Coptic," In *Proceedings of the 10th ACL SIGHUM Workshop on Language Technology for Cultural Heritage, Social Sciences, and Humanities, eds.* Nils Reiter, Beatrice Alex, and Kalliopi A. Zervanou (Association for Computational Linguistics (ACL), 2016), 146–155. *(LaTeCH2016)* https://doi.org/10.18653/v1/W16-2119;
[8] See Frank Feder, et al. *Coptic Dictionary Online*, 2016–2019, https://corpling.uis.georgetown.edu/coptic-dictionary/about.cgi; Frank Feder, Maxim Kupreyev, Emma Manning, Caroline T. Schroeder, Amir Zeldes. *A Linked Coptic Dictionary Online*. Proceedings of LaTeCH 2018—The 11th SIGHUM Workshop at COLING2018, 12–21, Santa Fe, NM: 2018; Tonio Sebastian Richter, et al. *Database and Dictionary of Greek Loanwords in Coptic*, https://www.geschkult.fu-berlin.de/en/e/ddglc/index.html; D. M. Burns, F. Feder, K. John, M. Kupreyev. *Comprehensive Coptic Lexicon: Including Loanwords from Ancient Greek* (2019), http://dx.doi.org/10.17169/refubium-2333.

reading" of the Coptic Gospels and the Nag Hammadi Corpus.[9] Linguistics, loan words, and digital codicology have also been studied using Coptic Scriptorium.[10]

In what follows, we provide a case study using CS to research the writing style of the figure often cited as Coptic's most important author, Shenoute of Atripe. To provide readers with a sense of how the research may be conducted, we have attached an appendix, which includes the queries used in this paper, along with links to the online corpora queries and results. The digital philology undertaken in this essay is not exhaustive; we do not examine in depth all instances of terms under study. Rather, we seek to show how CS can provide the basis for such investigations and help develop understanding of Shenoute as writer and as monastic leader. Coptic literature, even the writings of Shenoute, is fairly understudied. Many important texts have not yet been published in print, and even more have not been translated into a modern language. One fragment of *Abraham Our Father* and many fragments of *I See Your Eagerness* have never been published before CS published them. The digital and computational methods used in this paper will be especially fruitful for further study of Coptic, since patterns in style and recurrent themes made evident in vocabulary have *not* been substantively studied in this literature.

Writerly Subjectivities in the Work of Shenoute of Atripe

Coptic Scriptorium, particularly with its capacity to query and visualize the multi-layered annotations, provides scholars of late antique monasticism with a resource that expands traditional philological analysis. This section focuses primarily on two works of Shenoute: a letter written to monks, *Abraham, Our Father*, and a sermon, *I See Your Eagerness*, that was delivered to a mixed audience of monks and lay people, who have sought Shenoute's advice on several mat-

---

[9] So Miyagawa, and Marco Büchler, "Computational Analysis of Text Reuse in Shenoute and Besa," and Paul C. Dilley, "Coptic Scriptorium Beyond the Manuscript: Towards a Distant Reading of Coptic Literature," papers presented at the 11th International Congress of Coptic Studies, Claremont Graduate University, 28 July 2016.
[10] On the latter see Amir Zeldes, "Duplicitous Diabolos: Parallel Witness Encoding in Quantitative Studies of Coptic Manuscripts," *Proceedings of the Symposium on Cultural Heritage Markup*, vol. 16, (2015), https://doi.org/10.4242/BalisageVol16.Zeldes01.

ters.¹¹ For the letter, we show how the deeply layered annotations from Coptic Scriptorium expand understanding of the rhetorical structures of the letter and so also the monastic issues that were under debate. The sermon, which is significantly longer, covers more topics, likely due to the different audience and to the change in genre. Like the letter, however, the sermon uses extensive quotation and allusion to Scripture, here to examine trustworthy leaders of Christians and to illuminate the general expectation of how to live in order to achieve salvation. For the latter, Shenoute includes a discussion of those who have taken on a monastic life and what that requires. Further, he debates the role of teaching and the question of prophetic performance, both important themes in *Abraham Our Father* as well. Throughout this essay, we cite from the online publications of the Coptic text of both works at Coptic Scriptorium; English translations come from the online publication of Krawiec's translation of *Abraham Our Father* and the print publication of David Brakke's and Andrew Crislip's translation of *I See Your Eagerness*.¹² Coptic Scriptorium allows comparison of the works in order to ask whether there are markers for what Derek Krueger has called a particular "writerly subjectivity," that Christians wrote in ways that were "part of their identities as disciple, monk, priest, deacon, devotee, pilgrim, prophet and evangelist, and even sinner".¹³

In *Abraham, Our Father*, Shenoute extensively discusses what constitutes correct monastic "labor," ϩⲓⲥⲉ (*hise*). This word can also mean "suffering" and it is central to Shenoute's understanding of what is necessary to achieve salvation.¹⁴ Shenoute locates such labor in biblical examples of barren couples longing for children, which he then reinterprets in a monastic context. The monastery serves as a new family, but since there is no sexual reproduction, the parent-

---

11 David Brakke, and Andrew Crislip, *Selected Discourses of Shenoute the Great: Community, Theology, and Social Conflict in Late Antique Egypt* (Cambridge: Cambridge University Press, 2015), 83–85.
12 Rebecca Krawiec, Caroline T. Schroeder, and Amir Zeldes Shenoute, eds. *Abraham Our Father*. Translated by Rebecca Krawiec and Heike Behlmer. Coptic SCRIPTORIUM. urn:cts:copticLit:shenoute.abraham. (v. 1.3.0, 8 September 2015); Rebecca Krawiec, Caroline T. Schroeder, David Sriboonreuang, and Amir Zeldes Shenoute, eds. *I See Your Eagerness*. Coptic SCRIPTORIUM. urn:cts:copticLit:shenoute.eagerness. (v. 2.3.1, 5 April 2017); Brakke, and Crislip, *Selected Discourses of Shenoute the Great*, 91–105.
13 Derek Krueger, *Writing and Holiness: The Practice of Authorship in the Early Christian East* (Philadelphia: University of Pennsylvania Press, 2004), 191.
14 David Brakke, *Demons and the Making of the Monk: Spiritual Combat in Early Christianity* (Cambridge, MA: Harvard University Press, 2006), 100–114; Rebecca Krawiec, *Shenoute and the Women of the White Monastery: Egyptian Monasticism in Late Antiquity* (New York: Oxford, 2002), 52–72.

child relationship is now based on monastic rank. Because Shenoute links biblical sources to his monastic re-definition, a list of his variation in citations, which encompass a variety of grammatical phrases, helps to discern the range of rhetorical structures for the use of Scripture as an authoritative source. Janet Timbie argues for a particular set of markers for citations in Shenoute's *Discourses*, and these provide a template that maps onto Scriptorium's tools.[15] Three main searches in *Abraham, Our Father* (Appendix, queries 1–3) produce a visualization of how he refers to his biblical sources: according to "scripture" (ⲅⲣⲁⲫⲏ *graphē*), "written" (ⲥⲏϩ *sēh*), and simply ϫⲉ (*je*), a conjunction often used to indicate speech; *je* is not used exclusively for quotations, but including *je* in the query allowed a check to make sure all quotations were located, even if not linked to Scripture. While it remains to be seen, with an increase in the texts in Scriptorium's corpora, whether the patterns that appear here are characteristic of his texts written to monks as a whole, for *Abraham, Our Father* Krawiec has used these three searches to argue for several potential conclusions.[16] First, at certain points in his argument, Shenoute moves from quotation to quotation, including some unidentified texts. This particular use of Scripture is important because Shenoute's successor, Besa, has often been described as an inferior monastic writer for stringing together quotations. Instead it is possible that such a use of quotations assumes an audience would be able to make the connections and discern the correct teachings. Second, searching the various citations digitally creates a visual contextualization. The list of search results provides the key word searched in context, enabling a visual comparison between when Shenoute makes a general allusion to the authority of Scripture and when he cites a specific passage. Both uses provide insights into the role of Scripture in Shenoute's arguments and thus its association with the formation of monasticism as a social institution. Krawiec has argued, based on these lists of quotations and allusions, that when Shenoute wants to combine a variety of biblical references he tends to use a general allusion. The use of specific quotations has a wider range of uses, including identification of a biblical figure as the speaker of the passage and hence as the authority being referenced.

More significantly, digital querying of these terms in *Abraham Our Father* reveals more interconnections among the use of Scripture and some of the key

---

15 Janet A. Timbie, "Non-Canonical Scriptural Citation in Shenoute," In *Actes du huitième congrès international d'études coptes* (Orientalia Lovaniensia Analecta 163), eds, Nathalie Bosson, and Anne Bouvarel-Boud'hors (Leuven: Peeters, 2007), 625–634.
16 Rebecca Krawiec, "Reading Abraham in the White Monastery: Fathers, Sources, and History," In *From Gnostics to Monastics: Studies in Coptic and Early Christianity,* eds. David Brakke, Stephen J. Davis, and Stephen Emmel (Leuven: Peeters, 2017), 455–73.

terms for the letter—interconnections that had not been evident in earlier traditional analyses of the letter.[17] Queries for ⲅⲣⲁⲫⲏ link it to other terms that create the rhetorical structures of Shenoute's argument and definition of monasticism. Figure 1 shows a screenshot of search results as they appear in Coptic Scriptorium, with the search term offset in red and additional phrasing to a set number of characters on either side. Each result includes links to be able to see larger contextualization within the work or for linguistic analysis. Exploring these connections leads to additional queries of these key terms, such as "teaching" (ⲥⲃⲱ) which can be either a noun or part of a compound verb "to teach" (ϯⲥⲃⲱ). "Teaching" is a more prominent term and theme in *Abraham Our Father* than in *I See Your Eagerness*. "Teach" and related terms appear thrice as many times in the former as in the latter (12 vs 4), even though the surviving text of the latter is over 25% longer (see Appendix, query 4).

Examining these query results more closely, both independently and in relationship to each other, Krawiec showed that "teaching of Scriptures" is one of the works monks are to undertake as part of their relationship to each other. Further, it also becomes an area of contestation, specifically about proper interpretation of Scriptural passages as the basis for monastic identity. Shenoute's arguments with monks in his monastery about correct monastic practice are well-known. Examining the variety of references and uses of Scripture in the letter as a whole, a process made more efficient by using Scriptorium, extends this understanding by making it evident that Shenoute places himself in the role of correct heir to the biblical and monastic fathers because he understands Scripture correctly. The queries for "teach" reveal that this instruction is also contested, with Shenoute also expressing concern about people who refuse teaching (*mntatsbō*/ⲙⲛⲧⲁⲧⲥⲃⲱ and *atsbō*/ⲁⲧⲥⲃⲱ).[18]

This summation of Krawiec's earlier analysis contributes to the current notion of a "writerly subjectivity." In this letter Shenoute's "writerly subjectivity" includes being a correct teacher of Scripture. CS and the textual investigations it provides thus lead to a depth of understanding of how Shenoute created a monastic language. It allows scholars to track the complexity of Shenoute's language that reveals the connections between his terminology and his use of Scripture to define proper monasticism in the White Monastery.

---

[17] Rebecca Krawiec, *Shenoute and the Women of the White Monastery: Egyptian Monasticism in Late Antiquity* (New York: Oxford, 2002).
[18] Rebecca Krawiec, "Reading Abraham in the White Monastery: Fathers, Sources, and History," In *From Gnostics to Monastics: Studies in Coptic and Early Christianity*, eds. David Brakke, Stephen J. Davis, and Stephen Emmel (Leuven: Peeters, 2017), 455–73.

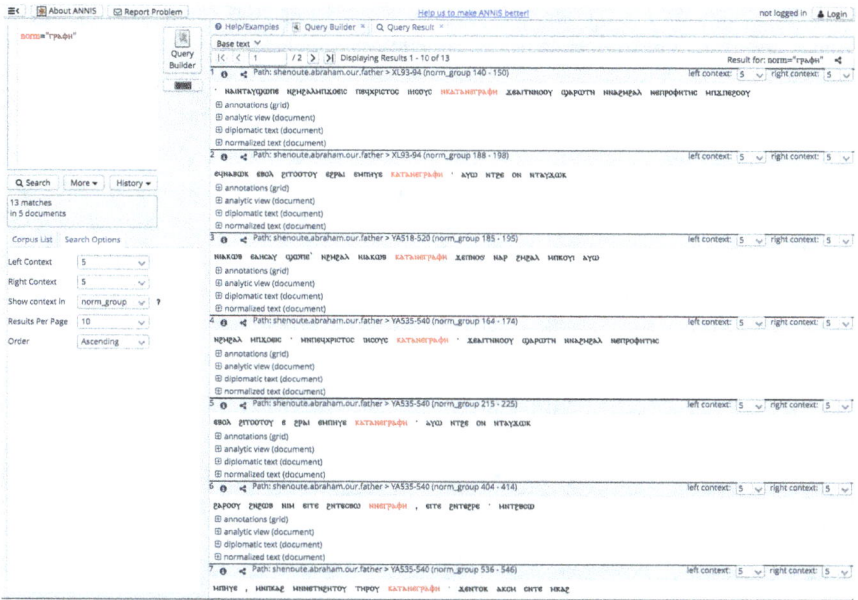

Figure 1: Results in the ANNIS search and visualization tool of the query for the Coptic term for "writing" (or "scripture") in Shenoute's *Abraham Our Father*. The query appears in the upper left; the "hits" appear on the right as key words in context. Hit #6 includes "teaching of scriptures".

Using these same digital strategies for *I See Your Eagerness* reveals both additional evidence of writerly subjectivities as well as exploration of monastic terminology. It also provides a comparison between a letter aimed at an internal audience, that of monks in the White Monastery, versus a sermon to an external one, albeit one that still has monks (perhaps from the White Monastery) present. At first, searches for these scriptural markers indicate simply that these citation techniques are representative of Shenoute's rhetorical style, particularly when his oratorical or written performance has a Christian audience. In contrast, for example, Shenoute directs *Not Because a Fox Barks* to an elite man in his region named Gesios, a man who has identified as Christian but whom Shenoute accused of still harboring traditional Egyptian idols to be worshipped. *Not Because a Fox Barks* contains allusions and references to biblical works, but it lacks the term "scripture" (Appendix, query 53) and the phrase "as it is written" (Appendix, queries 14 and 34). Here instead Shenoute favors the term "words" to refer to passages from the prophets or Jesus (Appendix, query 22). Digital searching of morphologically tagged text makes more visible the shifting rhetorical moves of one author in texts for different audiences.

For *I See Your Eagerness*, Scriptorium's queries and visualizations again make possible exploration of the connections between Scriptural citation, or allusion, and key elements of the sermon. A query for the phrase "as it is written" (ⲕⲁⲧⲁⲑⲉ ⲉⲧⲥⲏϩ/*katathe etsēh*), which appeared as a distinct phrase three times in *Abraham Our Father*, reveals no hits in *I See Your Eagerness* (Appendix, query 6). A query just for "written," however, reveals first, that the word appears in a different formulation than it appeared in *Abraham, Our Father*, as ⲛⲑⲉ ⲉⲧⲥⲏϩ (*nthe etsēh*); and second, that one use of the term "written" specifically links to the Gospel as a written source.

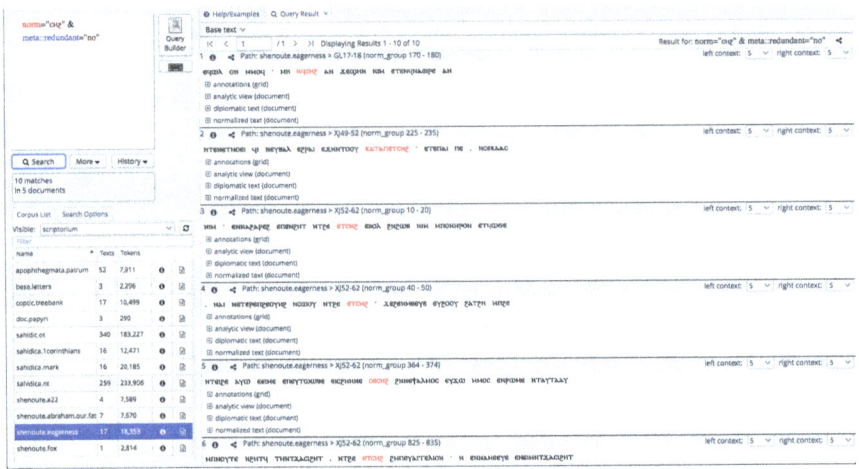

Figure 2: results of search for the Coptic term for "written" (sēh) in Shenoute's *I See Your Eagerness* in ANNIS. The upper left shows the query, and the lower left shows the corpus searched/queried. The "hits" from the query in that corpus appear on the right, showing repetition of the phrase "in the way that it is written (nthe etsēh)." Hit #6 includes a reference to "the Gospel". Appendix, query 13.

Shenoute challenges anyone who might disagree with what he is preaching as having "the abomination of God, arrogance" (ⲙⲛⲧϫⲁⲥⲓϩⲏⲧ/*mntjasihēt*) within him "as it is written in the Gospel". He then gives an example of such arrogance: for his audience ("you"), questioning that Shenoute would teach (ϯⲥⲃⲱ/*tisbō*) them is arrogance (*mntjasihēt*). Shenoute admits to his own arrogance, as well. When he does, the source of arrogance is the same—refusing to be taught—but he uses a different term for his arrogance, ⲕⲁⲑⲏⲅⲉⲓ (*kathēgei*), which appears only once elsewhere in Coptic Scriptorium's digitized corpus (1 Cor 14:19) (Appendix, queries 8–10). In contrast, *mntjasihēt* is a far more common term, appearing at least once in several works in Coptic Scriptorium (appendix, queries 36–39). This passage is important in pointing to some sort of dispute about teaching, specifically submitting to teaching authority. Both Shenoute and his audience are at fault for their arrogant refusal of this submission.[19] The sinfulness of this arrogance then leads Shenoute to imagine how people will account for themselves at their judgement before God. He lists the various acts that will be received positively, one of which is that "in your name we proclaim the Scriptures." This use of "Scripture" is not his usual citation method for Scripture; here he emphasizes an action about Scripture, namely proclaiming or possibly preaching, just as in *Abraham, Our Father* where "teaching of Scripture" is a deed that monks are to take on. Shenoute's argument in this section of the sermon is complex, as is apparent through the intersection of various key terms, "it is written" and "Scripture," and the appearance of unusual terms, "Gospel" (Appendix, queries 29–31) and the Greek loan word for "arrogance." This complexity raises the possibility that this section uses language that establishes Shenoute's writerly subjectivities as both preacher and teacher.

Computational methods assist in investigating the ways in which particular ideas and concepts recur at various points throughout a work. Although fragmentary at its beginning, *I See Your Eagerness* opens with the idea that some people have received (ϫⲓ/*ji*) the title (ⲣⲁⲛ/*ran*) "leader" (ⲁⲣⲭⲏ/*arxē*); this same claim occurs later in the text, in the passage we just discussed about declarations on judgment day. These two occurrences are easily found using the query syntax of Coptic Scriptorium's database (Appendix, query 32). When Shenoute lists the things that one can say to God at the hour of judgment, the first is "In your name (*ran*) we accepted (*ji*) positions of leadership (*arxē*)". The reoccurrence of this language functions as a rhetorical call back to the opening of the sermon. Digital methods allow us to trace uses of terms and phrases to see how they are connected to larger themes in the sermon.

Another writerly subjectivity we can discern using digital queries is Shenoute's prophetic persona. Shenoute references the prophets, along with other figures like the apostles, to create generic appeals to sacred models. This rhetorical technique is the basis of *Abraham Our Father*, as the very opening referencing the biblical patriarch suggests, and which he then develops by positing that these figures create the "pattern" (ⲥⲙⲟⲧ/*smot*) for monastic life.[20] This motif is less significant in *I See Your Eagerness*; despite the longer length of the text, it contains four usages of the word "prophet" (Appendix, queries 25–6) as opposed

---

19 See Stephen Emmel discussion of a teaching dispute in an undigitized fragment of the sermon in *Shenoute's Literary Corpus*, CSCO 599–600, Subsidia 111–12 (Leuven: Peeters, 2004), 631.
20 Rebecca Krawiec, *Shenoute and the Women of the White Monastery: Egyptian Monasticism in Late Antiquity* (New York: Oxford, 2002).

to twenty-six in *Abraham* (Appendix, query 27). Two of the four instances in *I See Your Eagerness* involve mentions of the prophets along with other figures (e. g., "fathers" and "apostles") to create a collective authority. One reference links an unnamed "prophet" to a particular saying (ϣⲁϫⲉ *šaje*). The last usage attaches the title "prophet" with the name of an individual, here Moses. The occurrences of the term "prophet" in *Abraham* can be classified into these three types of usages, as well. The difference is thematic rather than rhetorical or stylistic. *Abraham* examines the distinction between prophets and false prophets. *I See Your Eagerness*, however, examines the actions of prophets to determine whether they ever revealed the location of hidden, particularly stolen, items.[21] Shenoute decries *this* definition of being a prophet, noting that Moses never did such a thing, even as God himself kept nothing hidden from Moses. Shenoute then uses more biblical quotations, without using the term "prophet," to argue against the definition of prophetic actions as revealing hidden things. This debate focuses on interpretation of Scripture, which Shenoute concludes by listing several biblical passages with only the introduction "and also." Since this listing technique is one that appears in *Abraham, Our Father* (above), it creates the opportunity to examine how lists work in Shenoute's overall rhetorical reliance on Scriptural passages. Further, a query for "to prophesy" (Appendix, queries 40–1) shows that this verb is not used to describe any of the actions of the prophets; rather it appears in the list of actions, from earlier in *I See Your Eagerness*, that some might claim at judgment day. Whereas "in your name we proclaim the Scriptures" would lead to salvation (above), "prophesying" "in your name" leads to condemnation. The need to determine what are proper, and salvific, actions surrounding prophecy connects to another writerly subjectivity for Shenoute, that of prophet.

One last area of comparison between the two works is their shared use of a specific biblical passage, "Let him who can receive, receive" (Matthew 19:12). Shenoute cites the passage in *Abraham Our Father* in service to his larger project of constructing a monastic orthodoxy by defining terms and constructing biblical models. In *I See Your Eagerness*, Shenoute uses the reference as he develops language that defines different types of Christian religiosity (including lay piety). In *Abraham, Our Father*, Mt 19:12 describes those who were able to "receive" the call to be "eunuchs and virgins"—this could include a range of people: monks, but also those who lived celibate lives outside the monastery, even if such celibacy

---

[21] See also the discussion in David Brakke, "Shenoute, Weber, and the Monastic Prophet: Ancient and Modern Articulations of Ascetic Authority." In *Foundations of Power and Conflicts of Authority in Late-Antique Monasticism*, ed. Camplani. A. and Filoramo, (G. Leuven: Peeters, 2007), 47–73.

was embraced after marriage. This call also required a renunciation of other family ties, and possessions. In *I See Your Eagerness*, Shenoute distinguishes between those things all Christians must do, even if they have not specifically promised (ⲉⲣⲏⲧ) to do them, and those things which only need to be done if promised. Thus he creates a category of universal Christian expectations, and a separate category of promises. He links the biblical passage to this latter category: only those who can bear, or receive, this level of piety need take it on. The piety involving promises includes: "to practice virginity, to take up your cross and follow God (Lk 9:23/Mt 10:38), to be a priest in his house, and to be a monk". Violating these promises is the same as apostasy.

The pledge of virginity/celibacy applies to both genders, as Shenoute goes on to explain. In the Coptic of this passage, practicing virginity is a compound word containing the verb for "to do" (Coptic verb *eire* in the form *r*) combined with the noun "virgin" (*parthenos*): ⲣ-ⲡⲁⲣⲑⲉⲛⲟⲥ (*r-parthenos*). Syntactically in Coptic, this compound verb is one word, one verb, and so in Scriptorium's data model, it is annotated as one linguistic unit. To enable computational study of particularly Egyptian and Coptic compound word forms (such as verbs like *r-parthenos* formed by joining the morpheme "*r*" of "to do" with any noun to create a new verb), we have another layer of annotation that annotates "*r*" and "*parthenos*" as individual morphemes; anyone wishing to study the linguistics of the morpheme "*r*" may do so easily in this data model. In the texts we have digitized so far, including the Coptic bible, *I See Your Eagerness* contains the only use of *parthenos* as a part of a verb (Appendix, queries 16–19). Men are here told to "be virgin" whereas women simply claim the status of "virgin," saying, "I am a virgin" (ⲁⲛⲅⲟⲩⲡⲁⲣⲑⲉⲛⲟⲥ/*angouparthenos*). It remains to be seen, as more texts are added to the corpus, how the various uses of "virgin" appear in Coptic linguistic constructions as part of the language of monasticism in late antiquity.

The third and fourth vows Shenoute lists—being a priest and a monk—are recognizable offices, perhaps the sort of "leader" (*arxē*) Shenoute references elsewhere in the sermon. The second category, however, turns a biblical passage ("take up your cross and follow") into a "promise." Classifying the biblical command of Mt 10:38 as a vow turns imitation of Christ into a social rank, albeit without a clear title; these people are those who vow to "take up your cross and follow God." Shenoute does not explain what such a vow entails in *I See Your Eagerness*; but he uses the same biblical citation in *Abraham Our Father* to describe those who are "eunuchs," who either did not have biological children or renounced those they did have. In short, Scriptorium provides tools that allow for the rapid and detailed investigation of the use of particular terms in different texts, including patterns of terms and combinations of terms, even in texts previously never before published or studied to this extent. Its various layers of annotation also allow exploration of the use of these terms alongside other rhetorical features of Shenoute's writing, leading to an increased understanding of the development of the language and definitions of monasticism.

Two other examples show the exploratory potential of Scriptorium: examining the term "written" in works of Shenoute beyond these two, and searching the annotation of language tags for so-called "loan words," that is, words whose origin is non-Coptic. The first example recalls the searches for the term "written" which is often used as part of an overall construction to refer to Scripture as a written source.[22] As noted above, Shenoute almost exclusively uses the construction "the way it is written" (*nthe etsēh*) in *I See Your Eagerness*.[23] In *Not Because a Fox Barks*, Shenoute does not use this full phrase (Appendix, query 34) and the term "written" only appears in reference to things being written on papyri, apparently as a public announcement (Appendix, query 14). Further, Coptic Scriptorim has another Shenoute text, an untitled text that Shenoute grouped together with *Abraham Our Father* of writings for his monks to read as instruction. In the section of that text that has been digitized, ⲕⲁⲧⲁⲑⲉ ⲉⲧⲥⲏϩ appears but in addition there are varied references to things being "written" (Appendix, query 35). Things written are the "words of the Lord Jesus" in the Scriptures; are "in the papyri" "from the beginning"; in the "book"; and in the "letters" or "this letter." An area of debate within scholarship on Shenoute currently is what sort of written sources he used, their origins, and whether we can determine the extent to which he treated written sources as fixed or mutable.[24] Coptic Scriptorium, here specifically being able to combine the various grammatical combinations of referring to written sources, provides not just an efficient but a multi-layered approach to cataloguing the variety of citations that can help contribute to investigating this area of Shenoutean studies.

Finally, because Coptic Scriptorium includes tags for "loan words," terms with origins another language—i.e., Greek, Latin, Arabic—one can engage in computational research of the multilingual character of Coptic. Many of these tagged terms are biblical names. Since Scriptorium also includes a tag for proper names in its annotations for part of speech, one can construct a query for all

---

[22] Bentley Layton, "Some Observations on Shenoute's Sources: Who Are Our Fathers," *Journal of Coptic Studies* 11 (2009): 45–59.
[23] There is one use of ⲕⲁⲧⲁⲡⲉⲧⲥⲏϩ (*katapetsēh*)—a slightly different construction than ⲕⲁⲧⲁⲑⲉ ⲉⲧⲥⲏϩ— in *I See Your Eagerness* (Appendix, query 13).
[24] Rebecca Krawiec, "Reading Abraham in the White Monastery: Fathers, Sources, and History," In *From Gnostics to Monastics: Studies in Coptic and Early Christianity*, eds. David Brakke, Stephen J. Davis, and Stephen Emmel (Leuven: Peeters, 2017), 455–73.

loan words, only loan words from a particular language, or loan words from any/ all language(s) *excluding* proper names. Coptic linguists have examined the interrelationship of Greek and Coptic, particularly how Greek influenced Coptic syntax. Papyrologist Sarah Clackson argues for the bilingualism of Greek and Coptic in late antique Egypt, and for resistance in separating of separating the two languages.[25] In particular, she rejects theories of a "monolingual" Coptic monastic population or of Coptic as a "nationalist" Egyptian language. She and linguist Chris Reintges agree that one can trace Greek influence on Coptic syntax, and Clarkson provides a list of morphology of "Copto-Greek" words.[26] Here Scriptorium provides multiple tools that would prove useful to advance such linguistic analysis. One can query for patterns or combinations of part of speech tags, to investigate syntax. One can add into the analysis language of origin, to see if certain combinations or patterns co-occur more frequently with "loan word" vocabulary (or not). We can compare the simple number of Greek tags in various works; thus, for example, there is not a significant difference in the percentage of loan word tags in *Not Because a Fox Barks*, a public letter addressed to an educated local governor (Appendix, query 47), and *I See Your Eagerness*, a public sermon (Appendix, query 50). When proper names are excluded (a query easily made possible since both language of origin and part of speech are annotated in Coptic Scriptorium), loan words appear 57.97 times per 1,000 words in *Not Because a Fox Barks* (Appendix, query 48) and 59.59 times per 1,000 words in *I See Your Eagerness* (Appendix, query 51). *Abraham Our Father,* with a monastic audience, has a similar ratio of loan words, excluding proper names (Appendix, query 44). As we digitize and annotate more texts, we can study this pattern of loan word use further in Shenoute's corpus, to see if 44–60 loan words per 1,000 is his general range, if the variance in these three texts is a result of audience (monastic versus public) or something else. The part of speech tags combined with language of origin tags also enable research into what Reintges calls "the 'Hellenisation' of Egyptian syntax, by which I mean the restructuring of the native sentence patterns according to a Greek model."[27] Since Scriptorium includes

---

[25] Sarah Clackson, "Coptic or Greek? Bilingualism in the Papyri." In *The Multilingual Experience in Egypt, from the Ptolomies to the Abbasids,* ed. Arietta Papaconstantinou (Farnham: Ashgate, 2010), 73–104.
[26] Chris H. Reintges, "Coptic Egyptian as a Bilingual Language Variety," In *Lenguas en contacto: el testimonio escrito,* edited by Pedro Badenas, et. al. (Madrid: Consejo Superior de Investigaciones Cientificas, 2004), 69–86.
[27] Chris H. Reintges, "Coptic Egyptian as a Bilingual Language Variety," In *Lenguas en contacto: el testimonio escrito,* edited by Pedro Badenas, et. al. (Madrid: Consejo Superior de Investigaciones Cientificas, 2004), 69–86.

lemmas and normalized text as *annotations* of the original text, when one searches for a normalized word form or for a lemma, the search results can visualize each hit of the normalized text or lemma within an annotation grid that also displays in parallel the original text—including original orthography, spelling, punctuation, abbreviations, etc., to enable study of dialect and morphology across space and time.

These case studies are meant to show the diverse ways in which Coptic Scriptorium can be used to advance Coptic studies, which encompasses both linguistics and the production of religious literature in Coptic. In this section, the main area of interest explored has been the potency of Scriptorium for sorts of historical analysis of the development of monasticism as a social movement in late antique Egypt. While Shenoute's works can be, and have been, analyzed with traditional methods of linguistic and historical research, Scriptorium provides exploratory tools that allow researchers the ability to create combinations of words, word and part of speech, and frequency of Greek "loan words" in ways that encourage creativity in research. Ultimately, with the expansion of the corpora and an increase in usage, deeper understandings of the connections and choices of Shenoute's rhetorical structures will help us better understand the formation of his monastic literacy.

Conclusion

Coptic Scriptorium shows the potential for furthering religious studies by created a richly annotated corpus of late antique Christian texts particularly related to the development of monasticism. For some scholars, Coptic Scriptorium will allow a more efficient tool for asking traditional questions about the development of monasticism for which these texts serve as evidence. The digitized text and basic annotations such as normalization and lemmatization will enable simple search of researchers' primary sources. In addition, however, our case study has also shown the creative potential to see new questions by visualizing these texts through digitization. Here this creativity, and its results, are in regard to two works of Shenoute, but they apply to Egyptian monastic texts as a whole. As more texts are digitized, questions regarding a shared, or perhaps differentiated, monastic language and terminology among the texts can contribute to our understanding of a diversity of meanings of "monasticism" among late antique authors. The linguistic annotations can enable research into writing styles across genres of monastic texts (letters, sermons, rules, hagiograpy, etc.) For texts that have received little scholarly study, our research environment can facilitate rapid searches for textual reuse: quotations, allusions, and citations of other works (including but not limited to the bible).

Moreover, additional research questions beyond the history of monasticism can also be examined. For example, because Coptic Scriptorium includes works translated from Greek (e.g. the Bible), those composed in Coptic, and those about whose original language is a matter of debate, the linguistic tools it provides can help explore issues of bilingualism and its social role in late antiquity. As texts of questionable authorship are digitized, and previously unpublished texts are digitized, a larger corpus can enable research into authorship—who wrote these unattributed or questionably attributed texts? Thus this project shares objectives with other projects examined in this volume, especially the contributions by Bingenheimer and Elwert, while also enabling research into issues specific to late antique Egypt or Coptic linguistics. Linking our data to other digital projects in late antiquity through the Pleidaes gazetteer and Pelagios online portal for ancient places enables research into space and place across the Mediterranean (https://pelagios.org/; https://pelagios.org/). The possibilities of what we can learn remains to be seen as the corpus, and so its collaborators and users, increase.

Selected References

Almas, Bridget, and Caroline T. Schroeder. "Applying the Canonical Text Services Model to the Coptic SCRIPTORIUM." *Data Science Journal* 15, (2016): 13. http://doi.org/10.5334/dsj-2016-01315, (2016): 13. http://doi.org/10.5334/dsj-2016-013.

Brakke, David. *Demons and the Making of the Monk: Spiritual Combat in Early Christianity.* Cambridge, MA: Harvard University Press, 2006.

Brakke, David. "Shenoute, Weber, and the Monastic Prophet: Ancient and Modern Articulations of Ascetic Authority." In *Foundations of Power and Conflicts of Authority in Late-Antique Monasticism,* edited by Camplani. A. and Filoramo, 47–73. G. Leuven: Peeters, 2007.

Brakke, David, and Andrew Crislip. *Selected Discourses of Shenoute the Great: Community, Theology, and Social Conflict in Late Antique Egypt.* Cambridge: Cambridge University Press, 2015.

Buzi, Paola, and Stephen Emmel. "Coptic Codicology." In *Comparative Oriental Manuscript Studies: An Introduction*, edited by Alessandro Bausi et al., 137–53. Hamburg: Comparative Oriental Manuscript Studies, 2015.

Clackson, Sarah. "Coptic or Greek? Bilingualism in the Papyri." In *The Multilingual Experience in Egypt, from the Ptolomies to the Abbasids,* edited by Arietta Papaconstantinou, 73–104. Farnham: Ashgate, 2010.

Emmel, Stephen. *Shenoute's Literary Corpus.* CSCO 599–600, Subsidia 111–12. Leuven: Peeters, 2004.

Krause, Thomas, and Amir Zeldes. "ANNIS3: A New Architecture for Generic Corpus Query and Visualization." *Literary and Linguistic Computing* (October 2014): fqu057. *llc.oxfordjournals.org*, doi:10.1093/llc/fqu057.

Krawiec, Rebecca. *Shenoute and the Women of the White Monastery: Egyptian Monasticism in Late Antiquity.* New York: Oxford, 2002.

Krawiec, Rebecca. "Reading Abraham in the White Monastery: Fathers, Sources, and History." In *From Gnostics to Monastics: Studies in Coptic and Early Christianity,* edited by David Brakke, Stephen J. Davis, and Stephen Emmel, 455–73. Leuven: Peeters, 2017.

Krueger, Derek. *Writing and Holiness: The Practice of Authorship in the Early Christian East.* Philadelphia: University of Pennsylvania Press, 2004.

Layton, Bentley. "Some Observations on Shenoute's Sources: Who Are Our Fathers." *Journal of Coptic Studies* 11 (2009): 45–59.

Orlandi, Tito. "The Library of the Monastery of Saint Shenute at Atripe." In *Perspectives on Panopolis: An Egyptian Town from Alexander the Great to the Arab Conquest,* edited by A. Egberts, B.P. Muhs, and J. van der Vliet, 211–31. Leiden: Brill, 2002.

Reintges, Chris H. "Coptic Egyptian as a Bilingual Language Variety." In *Lenguas en contacto: el testimonio escrito,* edited by Pedro Badenas, et. al., 69–86. Madrid: Consejo Superior de Investigaciones Cientificas, 2004.

Schroeder, Caroline T. "Cultural Heritage Preservation and Canon Formation: What Syriac and Coptic Can Teach Us about the Historiography of the Digital Humanities." In *Garb of Being,* edited by Susan Holman, Georgia Frank, and Andrew Jacobs, 318–345. New York: Fordham University Press, 2019.

Schroeder, Caroline T., and Amir Zeldes. "Raiders of the Lost Corpus." *Digital Humanities Quarterly* 10, no. 2 (2016). http://digitalhumanities.org/dhq/vol/10/2/000247/000247.html.

Timbie, Janet A. "Non-Canonical Scriptural Citation in Shenoute." In *Actes du huitième congrès international d'études coptes* (Orientalia Lovaniensia Analecta 163), edited by Nathalie Bosson, and Anne Bouvarel-Boud'hors, 625–634. Leuven: Peeters, 2007.

Zeldes, Amir. "Duplicitous Diabolos: Parallel Witness Encoding in Quantitative Studies of Coptic Manuscripts." *Proceedings of the Symposium on Cultural Heritage Markup,* vol. 16, (2015). https://doi.org/10.4242/BalisageVol16.Zeldes01.

Zeldes, Amir, and Caroline T. Schroeder. "Computational Methods for Coptic: Developing and Using Part-of-Speech Tagging for Digital Scholarship in the Humanities." *Digital Scholarship in the Humanities,* vol. 30, no. suppl 1, (Dec. 2015): i164-76. https://doi.org/10.1093/llc/fqv043.

Zeldes, Amir, and Caroline T. Schroeder. "An NLP Pipeline for Coptic." In *Proceedings of the 10th ACL SIGHUM Workshop on Language Technology for Cultural Heritage, Social Sciences, and Humanities,* edited by Nils Reiter, Beatrice Alex, and Kalliopi A. Zervanou, 146–155. Association for Computational Linguistics (ACL), 2016. *(LaTeCH2016)* https://doi.org/10.18653/v1/W16-2119.

Appendix: Coptic Scriptorium Data and List of Queries/Searches

All queries derive from the following corpora. URLs link to *current* versions of the data for reading and querying, which may differ from the data and annotations used when this article was written. All data, including past versions of the data, are regularly archived for download on our GitHub site (https://github.com/CopticScriptorium/corpora); the version of data used for this paper is at https://github.com/CopticScriptorium/corpora/releases/tag/v2.4.0.

Coptic SCRIPTORIUM, *apophthegmata.patrum*, v. 2.4.0, 9 November 2017. http://data.copticscriptorium.org/urn:cts:copticLit:ap.
Coptic SCRIPTORIUM, *besa.letters*, v. 2.2.0, 8 December 2016. http://data.copticscriptorium.org/urn:cts:copticLit:besa.
Coptic SCRIPTORIUM, *sahidica.nt*, v. 2.1.0, 12 May 2016. http://data.copticscriptorium.org/urn:cts:copticLit:nt.
Coptic SCRIPTORIUM, *sahidic.ot*, v. 1.0, 8 June 2016. http://data.copticscriptorium.org/urn:cts:copticLit:ot.
Coptic SCRIPTORIUM, *shenoute.abraham*, v. 1.6.1, 29 January 2016. http://data.copticscriptorium.org/urn:cts:copticLit:shenoute.abraham.
Coptic SCRIPTORIUM, *shenoute.a22*, v. 1.6.1, 29 January 2016. http://data.copticscriptorium.org/urn:cts:copticLit:shenoute.a22.
Coptic SCRIPTORIUM, *shenoute.eagerness*, v. 2.3.1, 5 April 2017. http://data.copticscriptorium.org/urn:cts:copticLit:shenoute.eagerness.
Coptic SCRIPTORIUM, *shenoute.fox*, v. 2.2.0, 8 December 2016. http://data.copticscriptorium.org/urn:cts:copticLit:shenoute.fox.

The following listed *queries* of our corpora were used to conduct research for this article. Links to each query provided below will take you to the *current* results in our ANNIS database using *current* data.

1. "Writing" (ⲅⲣⲁⲫⲏ) in *Abraham, Our Father:* https://corpling.uis.georgetown.edu/annis/?id=e58363aa-6f9f-4fb5-88d5-f596a7406341
2. ⲥϩⲁⲓ ("written") in *Abraham, Our Father:* https://corpling.uis.georgetown.edu/annis/?id=941296f0-964e-46ba-91f9-d0a431748f6c
3. The word ϫⲉ in *Abraham, Our Father:* https://corpling.uis.georgetown.edu/annis/?id=4ab1fe34-2002-4528-903a-f9b51a3f9f94
4. "Teach" and related terms in *Abraham Our Father* and *I See Your Eagerness;* double hits in parallel manuscript witnesses have been eliminated in the query syntax: https://corpling.uis.georgetown.edu/annis/?id=2cbc241f-cf23-4886-aa70-8124ef8d1338 https://corpling.uis.georgetown.edu/annis/?id=37dd72c1-047e-4cbd-9741-4f1bb9083715
5. "As it is written" (ⲕⲁⲧⲁⲑⲉ/ⲕⲁⲧⲁⲧϩⲉ ⲉⲧⲥϩⲁⲓ) in *Abraham, Our Father:* https://corpling.uis.georgetown.edu/annis/?id=22557331-be18-4804-b0cc-6efc6de4570a
6. "As it is written" (ⲕⲁⲧⲁⲑⲉ/ⲕⲁⲧⲁⲧϩⲉ ⲉⲧⲥϩⲁⲓ) in *I See Your Eagerness*; double hits in parallel manuscript witnesses have been eliminated in the query syntax (no hits): https://corpling.uis.georgetown.edu/annis/?id=9d87b60c-b37a-4695-994c-f204eabe44e9

7. "Arrogance" (ⲕⲁⲑⲏⲅⲉⲓ) in Shenoute's *I See Your Eagerness*; double hits in parallel manuscript witnesses have been eliminated in the query syntax: https://corpling.uis.georgetown.edu/annis/?id=b4032f28-b9a9-4ff5-a446-2dae33008b56
8. "Arrogance" (ⲕⲁⲑⲏⲅⲉⲓ) in Shenoute's *Abraham, Our Father, Not Because a Fox Barks, A22*, and *I See Your Eagerness*: https://corpling.uis.georgetown.edu/annis/?id=54f51c6c-1928-454d-89b9-33f4cafbbe71
9. "Arrogance" (ⲕⲁⲑⲏⲅⲉⲓ) in digitized selections of the *Sayings of the Desert Fathers* (hereafter digitized *AP*), and *Letters* of Besa (no hits as of this writing): https://corpling.uis.georgetown.edu/annis/?id=b32dc4e9-f317-43a2-aaa2-ac74b34433a2
10. "Arrogance" (ⲕⲁⲑⲏⲅⲉⲓ) in the Sahidic New Testament, digitized Sahidic Coptic Old Testament books: https://corpling.uis.georgetown.edu/annis/?id=aa209b4d-756f-4e80-8e0e-b84fdf8d95f6
11. "Vow"/"promise" (ⲉⲣⲏⲧ) in Shenoute's *I See Your Eagerness*; double hits in parallel manuscript witnesses have been eliminated in the query syntax: https://corpling.uis.georgetown.edu/annis/?id=3284f91e-26d3-49a8-a43b-5e34e1fa9eab
12. "Writing"/"scriptures" (ⲅⲣⲁⲫⲏ) in Shenoute's *I See Your Eagerness*; double hits in parallel manuscript witnesses have been eliminated in the query syntax: https://corpling.uis.georgetown.edu/annis/?id=dacf39bc-6a58-49ed-aac1-dcb71b782632
13. "Written" (ⲥⲏϨ) in I See Your Eagerness; double hits in parallel manuscript witnesses have been eliminated in the query syntax: https://corpling.uis.georgetown.edu/annis/?id=401ec056-bfc5-4cd3-978f-dffc00fb0f5e
14. "Written" (ⲥⲏϨ) in Shenoute's *Not Because a Fox Barks*: https://corpling.uis.georgetown.edu/annis/?id=be027de5-218c-49a0-a03b-c7690b5e329e
15. "Written" (ⲥⲏϨ) in Shenoute's *Acephalous Work #22* (*A22*): https://corpling.uis.georgetown.edu/annis/?id=32e8e147-4dd6-42a7-9dd1-a782e82a6186
16. "Virgin" (ⲡⲁⲣⲑⲉⲛⲟⲥ) as a word or as a term within compound words in Shenoute's *I See Your Eagerness*; double hits in parallel manuscript witnesses have been eliminated in the query syntax: https://corpling.uis.georgetown.edu/annis/?id=abd42888-fd79-4941-860f-1d40ded854c7
17. "Virgin" (ⲡⲁⲣⲑⲉⲛⲟⲥ) as a word or as a term within compound words in Shenoute's *Abraham, Our Father* (no hits for compounds as of this writing): https://corpling.uis.georgetown.edu/annis/?id=1902011b-3c25-406b-8b28-99d1f8401ebd
18. "Virgin" (ⲡⲁⲣⲑⲉⲛⲟⲥ) as a term within compound words in Shenoute's *A22*: https://corpling.uis.georgetown.edu/annis/?id=2e372570-579a-4994-aa61-c2db80afa77c
19. "Virgin" (ⲡⲁⲣⲑⲉⲛⲟⲥ) as a term within compound words in the Sahidic New Testament, digitized Sahidic Coptic Old Testament books, digitized *AP*, and *Letters* of Besa (no hits as of this writing): https://corpling.uis.georgetown.edu/annis/?id=d5ae80e7-3abb-4daa-8834-071f329a2bde
20. "Speak" (ϣⲁϫⲉ) in Shenoute's *I See Your Eagerness*: https://corpling.uis.georgetown.edu/annis/?id=1fd6f58a-0a65-4f38-81d6-7bd3b8fd3a7a
21. "Speak" (ϣⲁϫⲉ) in Shenoute's *Abraham, Our Father*: https://corpling.uis.georgetown.edu/annis/?id=27f8b234-95e0-4872-9bfb-26d590dfe2cb
22. "Speak" (ϣⲁϫⲉ) in Shenoute's *Not Because a Fox Barks*: https://corpling.uis.georgetown.edu/annis/?id=12df40a8-e718-4365-8ae9-fab55210d150
23. "Speak" (ϣⲁϫⲉ) in Shenoute's *A22*: https://corpling.uis.georgetown.edu/annis/?id=94ca8272-f0f5-487b-8ce5-afb246a0582f

24. "Speak" (ϢⲀⲬⲈ) followed by the converter "ⲚⲦⲀ" in Shenoute's *I See Your Eagerness*; double hits in parallel manuscript witnesses have been eliminated in the query syntax: https://corpling.uis.georgetown.edu/annis/?id=c3861813-77e5-419d-b501-d7aa142a2ef3
25. "Prophet" (ⲠⲢⲞⲪⲎⲦⲎⲤ) in Shenoute's *I See Your Eagerness*; double hits in parallel manuscript witnesses have *NOT* been eliminated in the query syntax: https://corpling.uis.georgetown.edu/annis/?id=53cca8a1-81ab-40a2-aa81-1e854f47ee2a
26. "Prophet" (ⲠⲢⲞⲪⲎⲦⲎⲤ) in Shenoute's *I See Your Eagerness*; double hits in parallel manuscript witnesses have been eliminated in the query syntax: https://corpling.uis.georgetown.edu/annis/?id=a8546f66-324b-476e-80d1-e0a3b74296c6
27. "Prophet" (ⲠⲢⲞⲪⲎⲦⲎⲤ) in Shenoute's *Abraham, Our Father* (twenty-nine hits but three are in a parallel witness in manuscript MONB.XL 93–94): https://corpling.uis.georgetown.edu/annis/?id=d330069e-67cd-4a9e-ac55-a15ab9cb6f9a
28. "Pattern" (ⲤⲘⲞⲦ) in Shenoute's *Abraham, Our Father*: https://corpling.uis.georgetown.edu/annis/?id=59aca655-f0f4-4aa2-8c67-59562d6bef50
29. "Gospel" (ⲈⲨⲀⲄⲄⲈⲖⲒⲞⲚ) in Shenoute's *I See Your Eagerness*; double hits in parallel manuscript witnesses have been eliminated in the query syntax: https://corpling.uis.georgetown.edu/annis/?id=f2910fd8-2c6a-42bd-bb25-f39d4fd670a2
30. "Gospel" (ⲈⲨⲀⲄⲄⲈⲖⲒⲞⲚ) in Shenoute's *Abraham, Our Father* (no hits): https://corpling.uis.georgetown.edu/annis/?id=761209fe-6669-424a-a6b7-104b6d08aaf0
31. "Gospel" (ⲈⲨⲀⲄⲄⲈⲖⲒⲞⲚ) in all the digitized works of Shenoute: https://corpling.uis.georgetown.edu/annis/?id=ccd200a7-500a-4375-a2e1-e6f0f663c4f9
32. "Leader" (ⲀⲢⲬⲎ) in Shenoute's *I See Your Eagerness*; double hits in parallel manuscript witnesses have been eliminated in the query syntax: https://corpling.uis.georgetown.edu/annis/?id=4510050d-8612-4132-84dc-d8ccfccad8b5
33. "False prophet" (ⲠⲢⲞⲪⲎⲦⲎⲤ ⲚⲚⲞⲨⲬ) in *Abraham, Our Father*: https://corpling.uis.georgetown.edu/annis/?id=bcf028e9-5925-4012-931e-3f7933e22052
34. "…which is written" (ⲈⲦⲤⲎϨ) in Shenoute's *Not Because a Fox Barks* (no hits): https://corpling.uis.georgetown.edu/annis/?id=b4d55664-b450-45e0-a084-4fcb5535c933
35. "…which is written" (ⲈⲦⲤⲎϨ) in Shenoute's *A22*: https://corpling.uis.georgetown.edu/annis/?id=85f50f32-f238-4a6a-80b8-54df6ead23f1
36. "Vanity"/"arrogance" (ⲘⲚⲦϪⲀⲤⲒϨⲎⲦ) in Shenoute's *I See Your Eagerness*; double hits in parallel manuscript witnesses have been eliminated in the query syntax: https://corpling.uis.georgetown.edu/annis/?id=a0ba9fa0-c565-4f80-bf5c-6a5ad8e11455
37. "Vanity"/"arrogance" (ⲘⲚⲦϪⲀⲤⲒϨⲎⲦ) in all the digitized works of Shenoute: https://corpling.uis.georgetown.edu/annis/?id=a68c596c-796d-43ef-ab3a-258da44be361
38. "Vanity"/"arrogance" (ⲘⲚⲦϪⲀⲤⲒϨⲎⲦ) in the digitized *AP* and *Letters* of Besa: https://corpling.uis.georgetown.edu/annis/?id=10a83d74-acd6-4e62-83ac-0b4ddd970a35
39. "Vanity"/"arrogance" (ⲘⲚⲦϪⲀⲤⲒϨⲎⲦ) in Sahidic New Testament and digitized Sahidic Coptic Old Testament books: https://corpling.uis.georgetown.edu/annis/?id=e385cb74-22ce-4de7-9880-634bc65065a2
40. "Prophesy" (ⲠⲢⲞⲪⲎⲦⲈⲨⲈ) in Shenoute's *I See Your Eagerness*; double hits in parallel manuscript witnesses have been eliminated in the query syntax: https://corpling.uis.georgetown.edu/annis/?id=10910f08-833c-4791-a3b9-67289ac5814b
41. "Prophesy" (ⲠⲢⲞⲪⲎⲦⲈⲨⲈ) in Shenoute's *Abraham, Our Father*, *A22*, and *Not Because a Fox Barks*; *Letters* of Besa; *AP* (no hits): https://corpling.uis.georgetown.edu/annis/?id=f28e42d5-fec6-4707-bed1-d515d9b9c50a

42. Greek language tags in Shenoute's *Abraham, Our Father*, not including parallel witness in manuscript MONB.XL 93–94): https://corpling.uis.georgetown.edu/annis/?id=dd44ce72-610e-4dd9-b271-038a11d4a82a
43. All loan words/language tags in Shenoute's *Abraham, Our Father* (not including parallel witness in manuscript MONB.XL 93–94): https://corpling.uis.georgetown.edu/annis/?id=035b529c-e311-4ea0-afa3-d6899140fae0
44. All loan words/language tags in Shenoute's *Abraham, Our Father*, excluding proper names (not including parallel witness in manuscript MONB.XL 93–94): https://corpling.uis.georgetown.edu/annis/?id=804ddd04-4ace-4d6e-b73b-bc76488db726
45. All words in Shenoute's *Abraham, Our Father* (not including parallel witness in manuscript MONB.XL 93–94): https://corpling.uis.georgetown.edu/annis/?id=769302d7-bd6d-435e-9b2a-652fb9a0b6c3
46. Greek language tags in Shenoute's *Not Because a Fox Barks* = 150 hits (58.75 per 1000 words): https://corpling.uis.georgetown.edu/annis/?id=0d513875-6595-411a-9ecf-f29dbc3d19c1
47. All loan words/language tags in in Shenoute's *Not Because a Fox Barks* = 162 (63.45 per 1000 words): https://corpling.uis.georgetown.edu/annis/?id=99bf848d-29d0-4699-9fb5-9488dc9bcaf3
48. All loan words/language tags in Shenoute's *Not Because a Fox Barks*, excluding proper names= 148 hits (57.97 occurrences per 1000 words): https://corpling.uis.georgetown.edu/annis/?id=640e502e-2348-4a67-b061-b1cb64980085
49. All words in Shenoute's *Not Because a Fox Barks* = 2553: https://corpling.uis.georgetown.edu/annis/?id=0103916e-7139-4738-94ba-c9e335aa8e42
50. All loan words in Shenoute's *I See Your Eagerness* (double hits in parallel manuscript witnesses have been eliminated in the query syntax) = 628 hits (64.41 per 1000 words): https://corpling.uis.georgetown.edu/annis/?id=2d1f9453-fc50-4fef-87ff-2e56fbfeba20
51. All loan words in Shenoute's *I See Your Eagerness* (double hits in parallel manuscript witnesses have been eliminated in the query syntax) excluding proper names = 581 hits (59.59 per 1000 words): https://corpling.uis.georgetown.edu/annis/?id=0eb858ce-8787-402d-bb0f-0ba5844ff6ac
52. All words in Shenoute's *I See Your Eagerness* (double hits in parallel manuscript witnesses have been eliminated in the query syntax) = 9750: https://corpling.uis.georgetown.edu/annis/?id=6f9a4c01-8e86-4493-a361-123eeb585e6d
53. "Writing"/"scriptures" (ⲅⲣⲁⲫⲏ) in Shenoute's *Not Because a Fox Barks:* https://corpling.uis.georgetown.edu/annis/?id=dff453ae-8697-4507-90e1-b6b60545cf1f

Part II: **Images**

Andrew Quintman and Kurtis R. Schaeffer
Synthesizing Image and Text in the Life of the Buddha

Introduction

The challenge of studying visual art, literature, and their institutional contexts in a synthetic fashion is acute throughout the humanities today. The Life of the Buddha (LOTB) project addresses this challenge by providing a digital platform for presenting and analyzing for the first time monumental Tibetan murals depicting the Buddha's life and their related literature within their architectural and historical settings: www.lifeofthebuddha.org. LOTB offers scholarly and learning communities the first tool to research and engage Buddhist images, texts, architecture, and history as an integrated and meaning-rich whole. The project's impact for the humanities and the study of Buddhism is thus twofold: it is the largest study to date on visual and textual Buddha narratives in Tibet, and it is also a new digital tool for synthetic teaching and research of Buddhist images and texts in context.

Such work reflects changing interests and research paradigms in the fields of Buddhist and Tibetan studies. In the past two decades there has been a "cultural turn" in Tibetan literary and historical studies, in which researchers are interested in the ramifications of text beyond the written page, as manifested in visual, material, and performance cultures. Where once scholarship in these areas was siloed into such sub-fields as art history, literary studies, ritual studies, institutional history, and so forth, advances in digital technology have both allowed and encouraged researchers to transcend these boundaries.

LOTB now serves as a major repository for materials pertaining to the Buddha's life as rendered in Tibet. The project's value further lies in its capacity to correlate key components of cultural production in complex ways that has remained a challenge. From the mid-nineteenth through the mid- to late-twentieth centuries, philology and literary scholarship formed the core of "Buddhist studies" as a field of enquiry. Other disciplines such as anthropology or art history treated Buddhist subjects, of course, though often did so in parallel to rather than in conjunction with textual studies. The turn toward visual and material cultures in recent decades has led to a more methodologically plural field of research, in which texts and image are considered equally important and mutually informing sources for understanding Buddhist cultural practices and products. Digital tools allow for increasingly complex forms of analysis across multiple media simultaneously. They do so by providing means for evaluating text and image side by side in a flexible fashion, and for collecting, structuring, and relating data across large collections of primary sources, regardless of media type. In a digital environment, text and image are drawn together within the same analytical and interpretive space.

LOTB is based on a literary and visual corpus produced at the monastery of Takden Puntsokling (Rtag brtan Phun tshogs gling), seat of the Jonang tradition of Tibetan Buddhism in the Tsang (Gtsang) region of the Tibet Autonomous Region. Created by the monastery's founder Tāranātha (1575–1634), these materials include an extraordinary set of narrative murals depicting the life of Śākyamuni Buddha. The murals date from the first decades of the seventeenth century and are among only a handful of fully preserved narrative paintings in Central Tibet. Practically nothing has been written about the Jonang murals, and no complete visual documentation of them has been undertaken prior to this project.

The LOTB project provides a multimodal framework for reading and analyzing the visual narratives of the Jonang murals in conjunction with their primary literary source, Tāranātha's extensive treatment of the Buddha's life story entitled the *Sun of Faith (Dad pa'i nyin byed)*, and related materials. To do so, we have adapted and extended a suite of tools in the Mirador viewer, a configurable and extensible environment for displaying and annotating images compliant with the International Image Interoperability Framework (IIIF), which has become the international standard for digital image archives.[1]

We chose Mirador over other possible platforms (including WordPress, StoryMapJS, and Media Thread) because of its flexibility and extensibility.[2] The LOTB team valued Mirador for its ability to link texts to specific points and regions in large images (even though other early projects employing Mirador did not exploit this potential), and its flexibility in handling multiple layers of annotations. Media Thread, a platform for collaborative multimedia presentation, annotation, and evaluation developed by Columbia University's Center for New Media Teaching and Learning, was a promising option early on in the project. We worked closely with the Media Thread team to develop an instance for working with LOTB materials. Although it provides a robust and intuitive work environment, it was unable to accommodate our large image files. In the end, Mirador provided three key features in a single product: (1) image delivery and navigation; (2) 2D freeform shapes employed as annotations upon an image; and (3) multiple

---

1 For an overview and history of Mirador see Mirador, http://projectmirador.org/. Further details on IIIF technical details and community resources at IIIF, http://iiif.io.
2 See StoryMap, https://storymap.knightlab.com and Mediathread http://mediathread.info.

hierarchical layers of text annotation. At the time of development, we were not able to locate another tool that integrated these features. Finally, Mirador had, and continues to have, an active international community of developers and users who were instrumental to our first technical team's ability to modify the tool's functionality for the project needs.

Background

LOTB is the first full-scale study of the life of the Buddha in a Tibetan setting. Scholarship on Buddhism throughout Asia has analyzed relationships between temple and cave murals, ritual and narrative literature, and the architectural contexts and institutional settings in which they were produced. Ajanta in India, Dunhuang in China, and Borobudur in Indonesia all have been the focus of extensive research.[3] The present case afforded a distinctive opportunity: the combination of a broad Tibetan literary corpus and a parallel set of detailed murals allows us to explore issues such as the planning and design of visual narratives, relationships between written and painted life stories, as well as other social, political, and economic perspectives on art and literature in place.[4]

LOTB brings to light what we have identified as a "Buddha Program" at Tāranātha's monastic seat of Puntsokling: a broad organizational principle consisting of a large body of Tāranātha's writings, including narrative, poetic, ritual, and technical painting literature about the Buddha, and its related religious iconography, statuary, and narrative artwork. The Buddha Program formed a central theme for the monastery, affording it a high degree of institutional prestige in the eyes of his patrons and their potential rivals.

The emphasis on Śākyamuni Buddha at Jonang was the result of competition among major monasteries in Central Tibet, where religious leaders had re-

---

[3] See, for example, Luis O. Gómez and Hiram W. Woodward, Jr., *Barabudur: History and Significance of a Buddhist Monument* (Berkeley: Univ. of California.Woodward, 1981); Dieter Schlingloff, *Ajantā: Handbuch der Malereien = Handbook of the Paintings* (Wiesbaden: Harrasowitz, 2000); Roderick Whitfield, *Caves of the Thousand Buddhas: Chinese Art from the Silk Route* (New York: George Braziller, 1990); Roderick Whitfield, Susan Whitfield, and Neville Agnew, *Cave Temples of Magao: Art and History on the Silk Road, Conservation and Cultural Heritage* (Los Angeles: Getty Conservation Institute and the J. Getty Museum, 2000).

[4] See Andrew Quintman and Kurtis R. Schaeffer, "The Life of the Buddha at Rtag brtan Phun tshogs gling Monastery in Text, Image, and Institution: A Preliminary Overview," *Zangxue xuekan /Journal of Tibetology* 13 (Summer 2016): 31–71; Andrew Quintman, "Putting the Buddha to Work: Śākyamuni in the Service of Monastic Identity," *Journal of the International Association of Buddhist Studies* 40 (2017), 111–156.

course to a number of mythoi around which to construct a symbolically rich institution. The Buddhist bodhisattvas, or celestial beings, such as Avalokiteśvara or Maitreya, were already in use by other leading figures in Central Tibet (Avalokiteśvara by the Dalai Lamas; Maitreya at Tashilhunpo Monastery). The choice of Śākyamuni, the buddha of our age, was an underutilized but undeniably authoritative figure across Tibet. As a guiding mythos, it thus made good strategic sense. On the local level, the life of the Buddha could be employed as a model for emulation in the education of the monastic population of Jonang, with textual resources available for the relatively small group of highly literate monks and visual resources available for the general monastic population and lay visitors. On a regional level, the broad and varied deployment of the figure of Śākyamuni—especially as the institution's principal Buddha icon of miraculous origins—distinguishes the monastery from its neighboring peers even as it suggests comparison with such famous Buddha images as the Jowo Śākyamuni statue in Lhasa.

The literary source for the visual narrative—Tāranātha's composition entitled the *Sun of Faith*, often referred to as the *Hundred Acts of the Buddha (Ston pa Shākya'i dbang po'i mdzad pa brgya pa)*—is one of the most extensive Tibetan treatments of the Buddha's life story. Tāranātha's narrative was innovative; he utilized little-known elements from the early literature of Buddhist monastic law known as the Vinaya, rather than standard and much later Mahāyāna sūtras favored by most other Tibetan writers. Jonang thus serves as a rare, and perhaps unique, example of an extant Buddha life mural in Tibet drawn from Vinaya sources. A related text, Tāranātha's *Painting Manual for the Hundred Acts of the Buddha (Mdzad pa brgya pa'i bris yig)* composed sometime between 1618 and 1620, serves to translate the text into images, presenting a frame-by-frame discussion of the iconography, composition, and symbolism in the Jonang murals. A further work, the *Guide to Jonang Monastery (Dga' ldan phun tshogs gling gyi gnas bshad)* (which is typically included in Tāranātha's collected works, but is likely not authored by him) together with other related texts, maps out a plan of the murals' architectural setting within the monastic complex including the assembly hall and upper gallery, and presents a catalogue of religious objects, statues, and other materials in situ at the time of their construction. Tāranātha's massive autobiography provides additional evidence for the historical, political and economic contexts in which these materials were produced.

Two major mural cycles dedicated to Śākyamuni Buddha are extant at Puntsokling Monastery: (1) the main shrine room and assembly hall, which houses images of the Buddha teaching the texts associated with Jonang's doctrinal traditions, select scenes from the life of Buddha, as well as illustrations of other narratives. These murals are approximately 10 feet in height and are dominated by 40 iconographically and stylistically distinct representations of the buddha; and (2) the upper gallery, which houses a complete life story of the Buddha. This continuous mural is approximately 5.5 feet in height and runs along all four walls for over 300 linear feet in fifteen discrete panels. (See Figures 1–2)

Figure 1: Location of the Life of the Buddha Murals.

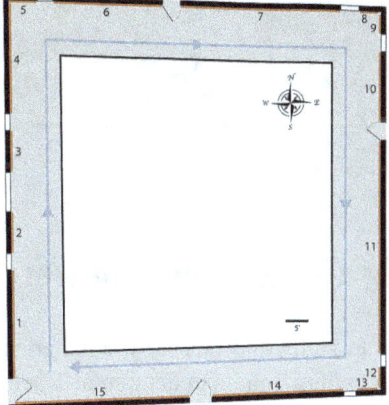

Figure 2: LOTB Mural Panel Locations.

The upper gallery is perhaps the more famous of the two, known by Tibetan authors and scholars of Tibet for its dynamic portrayal of narrative events through the juxtaposition of active human figures, diverse images of landscape and architectural settings, and judicious inscriptions that bind Tāranātha's longer literary renderings to the visual narrative. This gallery visually depicts each of the more than one hundred acts of the Buddha as described in Tāranātha's account. Each vignette, moreover, may consist of anywhere from one to ten individual human or divine figures, so that the mural contains hundreds of human figures in a dizzying array of vibrant poses. (See Figure 3)

Figure 3: Detail from Panel 1, Prince Siddhartha's Archery Contest.

The LOTB Website: Adapting the Mirador Viewer

How did we use and adapt web-based tools to study and present the materials described above? LOTB focuses on the upper gallery murals to document Tibet's most significant literary and visual materials depicting the life of Śākyamuni Buddha (though it could, and we hope will, extend to the lower gallery). At present the LOTB website presents all fifteen mural panels in full resolution, with detailed annotations, corresponding to the Buddha's birth, life in the palace, renunciation, enlightenment, and early teaching. This covers approximately 12% of the total visual and literary narrative content, but it forms a cohesive arc of the Buddha's life story, and includes many of the best-known episodes. Our thinking about how a digital framework can help meet our aspirations for synthetically engaging, analyzing, and interpreting visual and literary narratives thus begins with a manageable sample of the total corpus. We worked with Yale University's IT and Software Development staff, in conjunction with student research assistants, over several years (as described below) to produce a basic Drupal website to provide background information about the project and its materials, labor that was supported by our general grant funding. The website also serves as the portal to LOTB digital assets and the major platform at the project's core: the Mirador viewer. While Mirador has been used internationally for a number of archival projects, LOTB has adapted the tool to meet new needs.

The Mirador viewer was initially developed at Stanford University with support from the Andrew W. Mellon Foundation as a "configurable, extensible, and easy-to-integrate image viewer, which enables image annotation and comparison of images."[5] Because it complies with IIIF standards of digital asset creation and management, the viewer is able to draw from digital repositories anywhere in the world. As originally developed, Mirador presents several basic features. It allows for the viewing of an image, or multiple images, selected from a larger gallery. It provides a tool for smoothly zooming and panning across the selected images. It also allows users to create simple shapes for outlining regions of an image and link an associated text annotation. Prior to the LOTB project, Mirador's robust annotation features remained underutilized and we found no evidence that its freeform annotation functionality was used at all. LOTB thus represents the first large scale, public-facing digital project to significantly expand the application of Mirador's annotation features.

Mirador was originally adopted by university libraries and museums as a tool for documenting and showcasing digital surrogates of text manuscripts and artwork. For example, the Wolsey Manuscript Project at Christ Church and Magdalen College deployed Mirador "to enable and encourage academic study and research around the Wolsey Gospel and Epistle Lectionaries."[6] These two lavishly illustrated sixteenth-century manuscripts were originally produced as a set but are currently housed in the collections of separate institutions. Mirador

---

5 Mirador, http://projectmirador.org.
6 The Wolsey Manuscripts, http://www.wolseymanuscripts.ac.uk/about.

thus allows a user to view the two volumes in conjunction with one another, side by side, as originally conceived centuries ago.

A similar instance of Mirador is found in The Archaeology of Reading project, a collaboration by the libraries of Johns Hopkins University, University College London, and Princeton University, and the Principal Investigators Professors Lisa Jardine and Anthony Grafton.[7] The project aims to create a corpus of "important and representative annotated texts with searchable transcriptions and translations" in order to "compare and fully analyze early modern reading, and place that mass of research material within a broader historical context."[8] In its current state, the Archaeology of Reading project primarily uses Mirador as a platform for viewing digital surrogates of rare book materials. Its directors expect that basic annotations with transcriptions and translations of the primary materials will soon follow.

The SAT Taishōzō image database, led by Mr. Tetsuei Tsuda, chief curator in the National Research Institute for Cultural Properties in Tokyo, uses the Mirador viewer to tag and annotate iconography preserved in the image section of the Taishō Tripitaka, the multivolume Japanese Buddhist canon.[9] This image database, currently the first two of twelve volumes, allows users to click on predefined regions of an image, which results in popup windows containing relevant details, including iconographic identification and descriptions. Users can also perform basic keyword searches from a predefined lexicon of terms.

A more robust application of Mirador's text markup features has been deployed in the Ten Thousand Rooms Project developed at Yale University by Tina Lu and Mick Hunter.[10] The project "allows users to upload images of manuscript, print, inscriptional, and other sources and then organize projects around their transcription, translation, and/or annotation. Both as a workspace for crowdsourcing core textual research and as a publishing venue for scholarly contributions that are less well suited to conventional book formats, the Ten Thousand Rooms Project aims to establish an international online community committed to making the East Asian textual heritage more accessible to a wider audience."[11] Ten Thousand Rooms thus allows users to create and share their own collections of digital texts together with complex layers of associated data. The project required significant modifications to Mirador's user interface

---

7 The Archaeology of Reading, https://archaeologyofreading.org.
8 "What is the Archaeology of Reading," The Archaeology of Reading, http://archaeologyofreading.org/what-is-the-archaeology-of-reading/.
9 SAT Taishōzō Image DB, https://dzkimgs.l.u-tokyo.ac.jp/SATi/images.php.
10 "The Ten Thousand Rooms Project," Yale University, https://tenthousandrooms.yale.edu.
11 "The Ten Thousand Rooms Project," Yale University, http://tenthousandrooms.yale.edu.

in order to accommodate multiple annotation streams and facilitate the incorporation of East Asian languages.

Whereas the Ten Thousand Rooms project allows readers to transcribe, edit, translate, and annotate multiple small images in an "annotation window" (in the language of Mirador) arranged side-by-side with an image of a text according to user preference, LOTB required additional functionality that led to three main advancements: (1) the incorporation of large high-resolution images; (2) the creation of freeform annotation outlines on those images; and (3) a structured hierarchy of annotation layers that relate back to the image outlines and a table of contents. A guiding principle in the adaptation of Mirador was to make the technology frameworks as transparent as possible, allowing users to focus on content.

As discussed below in the section on Institutional Opportunities and Challenges, we developed early iterations of the LOTB project in the context of undergraduate classroom teaching with support from Yale's Instructional Technology Group. Together with a small team of technologists and student interns, we prepared basic visual and textual materials, and set up student work sites in WordPress and other digital platforms. Starting in 2015, a two-year collaborative research fellowship from the Robert H. N. Ho Family Foundation, who sponsored the project through the American Council of Learned Societies (ACLS), funded all major aspects of the project, including the development and adaptation of Mirador, digital image processing, database production and management, and website design. Two teams worked side by side, a team of technology experts and a team of research scholars in Tibetan and Buddhist Studies. The technical staff included a project manager; three software developers, one focused on Mirador, one on databases, and one on the Drupal site that serves as the portal to the Mirador program; a classroom technology consultant who helped with early WordPress iterations of the project and digital asset management; an image processing and graphic design specialist; and a user experience specialist. On the content side, a number of research assistants worked on the textual and visual materials: two Ph.Ds in art history and one graduate student worked on visual analysis of the mural; one Ph.D. student worked on Tibetan language transcription and translation; and one Ph.D. student researched the classical Indian literary background of the life of the Buddha narrative.

Unlike other Mirador instances, LOTB uses a small number of large high-resolution image files, each corresponding to one of fifteen mural panels that span approximately 7–30 linear feet (or 40–200 square feet) of painted surface. The two of us digitally photographed the mural panels in their entirety, producing approximately three thousand discrete RAW image files. (See Promey and Floyd's essay in this volume for some further methodological and ethical considerations about high-resolution site photography carried out in Peru.) We then employed a digital image specialist to stitch the images together and color correct the large files in Photoshop to replicate the mural panels in life-size detail. Significant analytical work was then necessary to identify elements in the mural panels that correspond to chapters in the *Sun of Faith*, as well as smaller narrative units within chapters we designate as "scenes."

A crucial innovation in the LOTB Mirador instance is the ability to create freeform visual markup as a way of demarcating chapters and scenes in the mural panels. Tāranātha divides his literary narrative of the Buddha's life into chapters (numbered 1–125) but not scenes. We therefore defined scenes as discrete instances of action, occurring in a unique time and place and located in a particular chapter of the story. While narrative units in the mural are often suggested by the placement of environmental features such as mountains, trees, and clouds, neither chapters nor scenes are explicitly noted.[12] Identifying and designating chapters and scenes therefore became a key task in our analysis of the murals. In our initial analysis, we used hand-drawn outlines superimposed upon printouts of mural sections to outline the scenes we defined. Mirador was originally built to allow only simple rectangular or circular outlines on an image. To our knowledge, prior to the LOTB project, Mirador did not allow users to define non-rectilineal spaces. Yet this was an indispensable feature for visually mapping out narrative elements that spread across the muralscape in irregular and disconnected ways. LOTB developers thus adapted Mirador to both portray non-rectilinear outlines upon the mural, and to allow content-creating users to draw freeform shapes within Mirador. What was initially a useful heuristic for "reading" the mural, namely drawing outlines around a given portion of a printed image in order to define a scene, is now available as a tool in Mirador. That being said, the drawing tool lacked the sensitivity to draw clean, clear, and elegant lines we required. As a workaround, we collaborated with a graphic design specialist to draw chapter and scene outlines in Adobe Illustrator. The technology team then imported outline coordinates into Mirador as machine-readable data capable of being linked to other annotation layers, such as the literary narrative, painting manual, and inscriptions.

The third innovation of the LOTB Mirador instance is the creation of a structured hierarchy of multiple layers of annotation data. This required an expansion of Mirador's capacity to portray discreet forms of data in a unified way. We began

---

[12] This is distinct from other examples of Tibetan narrative painting (both murals and portable hanging scrolls) in which discrete vignettes are inscribed with sequential numbers and occasionally descriptive text in order to facilitate reading.

by establishing a spreadsheet using Google Sheets for systematically documenting data and correspondences between types of data across multiple visual and textual resources including (1) mural panels, chapters, and scenes; (2) Tibetan and English versions of the *Sun of Faith*; (3) Tibetan and English versions of the *Painting Manual*; (4) Tibetan and English versions of the inscriptions; and (5) Buddhist canonical sources for the narratives. Structured correspondences are identified in the spreadsheet across all resources down to the level of sentences in the texts and scenes (or parts of scenes) in the murals. The spreadsheet now serves as the source for the content database behind the Mirador viewer. Chapter and scene outlines serve as annotation anchors, to which other annotation layers are linked. This allow users to interact with the various kinds of source materials as a seamless and integrated whole. A collapsible Table of Contents of chapter titles in English and Tibetan serves as a top-level organizing structure, allowing users to find and read both image and text in an ordered way, much like reading a book.

Reading the Life of the Buddha on LOTB

LOTB users can engage with the life of the Buddha in a variety of ways. First, they can approach the materials in a linear fashion, which emulates the traditional experience of reading the Buddha narrative. A brief walkthrough is the best way to understand how LOTB's Mirador instance works in this approach. Upon entering the Mirador viewer from the LOTB site, a "table of contents" appears screen-left, and the entire panel one mural appears screen-right. Clicking any chapter in the table of contents zooms the mural image to the first scene of the selected chapter, adds a non-rectilineal visual outline in the form of a yellow line around the scene in the center of the screen, and brings up our English translation of Tāranātha's *Sun of Faith*, which appears on screen-right (Figure 4). From there, the user can navigate the story through any of the three windows: table of contents, image window, or text window.

The table of contents is perhaps the easiest tool for navigating the story in a chronological fashion. Selecting a chapter title calls up the text and its associated part of the mural. As the user scrolls through subsequent scenes and chapters in the narrative text, the mural "follows along" as scenes zoom to the center of the image window. Selecting, dragging and scrolling the mural image, or navigating using the movement controls in the lower right of the image window allows one to explore the visual narrative more freely. As a user hovers the cursor over a given section of the mural, an outline appears around the scene, together with a

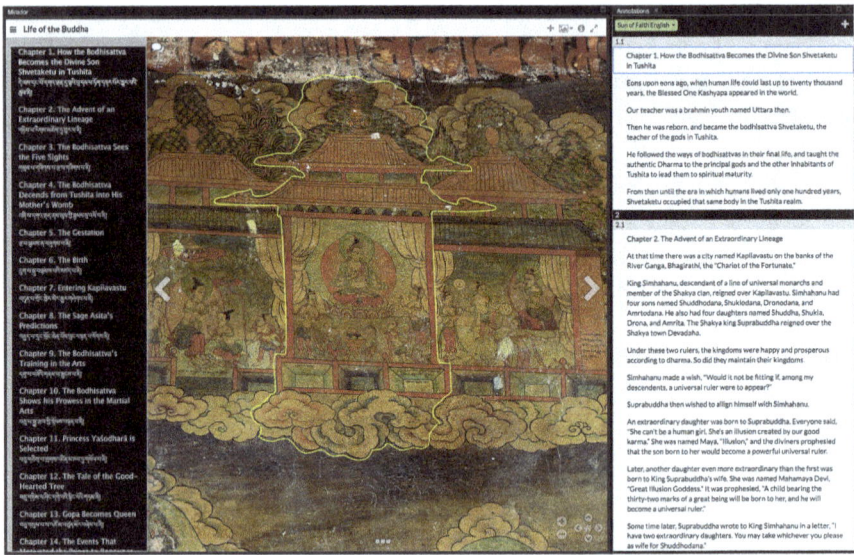

Figure 4: Chapter 1, Scene 1 (ie. 1.1) of the Life of the Buddha.

translucent pop-up box that names the outlined scene by chapter and scene numbers. (Figure 5)

Figure 5: Scene 1 (i. e., 1.1) of the Life of the Buddha, Image and Pop-up Box.

Users who want to read more than the translation of the *Sun of Faith* can click the "plus" (+) button in the upper right-hand corner of the open text window and the Tibetan-language text of the *Sun of Faith* will appear in an adjacent text window. Clicking the "plus" button again opens another text window, this time the English translation of Tāranātha's *Painting Manual*, again outlined

Synthesizing Image and Text in the Life of the Buddha — 109

and formatted in such a way that it synchronizes with the already open text windows so that one can read them all together. Clicking "plus" once more brings up the Tibetan text of the *Painting Manual*. (Figure 6) Seven text windows in total are available: *Sun of Faith* in English and Tibetan; *Painting Manual* in English and Tibetan; mural inscriptions in English and Tibetan; and canonical sources of the *Sun of Faith*. (Figure 7) Users may also select a particular text from the dropdown menu at the top of the window.

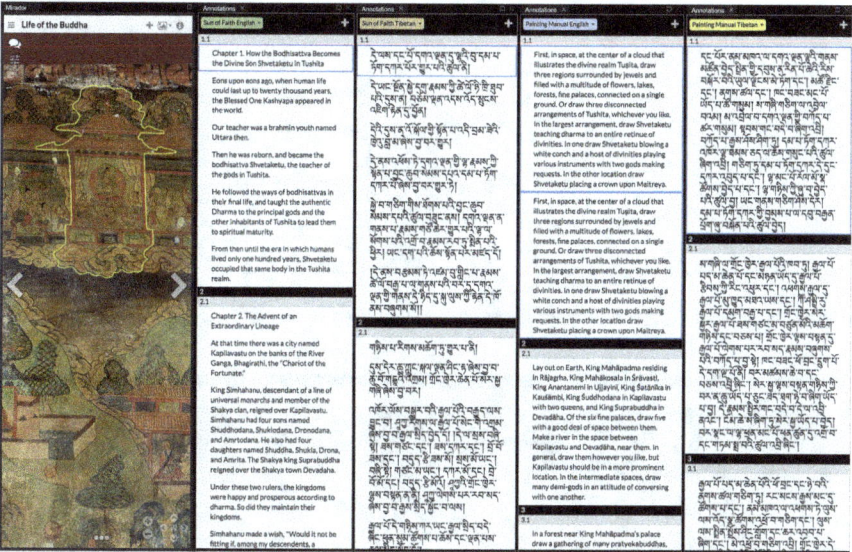

Figure 6: Chapter 1, Scene 1, with the *Sun of Faith* and the *Painting Manual* open in both English and Tibetan.

Mirador allows users to move and resize windows to suit specific needs and preferences, and LOTB takes full advantage of this to enhance user experience of the large mural panels and multiple text types. Users can, for instance, move text windows below the image window to maximize horizontal image space if the visual narrative calls for it (Figure 8).

Users can also prioritize the images over text and thus explore the life of the Buddha in a non-linear way. To do so, the user collapses the Table of Contents and removes the text windows entirely, which makes room for the largest possible view of the mural. The user can then pan and zoom across the visual space and as the cursor moves, it activates chapter and scene outlines and calls up links to the related text passages. (Figure 9).

Figure 7: Chapter 1, Scene 1, with all seven text windows open.

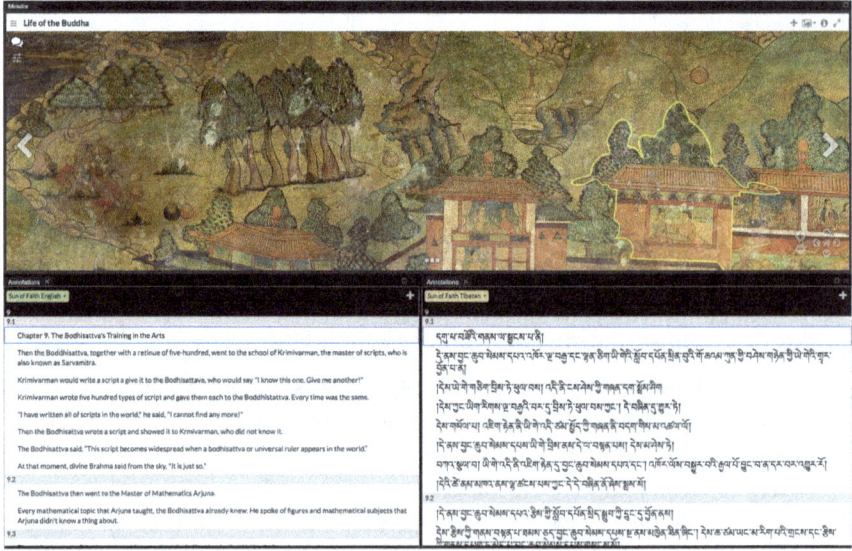

Figure 8: Chapter 9, Scene 1, with image window arranged horizontally above text windows.

Clicking on one of the links (a scene or chapter), recenters the image to that scene and calls up the associated text in an adjacent window.

Figure 9: Chapter 4, Scene 2, with image window maximized within Mirador.

As a final mode of engagement, users may search the entire LOTB corpus of materials for key terms such as proper names, places, and other text strings. From the homepage, the user enters a term into the search box and presses return. This results in a list of all instances in which the term appears, across all texts, together with thumbnail images of the associated mural section. It is then possible to filter the results by text source, panel, chapter, or scene.

Technical Opportunities and Challenges

The challenges we faced in working with a nascent open-source software package may be instructive for those considering similar options in a digital humanities (DH) project. Mirador is, in our estimation, a flexible, extensible application whose potential has yet to be fully tested. The Mirador viewer was designed primarily for images of relatively modest size. Early uses included delivery systems for those images, especially medieval manuscripts in various languages, and associated shallow metadata. Mirador's basic editing capability provided a convenient space in which to make notes, or "annotations," related to features of the manuscript page. For instance, one could easily transcribe the original text, add descriptive notes, or prepare side-by-side editions in the original language and in translation. Projects such as Ten Thousand Rooms make abundant use of these features by offering the reader a highlighted passage of Chinese text in an image of a manuscript page, and the ability to view electronically readable Chinese text side-by-side with English-language translation, references to commentarial literature, analytical notes, and so forth. In such cases, visual markup of the image is typically limited to rectangular outlines that bound a particular passage. However, we have found Mirador powerful enough to manage larger image files and navigate more complex annotation structures.

The large-scale problem we set before ourselves was to make engaging with the LOTB images and texts as intuitive and natural as reading a book. The book is an obvious metaphor for the technological means to present a narrative in which image and text sit side-by-side on the page. Book readers can move between image and text with the slightest movement of the eye. And depending on the useable area within a given page layout, secondary sources are easy to access. However, presenting large image and text collections using the book as a technological metaphor entails two primary challenges. The first is the challenge of designating precise relationships between a passage of text and a section of image without compromising the reader's engagement of either source on its own. Text superimposed on the image hinders visual appreciation and can make reading difficult. Text set on the margins of the image introduces ambiguity in the text-image relationship. Outlines superimposed on the image impedes the visual impact unless one prints the image twice.

The second challenge is the complete representation of a large and varied body of primary sources. In other words, if you have a large number of images (about 1,500 square feet of primary image assets in the case of LOTB), and many types of textual data to "line up" with them (nine types of text data for the LOTB project, including designations of chapters and scenes), then you will quickly exceed the capacity of all but the most elaborately and expensively produced book. Creating a system for synchronizing the various texts in two languages was further complicated by the differences in their order and organizational structure. The *Painting Manual* proved to be our most challenging source because it sharply diverges from the chapter and scene divisions of the Buddha's life story as presented in the *Sun of Faith*, which served as our benchmark for narrative sequencing.

While the LOTB website began with the book metaphor, we knew we needed to produce a dynamic book, one capable of connecting multiple layers of text with their many associated locations, all within the context of a series of massive images. We began with the basic Mirador viewer framework and gradually added modifications, including a table of contents, drop-down menus for choosing text types, the ability to hover over an image to reveal associated chapters and scenes in both text and image, and so forth. As the LOTB platform grew, we prioritized the creation of a seamless reading and viewing experience. This necessitated, among other things, changing the text window displays from using spreadsheet-like cells (as they appear in the basic Mirador instance) to a single continuous page of text with designated breaks between chapters and scenes. It also required maintaining visual links between parallel texts, such as the Tibetan and English versions of the *Sun of Faith*.

Developing Mirador with LOTB project goals in mind entailed significant programming effort. Our work was constrained by the state of Mirador's functionality at the time the project launched, and by the limits of our own capacity for software development in terms of financial and technological support. No one was more aware of the challenges these limitations presented than our team of technologists. After a period of preliminary planning and design work, technical development of LOTB progressed in three logical parts.[13]

The first aspect of technical development involved authoring our own application code (Yale-Mirador) for the HTML page that hosts the Mirador instance, which itself is pointed to from a descriptive Drupal website.[14] It included the creation of (1) text windows that can display annotations across different canvases (images), chapters, and scenes, and respond to chapter/scene selection by highlighting the relevant annotation; (2) a layout management system to coordinate the creation, removal, and relocation of image and annotation windows; (3) a tag-based annotation hierarchy system that organizes the annotations into chapters and scenes; and (4) a mechanism to translate annotation IDs contained in a search result into appropriate chapters and scenes and their corresponding locations on the canvas.

The second part entailed the creation of various plugin modules that communicate with Mirador, modify or override some of Mirador's default visual styles and behaviors, and provide additional functionality. These modules include (1) an override of CSS styles to safely embed the Mirador instance within the bigger HTML application, a necessity since Mirador was originally designed to take up the whole browser window space; (2) a side panel that can show the table of contents of the annotations; (3) a custom annotation editor that can attach an annotation layer to the appropriate annotation text; and (4) the ability to programmatically pan and zoom the image on view.

The third part required the rewriting of some elements of the Mirador code itself, creating a Mirador fork ("Mirador-y") that contains minimal modifications

---

13 The authors would like to thank Seong-June Kim for his assistance in writing this overview of LOTB's technical development.
14 Details about Yale-Mirador at "yale-web-technologies/yale-mirador," GitHub, https://github.com/yale-web-technologies/yale-mirador.

that could not be taken care of by plugins and overrides in the second part.[15] Modifications of the core Mirador code for LOTB were kept to a minimum in order to avoid creating a "brittle" software package as the program is updated.

Institutional Opportunities and Challenges: From Teaching to Research and Back Again

The process of building LOTB raised a number of practical questions early on. How could we leverage DH research and production for use in the classroom? What aspects of DH-inflected pedagogy could serve to support the core missions of academic research? LOTB served as a test case for the integration of research and teaching, a desirable outcome in theory but difficult to achieve in practice. This required us to navigate the complexities of university IT development and support, which often segregates research and teaching technologies. We found that staff, resources, and funding streams for research and teaching were frequently walled off from one another in a similar way. In-house university grants were typically designated for the support of either pedagogical or research development. Similarly, staff support for classroom and research applications typically came from different units of university IT. On several occasions, LOTB development and implementation was delayed due to reorganization of university IT departments and staffing transfers and turnover.

While we fully intended LOTB to serve as both a teaching and research tool, the project effectively began with the creation of a teaching platform, developed over several years. This decision was largely pragmatic, since modest institutional support and funding was most readily available for the creation of new digital resources to use in the classroom. First, we created a basic database of assets and some shallow metadata. Concurrently we created several presentations of the images and text in WordPress with support by Yale's Instructional Technology Group for use in the classroom. This allowed us to begin image processing, organization, and cataloguing, primarily because these materials were directly relevant to course instruction. Early support for the development of teaching resources also enabled us to create a digital asset management workflow for our image collection, at the time using in the Portfolio DAM system. Such preliminary work allowed us to begin analysing the visual narratives several years before the LOTB Mirador project was launched, even though some of the image process-

---

[15] See the technical description of Mirador-y at "yale-web-technologies/mirador-y," GitHub, https://github.com/yale-web-technologies/mirador-y.

ing needed to be repeated later on. In order to build up teaching resources for student work, classroom IT groups provided modest funding to hire graduate student research assistants, who in turn helped us undertake preliminary photo processing and stitching, visual analysis, and digital asset organization. The creation of digital resources for teaching also allowed us to test and evaluate a variety of existing digital platforms. With some technical assistance from our universities' classroom IT groups we devised class projects about the Buddha's life story by utilizing products such as WordPress, Zoomify, Google Slides, StoryMap JS, and Media Thread. These student projects in turn helped us test and evaluate these tools for possible use in LOTB, even though we eventually determined no existing "out of the box" platform would serve our needs. LOTB in the classroom thus helped build the foundation upon which the Mirador project would be established.

The scope of LOTB as a research project expanded significantly with the award of a two-year collaborative research fellowship and grant from the the Robert H. N. Ho Family Foundation and administered by the American Council of Learned Societies (ACLS). This additional support allowed the project to begin work in earnest with a team that included computer developers, a digital database manager, front end website and user experience specialists, and a digital image designer, as well as several graduate student research assistants who helped prepare both visual and textual materials and create a more robust Drupal project website. The grant also supported evaluation and testing of the Drupal and Mirador sites by an in-house team of User Interface (UI) and User Experience (UX) specialists.

By the conclusion of its initial two-year development period, we found the LOTB project and the Mirador website was already serving as a platform for successfully integrating teaching and research. The contents of the project, and the publications that have resulted from the related research, have now been successfully incorporated into the classroom, as instructional aids and as resources for individual and collaborative student projects. The Mirador framework has allowed us to make complex materials accessible to an undergraduate audience in a clear and manageable way. Classroom discussions and students' written work have likewise informed and augmented our ongoing research agenda. As the LOTB project moves forward, its additional resources and publications in turn serve to re-seed subsequent student projects. We have found a web-based research environment to be flexible enough to allow, and even to encourage, synergy between research, research reporting, and project-based learning for students at the graduate and undergraduate levels.

LOTB has the capacity to allow users to read the Buddha's life story across different kinds of media. It is also designed to facilitate interpretive analysis about the relationships between visual and literary expressions of the Buddha's life. (See the essay by Bielo and Vaugn in this volume on the pedagogical value of immersive environments.) It thus allows users to consider the forms and functions of Buddha life narratives by suggesting a series of interrelated questions. How are the murals arranged in the context of their architectural space? Who were the intended audiences for the text and murals, and how might they have been used or interpreted by those audiences? How might we understand the "visual logic" of the murals, or differences between visual and textual logics? Plans are currently in place to expand LOTB's functionality as a platform for collaborative research and learning. This expansion will allow users to upload personalized content (such as reading notes and observations), draw their own visual annotations (say, to highlight and identify all instances of serpent deities in the mural), save these materials, and then share them with a defined group of other users.

Conclusions

Once the primary sources were assembled and integrated within a digital platform for viewing and reading, the central challenge of the LOTB was working at the intersteces of computer science and the humanities to co-create a tool that enhanced the engagement with and study of a collection of rich cultural artifacts. As humanists we could imagine many types of tasks we might want a digital platform to perform. At one point, for example, we imagined that it would be productive to mark up every instance of a humanoid figure in the murals and then enter their coordinates and other metadata into a searchable database. This might afford researchers enhanced means to do fine-grained analysis of artistic style and technique, or to theorize how representations of bodily form or movement are related to certain narrative actions across the mural (Is the Buddha sitting during teachings? Is he standing? Walking? Flying? etc.). Yet such tasks were so far from the originally intended usage of our tool in theory, and so time-consuming to carry out in terms of data collection in practice, that they were simply impractical. Throughout the project we, as both humanists and project directors, found ourselves in conversation with the technology team about the gulf between what is imaginable, what is possible, and what is ultimately worth doing given constraints of time, funding, and technology. As one of our technology mentors once said of DH projects, "Almost anything is possible; few things are practical." So the collaborative work of defining and achieving the practical was, and remains, a challenge. And yet in the case of LOTB, the realistic establishment of limits was a profoundly productive exercise, one that led not only to the creation of the tools that make the life of the Buddha come alive in a new form, but also to their completion.

The LOTB Team 2015–2017

Many people worked on LOTB from 2015–2017. The project would not have happened at all without the generous support of the Robert H. N. Ho Family Foundation who sponsored the project through an ACLS Buddhist Studies collaborative research grant. This enabled us to assemble two teams working side by side: a team of technology experts and a team of research scholars in Tibetan and Buddhist Studies. The technical staff included: Michael Appleby (technology team manager 2015), Oren Kanner (project management 2017–2018), Seong-June Kim (software development), Roy Lechich (software development), Lec Maj (project management, 2016–2017), Pam Patterson (classroom technology, image processing, and database management), Mark Saba (image processing and graphic design), Harry Shyket (software development), and Heather White (user experience design). A number of research assistants worked on the textual and visual materials: Rebecca Bloom (visual analysis), Priyankar Chand (website design), Yong Cho (visual analysis), Rinchen Dorjé (text translation), Christopher Hiebert (canonical text analysis), Peter Jang (website design), Dr. Ariana Maki (mural analysis), and Dr. Sarah Richardson (mural analysis).

Development of the LOTB Mirador instance and associated digital frameworks now continues, with further support from the Robert H. N. Ho Family Foundation, and with the support of the University of Virginia's Institute for Advanced Technologies in the Humanities (IATH), including IATH Director Worthy Martin, Systems Administrator Shayne Brandon, and Scholarly and Technical Communications Officer Sarah Wells.

Selected References

Gómez, Luis O. and Hiram W. Woodward, Jr. *Barabudur: History and Significance of a Buddhist Monument.* Berkeley: University of California, 1981.
Quintman, Andrew. "Putting the Buddha to Work: Śākyamuni in the Service of Monastic Identity." *Journal of the International Association of Buddhist Studies* 40 (2017): 111–156.
Quintman, Andrew, and Kurtis R. Schaeffer. "The Life of the Buddha at Rtag brtan Phun tshogs gling Monastery in Text, Image, and Institution: A Preliminary Overview." *Zangxue xuekan /Journal of Tibetology* 13 (Summer 2016): 31–71.
Schlingloff, Dieter. *Ajantā: Handbuch der Malereien = Handbook of the Paintings.* Wiesbaden: Harrasowitz, 2000.

Whitfield, Roderick. *Caves of the Thousand Buddhas: Chinese Art from the Silk Route.* New York: George Braziller, 1990.

Whitfield, Roderick, Susan Whitfield, and Neville Agnew. *Cave Temples of Magao: Art and History on the Silk Road, Conservation and Cultural Heritage.* Los Angeles: Getty Conservation Institute and the J. Getty Museum, 2000.

James S. Bielo & Claire Vaughn
Materializing the Bible: A Digital Scholarship Project from the Anthropology of Religion

How do people transform written scriptures into experiential, choreographed environments? This question organizes the digital scholarship project, *Materializing the Bible* (*MB*), which is curated collaboratively by a faculty director and undergraduate research assistants. Launched in July 2015, *MB* is dedicated to understanding how Protestants, Catholics, Mormons, and Jews create places themed by biblical narratives. The project is multi-functional. It is curatorial, detailing nearly 500 sites around the world. It integrates ethnographic and archival data, fostering methodological transparency and exploring mixed-media representational strategies. Finally, it is designed for pedagogical use, providing resources for students in the anthropological and critical study of religion to engage questions of social theory, method, and analysis. In this chapter, the project founder (Bielo) and an undergraduate assistant who worked on the project for several years (Vaughn) examine the origins and development of *MB*, with a particular emphasis on how its theoretical grounding in material religion interacts with the affordances of a digital platform.

Global Phenomenon, Local Places

As a comparative project, *MB* emerged from an anthropological case study. From 2011–2016, I (Bielo) conducted ethnographic fieldwork on the cultural production of, and touristic engagement with, Ark Encounter: a creationist theme park in the U.S. state of Kentucky. The centerpiece of the ethnographic labor was two and a half years of fieldwork with the creative team who led the park's design. Most of this fieldwork took place in the team's design studio, sitting in cubicles with the artists as they created concept art, wrote exhibits scripts, and conceptualized how to choreograph visitor experiences. The centerpiece of their creative labor was a re-creation of Noah's ark, inspired by a literalist reading of Genesis as actual history. Opened to the public in July 2016, the re-created ark features more than 100,000 square feet of themed space. Onboard, displays use multi-sensory techniques to teach the creationist worldview.

When my ethnographic access to the creative team ended in June 2014, the research adapted in two ways. First, ethnographic attention was reframed from the team's creative process to the attraction's publicity strategies and creationist visitors' engagement with the park as religious tourists. Second, in collaboration with undergraduate research assistants, we asked how Ark Encounter fits among other kinds of sites that transform written scripture into experiential, choreographed environments. By contextualizing Ark Encounter in this way, we hoped to illustrate how something like a creationist theme park is not only revealing vis-à-vis the cultural and political ambitions of Protestant fundamentalism, but also with respect to a broader kind of Christian material practice and interdisciplinary dialogues about material religion[1]

After all, Ark Encounter is not the only token of its type. Selected sites have accrued significant attention. For example, The Holy Land Experience in Orlando, Florida has been variously interpreted as heritage entertainment detached from archaeological evidence;[2] as a spectacle past eliding the presence of Catholics, Orthodox Christians, and Muslims in Jerusalem in favor of a "purest apostolic form";[3] and, as a spatial enactment of Christian Zionism.[4] Scholars have also analyzed sites in comparative perspective. Burke Long's *Imagining the Holy Land* (2003) explores how biblical Palestine has been replicated numerous times throughout American history, typically remade in the theological and sociological image of its re-creators.[5] In *Sensational Devotion* (2013), Jill Stevenson explores how "evangelical dramaturgy" works to affectively instill core religious commitments at places like The Great Passion Play (Eureka Springs, Arkansas) and the Creation Museum.[6]

The most dedicated comparative analysis of biblically themed attractions is Timothy Beal's *Roadside Religion*.[7] Focusing on ten sites in the continental United States, Beal argues that these places are extensions of religious selves and

---

[1] Sally M. Promey, ed., *Sensational Religion: Sensory Cultures in Material Practice* (New Haven: Yale University Press, 2014).
[2] Y. Rowan, "Repacking the pilgrimage: Visiting the Holy Land in Orlando." In *Marketing Heritage: Archaeology and the consumption of the past*, ed. Y. Rowan and U. Baram (Walnut Creek, CA: Alta Mira Press, 2004), 249–266.
[3] Annabel Jane Wharton, *Selling Jerusalem: Relics, Replicas, Theme Parks,* (Chicago: University of Chicago Press. 2006).
[4] Ronald Lukens-Bull, and Mark Fafard, "Next Year in Orlando: (Re)creating Israel in Christian Zionism," *Journal of Religion and Society* 9 (2007): 1–20.
[5] Burke O. Long, *Imagining the Holy Land: Maps, Models, and Fantasy Travels* (Bloomington, IN: Indiana University Press, 2003).
[6] Jill Stevenson, *Sensational Devotion: Evangelical Performance in 21st Century America* (Ann Arbor, MI: University of Michigan Press, 2013).
[7] Timothy K. Beal, *Roadside Religion: In Search of the Sacred, the Strange, and the Substance of Faith* (Boston: Beacon, 2005).

ambitions. They exist as theological, biographical, and aesthetic imprints on local landscapes. Through their creations, designers and builders publicly display their stance toward scripture and their unique version of spirituality. Sara Patterson *Middle of Nowhere* a case study analysis of Salvation Mountain in the southern California desert.[8] The mountain is a several-story high, 100-yard long heap of adobe bricks (straw bale hay, water, clay), discarded car tires, and other scavenged and donated desert finds, all covered by thousands of gallons of paint. Leonard Knight began the project in 1984 after a born-again conversion experience, watched five years of work wash away in a massive rainstorm, and began again, working continuously until his death in 2014. A Sea of Galilee is the only direct biblical replication; the remainder materializes a singular biblical interpretation: "God is Love." Through his creative labor, Knight built a sacred space, an emplaced mirror of his theological conviction, and a popular destination for ecumenically-minded Christians, inter-faith activists, outsider artists, and other travelers seeking "weird" America.[9]

Scholars have expertly demonstrated how individual sites have fascinating histories and how idiosyncratic places can advance our understanding of key issues in the study of religion (from affect to embodiment, sacred space, and ritual creativity). What remains, and what we aim for with *MB*, is a genre-wide analysis, a comparative accounting of the material, ideological, and experiential forms that constitute and emerge from the seemingly disparate sites that materialize the Bible. As we demonstrate below, engaging this task on a digital platform opens valuable possibilities for analyzing and representing data.

Empirically, *MB* aims for both breadth and depth. For the former, the project has a global aspiration to identify and describe all sites that are arguably part of the phenomenon of materializing the Bible. With more than 250 sites, the United States hosts more than any other nation. However, we have also identified more than 230 extant sites in 43 other nations: Argentina, Australia, Austria, Bahamas, Bosnia, Brazil, Canada, China, Croatia, Czech Republic, Denmark, Egypt, England, Ethiopia, France, Germany, Hungary, India, Ireland, Israel, Italy, Japan, Kenya, Latvia, Lithuania, Malta, Mexico, New Zealand, the Netherlands, Northern Ireland, Pakistan, Philippines, Poland, Portugal, Russia, Scotland, Slovakia, South Africa, South Korea, Switzerland, Taiwan, Ukraine, and Wales.

Within this global range, the project differentiates among four kinds of attractions: re-creations, biblical gardens, creationist sites, and Bible history mu-

---

8 Suara Patterson, *Middle of Nowhere* (Santa Fe: University of New Mexico Press, 2016).
9 Greg Bishop, Joe Oesterie, and Mike Marinacci, *Weird California: Your Travel Guide to California's Local Legends and Best Kept Secrets* (New York: Sterling, 2009).

seums. The latter three are consistent aesthetically and discursively, while "re-creations" is more diverse and encompasses everything from miniature and "life-sized" replications of Holy Land sites to Passion Plays and other biblical theatre productions, "exact" models of various biblical references (e.g., Moses' Tabernacle), and others. Our hope is that users of *MB* will learn from this comparative arrangement, recognizing the diversity of media forms mobilized to materialize the Bible.

To complement this curatorial and analytical breadth, *MB* zooms in on specific attractions to explore how biblical texts are materialized in particular places by particular people with particular ambitions. Attractions draw from shared scripts (from archaeological data to Disney imagineering and ritual tradition), but they are always actualized in relation to local cultural contexts. A range of elements intersect to make and maintain each attraction, such as topography, interpretive and sensory annotations (e.g., onsite signage), props (e.g., guidebooks), architectural materiality, and the public circulation of images and sounds through social media and commodities. As we explore individual attractions we draw together these various elements through ethnographic and archival data. Joining ethnographic and archival materials is a key dynamic of *MB*, integrating thick descriptions of how sites are choreographed with their development over time. In this way, *MB* complements related digital projects, such as *MAVCOR*, which offers virtual immersion into religious attractions via 360-degree panorama photography (see Promey and Floyd, this volume).

In 2017, the American Academy of Religion published a set of guidelines for evaluating digital scholarship projects.[10] Based on their schema of "types or genres of scholarly digital work," *MB* integrates four types: archive; essay/exhibit/digital narrative; teaching resource; and gateway/clearinghouse. We engage these four types of digital work through multiple interactive features. (1) Each attraction is identified, briefly described, and hyperlinked to either an official webpage or web address with useful information. In addition to extant re-creations, biblical gardens, creationist sites, and Bible history museums, an additional page focuses on non-extant and proposed attractions. (2) A "Map" portal uses Google technology to display the attractions on a global geo-political map. Four colors—designating re-creations, biblical gardens, creationist sites, and Bible history museums—enable visitors to visually explore where attractions are located and clustered. (3) A "Tours" portal explores individual attractions,

---

[10] "AAR Guidelines for Evaluating Digital Scholarships," American Academy of Religion, rsn.aarweb.org/articles/draft-aar-guidelines-evaluating-digital-scholarship (accessed: July 25, 2019).

combining fieldwork photography and/or video; publicity materials (e.g., promotional videos); archival scans; hyperlinks to relevant data and/or scholarship; and, narrative description. (4) An index for a physical archive of project materials is maintained, including visitor materials and related religious ephemera dating from the early 1800s to the present. Selected archival materials are digitized for viewing and download, arranged by textual genre: attraction maps, brochures, guidebooks, postcards, and news stories. (5) Finally, a "Scholarship" section includes an interdisciplinary bibliography of scholarly publications that address specific attractions and discussion questions written for university courses.

Having outlined the project's origins, the empirical phenomenon it is dedicated to understanding, and the general scope of the digital curation, we can now examine how the project engages its orienting theoretical inspiration. The central question animating the following discussion is this: how do the project's analytical themes of material religion translate to a digital platform?

Project Argument and the Affordances of Digital Representation

The field of material religion emerged from a "media turn" in the study of religion, which recalibrates the center of gravity to the sensational forms and social processes of mediation.[11] The media turn has emphasized how religious actors use different kinds of materiality—from physical objects to technological apparatuses and the human sensorium—to construct and perform religious experience, learning, communication, and sociality.[12] The central theoretical conceit is that mediation is constitutive of religious worlds, not an incidental byproduct.[13] Religious actors use material forms to address the central problems that define their religious tradition(s), such as authority, belonging, and presence. The media turn puts to rest tired ideologies that resist, doubt, or deny the fact that religious life is fundamentally entangled in life's gritty and polished materialities.

In the case of materializing the Bible, human relations with scriptural texts are mediated by embodied experience, architecture, art, and the multi-sensory choreography of space. The attractions curated on *MB*—from creationist theme

---

11 Matthew Engelke, "Religion and the media turn: a review essay," *American Ethnologist* 37, no. 2 (2010): 371–379.
12 S. Brent Plate, ed., *Key Terms in Material Religion* (London: Bloomsbury, 2015).
13 Promey, ed., *Sensational Religion*.

parks to biblical gardens—make a promise to visitors. They promise that the power and meaning of sacred texts will be revealed or rediscovered. Visitors are invited to "experience," "engage," "interact with," "see," and "step into" the Bible. Through this physical encounter, attractions seek to persuade visitors that they should develop an affective intimacy with scripture.

The invocation of "affect" is pivotal. As a conceptual apparatus, affect is valuable because it draws our attention to the entanglements that develop among structural forces, sociality, ideology, materiality, and subjectivity.[14] This project begins with the premise that "the Bible" as a cultural category is not reducible to a printed text that people read, interpret, memorize, and discursively circulate. "The Bible" has historically been performed through a wide range of experiential registers: from stained glass and other artistic media to film, video games, and toy objects. We have the story of Noah's ark in Genesis 6–9, and we have (among many others) Edward Hicks' widely reprinted 1846 painting, the 1928 romantic melodrama film *Noah's Ark*, the 1991 *Bible Adventures* game for Nintendo, craft and mass-produced wooden playsets designed throughout the 19th and 20th centuries, and now a young earth creationist "theme park" centered on a "life-sized" replica of Noah's ark.

These transmedial performances have multiple functions—religious pedagogy, devotion, fun, and evangelism—but their capacity to be efficacious in any function is grounded by an affective relation. Any cognitive knowledge that religious actors develop about scripture is anchored by the development of intense bodily and emotional bonds with scripture. While these affective bonds can certainly support an authoritative view of scripture, they also engage religious actors in an ongoing authorizing process where the aura of scripture is internalized. The ambition is to experience scripture from as many angles as possible, in as many sensory configurations as possible, because just as "the Bible" is inexhaustible for readers it is also experientially inexhaustible.

Attractions like museums, gardens, and theme parks testify to the power of sensory affect. The arrangement of bodily experiences in choreographed space registers effects on and through the sensations of visitors. Ark Encounter, for example, is not merely about teaching creationist biblical history, it is about getting caught up in the multi-sensory presentation of that history. This argument echoes historian Vanessa Agnew's depiction of re-enactment as a form of "affective history," in which the past is imagined through the "physical and psychological

---

14 Jenna Supp-Montgomerie, "Affect and the Study of Religion," *Religion Compass* 9, no. 10 (2015): 335–345.

experience" of individuals.¹⁵ This mode of performing history aspires to provoke the body, for the body to respond in ways that may or may not have the consent of language or cognition. For example, Agnew (2019) describes the sensorial ambition of "gooseflesh," which affective history seeks to achieve as an involuntary aesthetic evaluation. Contrary to the fact that bodies are encultured to respond to particular stimuli in particular ways, gooseflesh promises the consumer of affective history that they are experiencing something timeless and unmediated.

Sites that materialize the Bible construct and elaborate this kind of affective intimacy with scripture through choreographed place. Before considering some examples from our curation, we review below some representational affordances of working within digital platforms.

Digital Affordances

Compared with print-based scholarship, digital scholarship entails authorial and user affordances that open new analytical and representational possibilities.[16] Consider three affordances that are particularly germane to the work of *MB*.

First, digital platforms are interactive, enabling users to navigate content in ways that are personalized, self-paced, and self-directed. This is expressed through a variety of features, from search functions to hyperlinks, downloadable files, and data visualizations (e. g., maps, timelines, animated graphics). Interactivity can also take the form of participatory media, as users respond to and circulate content across other digital platforms (e. g., Twitter). Functions such as liking, sharing, commenting, and (re)posting enable users to integrate a digital project into their online networks.

Second, digital platforms are multi-modal. In addition to written text, they can host audio, video with sound, digitized scans of written texts and 3-D objects, Virtual Reality simulation, and a wide variety of images (from basic photographs to 360 degree panoramas). Multi-modality enables project creators to present analyses in multi-sensory ways, and enables users to have an immersive experience of data and argument. Digital projects have massive storage capacities, which allows tremendous amounts of data to be integrated. The significance

---

15 Vanessa Agnew, "History's Affective Turn: Historical Reenactment and Its Work in the Present," *Rethinking History* 11, no. 2 (2007): 301.
16 Christopher D. Cantwell, and Hussein Rashid, *Religion, Media, and the Digital Turn: A Report for the Religion and the Public Sphere Program* (New York: Social Science Research Council, 2015).

of this should not be underestimated, particularly in the context of print publishers struggling to remain financially viable due to production costs.

Third, digital platforms are open ended, meaning that digital scholarship projects can function in a condition of "perpetual beta."[17] Existing content can continually be edited or removed, new content can continually be added. These adjustments can happen in response to critical peer review, ongoing data collection and analysis, the emergent nature of knowledge production, and innovations in digital technology. Moreover, content and format changes can be published with little lag, often in real time as a project's analysis advances. This open-ended capacity aligns with the function of data, resource, and knowledge aggregation, in which digital projects operate as curated portals for entry into specialized subject areas.

With these broad observations identified, we now pivot to this chapter's primary question: how have the affordances of working on a digital platform shaped *MB*'s curatorial, analytical, and representational work?

Connecting Ethnographic and Archival Data

An enduring methodological interest in both anthropology and religious studies is how to meaningfully integrate two forms of qualitative data: ethnographic and archival. How can we historicize the densely textured, localized nature of ethnographic data? And, how can we illustrate the contemporary resonance of historical phenomena? Digital platforms enable scholars to embed data into multimodal representations, which creates valuable opportunities for accentuating the interplay of past and present, ethnographic and archival. This embedding of data exceeds merely highlighting illustrative examples, it alters the reading/viewing/listening experience by heightening the transparency of data collection and analysis. For example, consider *MB*'s virtual tour of an attraction in northern Kentucky.

Covington is a small city in northern Kentucky, just across the Ohio River from Cincinnati. On its southwestern edge—tucked away on the backside of a residential section, at the end of a No Outlet road, set atop a steep rise that affords a gorgeously unobstructed view of the Cincinnati skyline—is a place called the Garden of Hope (see Figure 1).

The Garden opened to the public on Palm Sunday 1958, consummating the nearly 20-year vision of a Southern Baptist minister, Reverend Morris Coers. After

---

17 Ibid., 17.

Figure 1: Scene from the Garden of Hope (Covington, Kentucky). Photo by James S. Bielo.

three pastorates in Indiana, Illinois, and Michigan, and two terms in the Indiana state legislature, Coers accepted the senior pastor position at a Covington Baptist church in 1945. Coers became well-known in the city as the host of local radio and television ministries, but his ambition exceeded pulpits, politics, and mass media preaching. In 1938, he had made a pilgrimage to Jerusalem and was inspired to create a place that would be a beacon for all who could never make the trip themselves. In a 1956 interview with the *Cincinnati Times-Star*, Coers is quoted: "we expect to attract thousands of tourists from all parts of the world—persons who will never have the privilege of walking in the Holy Land." After four years of fundraising, he purchased the 2.5-acre hilltop plot.

The Garden of Hope's centerpiece is a 1:1 scale replica of Jerusalem's Garden Tomb. As a biblical site, the Garden Tomb is contested. Many Protestants (especially evangelicals and fundamentalists) claim that the tomb, located outside Jerusalem's Old City walls, is where Jesus was buried and resurrected. However, this claim is rejected by mainstream biblical archaeology and dismissed by the majority of the world's Christians who claim that the Church of the Holy Sepulcher inside the Old City houses both Golgotha and the burial-resurrection site.[18] Despite its controversial status, the Garden Tomb has come to be more satisfying for Protestant pilgrim-tourists who favor the open-air garden feel over the Old City Church, which is busy with Catholic and Eastern Orthodox ritual elements.[19]

---

18 Ruth Kark, and Seth J. Frantzman, "The Protestant Garden Tomb in Jerusalem, Englishwomen, and a land transaction in late Ottomon Palestine," *Palestine Exploration Quarterly* 142, no. 3 (2010): 199–216.

19 Amos S. Ron, and Jackie Feldman, "From spots to themed sites—the evolution of the Protestant Holy Land," *Journal of Heritage Tourism* 4, no. 3 (2009): 201–216; Hillary Kaell, *Walking Where Jesus Walked: American Christians and Holy Land Pilgrimage* (New York: NYU Press,

As a thoroughly Protestant attraction, Kentucky's Garden of Hope elides all traces of this disputed historicity. Neither the Church of the Holy Sepulchre nor archaeological arguments against the Garden Tomb's veracity are mentioned in Coers' late 1950s newspaper interviews, in the written text of onsite signage or visitor pamphlets, or in tour guide performances. The Garden of Hope's appeal to authenticity happens through this process of erasure, and through the immersive ambition of creating a surrogate experience of biblical land in Kentucky.

In addition to his own Holy Land pilgrimage, Coers' claim for the Garden of Hope's "exact" replication was his relationship with the Garden Tomb's warden. From 1953 until his untimely death in 1967, Solomon Mattar helped care for the Tomb in Jerusalem and guided pilgrim-tourists. Coers' noted in a December 1956 interview with the *Cincinnati Enquirer*, "I have been collaborating long distance with the warden of the garden in Jerusalem. He arranged for an architect to draw exact plans of the tomb of Christ. I have arranged to have these plans followed in the minutest detail." To accompany this architectural precision, Coers planted numerous botanical species indigenous to Jerusalem throughout the attraction. By mobilizing the sight, feel, and aroma of "cedars of Lebanon," juniper, and different flowers the Garden of Hope appealed to the natural world of scripture to help transport visitors.

Today, without Coers' devotional labor, few of these species remain and the botanical aspect of the experience is lost from his original vision. However, another claim to authenticity via the natural landscape continues. Scattered throughout the attraction are four stones, each associated with a biblical story and originating from the corresponding location in Israel-Palestine. For example, one of the stones is located inside the Garden's chapel, a replica of a 16th century Spanish mission church. Along with worship services, the chapel was primarily intended for weddings, a practice that Coers initiated in 1958 and that continues today. During the ceremony couples stand atop a pink stone embedded into the floor, which signage explains "is from the Horns of Hatton where Jesus delivered the Sermon on the Mount." The other stones, as signage onsite explains, come from the Western Wall, the Jordan River, and the Samaritan Inn.

Reverend Morris Coers realized his 20-year vision, but he enjoyed it for only two years. He died in his sleep in late February 1960 after multiple health difficulties. With Coers' passing, the Garden fell into a lengthy cycle of disrepair, sale, purchase, repair, disrepair, sale, and so on. Following a major restoration effort to re-open the Garden in 1998, a volunteer tour guide was hired who still guides

---

2014); Jackie Feldman, *A Jewish Guide in the Holy Land: How Christian Pilgrims Made Me Israeli* (Bloomington: Indiana University Press, 2016).

visitors today. Steve, a 70-year old Covington native, has "shown it to over eight thousand people" since he started in 2003. In summer 2016, the site entered a new, and newly popular, phase of its life. The opening of Ark Encounter, located 40 miles south of Covington, has sparked the organization of bus tours to the area. For example, the company Ohio Travel Treasures arranges for groups to take an Ohio River cruise featuring gospel music, visit the Ark, the Creation Museum (another creationist ministry open since 2007, located 20 miles west of Covington), the Cathedral Basilica of the Assumption (a replica of Paris' Notre Dame Cathedral) in downtown Covington, and the Garden of Hope.

Drawing together ethnographic and archival materials, we created a virtual tour of the Garden of Hope in July 2017. The tour integrates fieldwork images, video, and audio; narrative description; a 3-D model of one Holy Land stone; and, digitized scans of local newspaper stories and other archival texts. The aim is to connect the attraction's history with the contemporary guided tour, and ultimately offer visitors to *MB* a virtual glimpse of the embodied experience of being onsite.

To tell the story of the attraction's origin and development, we use three forms of archival data. First, courtesy of the Cincinnati Public Library's digitized collections, we integrate scans and downloadable copies of local newspaper stories reporting on the Garden of Hope. Second, courtesy of the Garden of Hope's own informal collection of materials donated by the Coers family, we integrate archival data from the process of building the Garden. This features a January 1958 calendar, with each day including Coers' handwritten notes detailing the construction's progress. Third, we acquired post cards (c.1965) produced for the attraction through eBay. We juxtapose these post card images with contemporary fieldwork images to help conjure a sense for how the appearance of the attraction has changed over 60+ years. Along with fieldwork photography that alternates between close-up and mid-range images, we integrate video recording clips that capture key moments from Steve's guiding performance.

Because the virtual tour is designed on an open-ended platform, we continue to adjust content over time. For example, throughout spring 2018 I (Vaughn) led the process of adding a 3-D model to the virtual tour. Using photogrammetry techniques, I started with one of the smallest objects onsite: a display of stones from the Samaritan Inn referenced in the Parable of the Good Samaritan (Luke 10:25–37). Starting from the bottom of the display, I photographed in levels circling the object, until reaching the top. This required about 120 photos. From here, I imported the photographs into Agisoft, a stand-alone software product that performs photogrammetric processing of digital images and generates 3D spatial data. I then grouped the images by level. Third, I aligned the images and built dense cloud point, mesh, and texture. The latter steps require not much more than clicking and waiting roughly one hour per step. Finally, I exported the model from Agisoft into Meshlab, an open source software which allows for viewing and manipulating 3-D spatial data. However, this first rendering encountered several problems (see Figure 2).

Figure 2: Initial Meshlab rendering of display; Samaritan stone display at Garden of Hope. Screen shot and Photo by Claire Vaughn.

For starters, I could not export the file as a 3-D model. I was only able to export the .obj file. I had the .jpg files, the uploaded photographic images, but I could not merge the two together. After exploring the program further, I realized that I needed all three files: .obj, .jpg, and .mtl. The .mtl is the file, which joins the .jpg files to the .obj. In other words, having all three together enables color and texture to be added to the 3-D model. The rendering problems owed to the lack of mapped spatial data, creating the empty spaces where the software could not recognize all of the photographic data points.

On a second attempt, I restarted the process with one of the Garden's other Holy Land stones: the 500-pound block from the Western Wall. I recorded the same number of pictures, about 120, but they were more spread out compared to the first display. They covered more of the stone's surface area and were also more precise in terms of my distance from the display. I repeated the steps in Agisoft, though I took more time grouping the photos so that they would align more precisely, thus creating more representative spatial data. After four hours of clicking and waiting, the 3-D model rendered; this time, with more data points. When I exported the model from Agisoft, the program produced three files in the same folder: .obj, jpg, and .mtl. I then imported those files into MeshLab, which allowed me to view and manipulate the model from all sides as well as zoom in and out. The second model is dramatically improved: more complete, more data points mapped, with color and texture added. In turn, this 3-D replication appears rather realistic and much closer to the stone display at the Garden of Hope compared to the first rendering (see Figure 3).

Figure 3: Second Meshlab rendering of display; Western Wall stone display at Garden of Hope. Screen shot by Claire Vaughn; Photo by James S. Bielo.

Our goal is to repeat this 3-D modeling process for each of the components at the Garden of Hope, ultimately joining them to build a cohesive 3-D model for the entire attraction. MeshLab includes a measurement tool that allows point-to-point measurement, which means that this whole attraction model would be to scale. MeshLab also allows the inclusion of textual and audio-visual annotations, which means that we are able to layer in various forms of ethnographic material (e. g., images of visitor interaction with the site; videos and transcripts of tour guide performances at different site locations).

From archival scans to fieldwork video and 3-D models, these data work as embedded material, meaning that users of the digital project can either capture or download them for their own analysis. Rather than simply following (and trusting!) our selected narrative representation of the Garden of Hope, *MB* users are able to interact with the primary sources. By embedding data in this way, there is increased transparency of both our decisions of research methodology and analytical decisions of interpreting archival and ethnographic data. Users can read the original news stories, select from a wide range of archival and fieldwork images, view/listen to unedited fieldwork video, and explore a 3-D object replica. Ultimately, embedding data also responds to some brute material realities of publishing research in print media. Whether the venue is a journal or a book, there are limitations that structure how representations can be composed. Word counts are enforced. The number of allowed images is monitored. Reproducing color images is monitored even more closely. Multi-modal, interactive material must be rendered in the two-dimensional form of written text accented with limited imagery. As an author, being able to share your data as embedded content in addition to your own analysis and representation opens new possibilities for contributing to comparative scholarship. As Promey and Floyd (this volume) discuss, such digital affordances offer particular potency for studies of material religion. If religious worlds are constituted through the media of bodies, technologies, and objects, then being able to share more of those media in a wider range of formats advances both methodological rigor and theory-building.

Embodied Presence

The interdisciplinary field of material religion is interested in how multiple media can create and re-create affectively dense cultural dispositions (e. g., identity, belonging, commitment). As a result, the field has focused intensely on how humans engage their religious worlds with a full-bodied presence. The concern with multi-sensory experience aims, in part, to counter historical ideologies that reproduce sensory hierarchies that elevate a single sensation—such as vision or sound. As David Morgan notes in his analysis of religious visual culture: "seeing is not disembodied or immaterial and vision should not be isolated from other forms of sensation and the social life of feeling."[20] In turn, material religion scholars insist on a model of integrated sensation, in which every sense always exists in dynamic relation with other senses as well as other materialities such as places and technologies.

Embodied presence is a foundational premise for both the empirical phenomenon of materializing the Bible and for *MB* as a digital scholarship project. Creators of, and visitors to, these attractions do not merely visually consume the sights of biblical replication. They walk through them. They experience them as ritual spaces where the outside world may still intrude, but where choreographed experiences of visual fields, sounds, smells, and felt textures are central to the experience. The effectiveness of these attractions as devotional, pedagogical, and enjoyable spaces is decided in part by how they immerse people as embodied visitors into a biblically themed world.

---

20 David Morgan, *The Embodied Eye: Religious Visual Culture and the Social Life of Feeling* (Berkeley: University of California Press, 2012), xvii.

Attractions that materialize the Bible situate visitors in physical space, function via an embodied encounter with that space, and require visitors to move through space. This movement is often guided narratively, in which physical progression aligns with progression through a theological or historical account. Digital techniques such as 3-D models, simulations, and audio-visual recordings of movement are strategies for representing emplaced experience. An additional technique *MB* uses is to collect the tourist artifact of "park maps," and allow visitors to comparatively explore attractions through this familiar textual device. Park maps are especially useful as a kind of visual technology that contributes to the embodied choreography of attractions.

For example, one kind of attraction replicates the Way of the Cross (or, Stations of the Cross), a mobile ritual that follows Jesus' path on the Via Dolorosa in Jerusalem, from condemnation by Pontius Pilate to Resurrection.[21] The modern form of the Stations as a devotional ritual was codified by Catholic Holy Land pilgrims in the 15th century, and replications began soon after in sites such as northwestern Italy's Sacred Mount of Varallo in 1486.[22] While it is exceedingly common for Catholic, Anglican, and other denominational church buildings to re-create the Stations as small art installations, *MB* focuses on outdoor replicas that feature life-sized three-dimensional statues arranged on a prayer path (see Figure 4). None of these attractions seek to re-create the Via Dolorosa as a physical place, and they vary widely in how elaborately the individual stations are designed and in how they make other claims to authenticity. For example, in 1936 a Catholic parish in southeastern Michigan completed an outdoor Stations and today entices visitors with this description: "The hill upon which [the Shrine of] St. Joseph stands very closely approximates in size and contour the Mount of Calvary. Historically, the distance from the crucifixion to the tomb is only 65 paces, the exact distance in the Irish Hills Way of the Cross."[23]

Among the park maps we have collected on *MB*, several feature Stations replications. Users can examine the maps to compare how this walking prayer ritual is spatially choreographed across attractions. For example, they can question how the walking paths are shaped; the respective distances between individual

---

[21] The number and order of the Stations of the Cross can vary slightly. Perhaps the most common variation is whether the 14th station (burial in the Tomb) is the final stop or whether a 15th station (Resurrection) is added.

[22] William Hood, "The *Sacro Monte* of Varallo: Renaissance Art and Popular Religion," In *Monasticism and the Arts*, ed. Timothy Gregory Verdon (Syracuse: Syracuse University Press, 1984), 291–311.

[23] Source: https://sites.google.com/site/stjosephbrooklynmi/stations-of-the-cross (accessed: July 25, 2019)

Figure 4: Example of outdoor, "life-sized" mobile Stations of the Cross attraction: The Shrine of Christ's Passion (St. John, Indiana). Photo by James S. Bielo.

stations; the landscape placement of the Stations on attraction grounds and local topography; and how devotional elements are integrated. Altogether, the park maps help reveal how the emplaced experience of movement through space, physically and narratively, works within this kind of attraction.

Another aspect of embodied presence that *MB* has sought to re-create is auditory visuality; that is, the ways in which sonic experience shapes visual experience. This is especially valuable for attractions where the soundscape is choreographed for visitors. Consider Ark Encounter. If this attraction is designed to teach the creationist worldview, then sound is a key sensorial resource designers have mobilized toward this ideological end.

As visitors approach the queue line to enter the re-created ark, they first encounter the park's baseline instrumental soundtrack. Composed uniquely for Ark Encounter, and written to connote a "Middle Eastern" musical aesthetic, the soundtrack runs on a constant loop throughout the three decks. Onboard, as you walk through and among selected exhibits, this baseline soundtrack is replaced by auditory annotations that stream down from overhead speakers located inside exhibits.

This auditory shift happens immediately upon entering Deck One. You are surrounded by animal cages stacked one on top another, and the narrow walkway turns sharply to wind among the cages. One soundtrack, playing at a lower volume but directed nearer to visitors, features an indiscernible mix of animal sounds. They are lively, even a bit unhinged by the storm. Then, there is the second soundtrack, playing louder but projecting from a further distance: a loud, unnerving mix of booming thunder, cracking lightning, and pouring rain.

Other auditory annotations range widely. On Deck Three, one of Noah's daughters-in-law works contentedly in a replication of the family's kitchen.

The orchestration streams down from overhead: she hums peacefully, even delightedly, while a knife chops vegetables methodically and scrapes them to the side. On Deck Two, the "Pre-Flood World" exhibit is a winding walkway that moves through five spaces: creation, Garden of Eden, the Fall, "Descent into Darkness" (i.e., the extravagant sinfulness of the generations preceding Noah), and the Flood. These spaces and themes are performed through a series of elaborate dioramas and colorful murals, framed along the way by minimal textual framing and scriptural citations. The striking visuality of the dioramas and murals is annotated aurally. Amid the "Descent into Darkness" section, a large wall mural depicts a dramatic scene of "pagan" ritual in which humans are sacrificed to appease a "false god." And, a diorama depicts a cheering stadium crowd as they gaze down upon humans fighting with each other, an armored Nephilim, and an attacking dinosaur. Overhead, a cacophonous arrangement plays on a two-minute loop: raging fires, a duel of colliding and sliding swords, raucous crowds, and human screams. There is little interpretive space to imagine anything except for death-by-combat staged for a blood-thirsty public.

Perhaps the most indelible auditory imprint in my experience of Ark Encounter comes from the "Fairy Tale Ark" exhibit on Deck Two. Approaching the exhibit, your visual field is immediately drawn upwards to a series of animals lining the top of the entrance. They are certainly cartoon-ish, but they somehow exceed that description. In my initial field notes I (Bielo) described them as looking "zany, even slightly imbalanced or crazed," signified by their eyes, facial expressions, and jumbled arrangement. The longer I stared at them, the more an unsettled affect sank in (see Figure 5).

Figure 5: Entrance to Fairy Tale Ark at Ark Encounter. Photo by James S. Bielo.

Unlike some other exhibits, where a wooden rail bars visitors from entering the space, you must step into the Fairy Tale Ark. Once inside the small room, there are two dominating features. The smaller of the two and positioned on the wall to your left, is a snake-encircled sign. It reads ominously, dialogically voiced as Satan himself: "If I can convince you that the Flood was not real then I can convince you that Heaven and Hell are not real." The primary display is positioned directly in front of you, covering the entire wall. It is a collection of nearly 100 Noah's ark themed books written for children. Most are in English, though a few are written in Spanish and French. They are arranged neatly on six rows, interspersed with other ark-themed kids' toys and games, housed directly behind glass panes.

Two textual annotations frame visitors' reading of the collection. First, three rows up from the bottom, a series of small books are lined across side-by-side. The features of their book form suggest an antique collection of fables, and they present the "7 D's of Deception." For example, "Destructive For All Ages" explains: "The cute fairy tale arks are not only marketed to children, thousands of items featuring whimsical arks have been made for adults too. The abundance of these fanciful objects attacks the truthfulness of Scripture." At the center of the display is a larger, again fable-looking book, that is voiced in a rhyming, fairy tale register. It begins: "Once Upon a Time, there was an old man of god. His name was Noah and his task was quite odd. One day, the Lord said to build a little boat, 'Make it nice and cute, but who cares if it will float ... '"

The cartoonish, fable-ish, and simultaneously playful-ominous aesthetic teaches a singular lesson. A literalist reading of Genesis—complete with an actual Flood, actual ark, and actual Noah and family—is lampooned every day by the ubiquitous circulation of "fairy tale" arks. This lampooning is no accident, but who is to blame? The snake-encircled sign suggests devilish agency. The bounds of responsibility widen in the text of "Discrediting the Truth," which identifies "many atheists and other skeptics" as directly culpable. The "abundance" of unrealistic ark representations targets children, impacts everyone, and is an orchestrated "attack" on the authority and historicity of literalist scripture. It is, in short, conspiracy.

The arresting effect of Fairy Tale Ark's visual and textual experience is heightened by a subtle, but powerful sound shift. When you enter the exhibit, the upbeat baseline soundtrack from the hallway shifts to a very different tune (again, playing on a two-minute loop). It reminded me of a dream sequence in a film, perhaps an animated film, where something terrible is about to happen. Interspersed throughout the dream-y instrumental is the sporadic sound of children laughing. The volume of their voices steadily increases and they transform from young children playfully giggling to teenagers laughing in mockery. The exhibit's ominous message of a secular conspiracy is enhanced by this eerily disturbing auditory annotation.

As we consider how to represent Ark Encounter on our digital platform, we are keen to capture the park's shifting soundscape as an integral feature of the emplaced experience. The first step for creating a virtual tour of this attraction is to make high quality audio-visual recordings of the auditory annotations inside exhibits. For examples like Fairy Tale Ark this involves a few seemingly minor, but actually important decisions. First, the recording needs to align with the two-minute loop. Second, the video recorder needs to be positioned nearest the overhead speaker, which is not the optimal filming location. Because the speaker is positioned above the center entrance, our recording spot is just right of the entrance, rather than the far-right corner, which would enlarge the visual field. Finally, we aim to record at a time when some visitors are present but that avoids peak traffic. Too many people will drown out the sound and be visually distracting. However, having a few people in the frame illustrates how the visitor experience includes others interacting with the exhibit. Whatever approach we take to the virtual tour's final design, it will prominently feature a series of audio-visual clips, offering users a partial re-creation of the shifting soundscape onboard the ark.

As a project grounded in material religion, *MB* is deeply concerned with how matters of embodied presence at particular attractions translates to what users encounter through our digital platform. How can we mobilize the affordances and techniques of digital representation to reveal and re-create the integrated sensorial experiences of being onsite? As we continue to curate the ethnographic and archival material collected for this project, we will continually revisit this question. Indeed, one of the exciting elements of developing this project digitally is the iterative process of collecting and analyzing data, experimenting with representational techniques, returning to data, discovering existing and emerging affordances, returning to data, and so on.

Conclusion

Materializing the Bible is a digital scholarship project, originating from the anthropology of religion and directed outward to other anthropologists, scholars from other disciplines, as well as other public audiences. Our primary aim is to curate a collection of ethnographic and archival materials that examine how religious actors transform written scripture into experiential, choreographed environments. We approach this empirical phenomenon with depth and breadth, exploring the global and historical diversity of this practice as well as zooming in on particular places. The curatorial, analytical, and representational work of this project is grounded in the interdisciplinary study of material religion, which means we seek to closely examine how people aesthetically engage ritualized spaces.

We do not approach digital scholarship as affording wholly new or fundamentally different possibilities in contrast with print scholarship. In many ways, what makes for meaningful digital scholarship is what makes for meaningful scholarship irrespective of the medium. However, a range of curatorial, analytical, and representational practices are renewed in a digital context. Some of the brute, and often stymying, limitations of print publishing are relieved. Vivid description need not be any less vivid, but it is greatly enhanced by the integration of embedded data and multi-modal techniques. The number of high quality color images need not be closely policed. And, engendered by the nature of perpetual beta, the results of rigorous analysis can continually be refined and the means of representation can continually be bolstered through open and inventive platforms.

One of this chapter's organizing claims is that key themes in material religion, such as embodied presence and emplaced experience, are especially well suited for representation on digital platforms. Averts and Counts and Promey and Floyd's essays in this volume offer similar views on this issue. Three affordances of digital scholarship are particularly resonant with *MB*. First, interactivity has guided our process. We hope that *MB* draws users into its empirical orbit and enjoy moving from one attraction to another. There is no single pathway for navigating the project, and we encourage this non-linearity. Through hyperlinks to attraction homepages (which themselves often contain a wealth of digital materials), embedded data, and visualizations (such as our global map of attractions), we invite users to develop their interests within the pedagogical frame we have set. Second, the multi-modal character of digital platforms enables users to engage the forms of integrated sensation that make the physical attractions compelling for visitors. It is one thing to represent the auditory annotations onboard Ark Encounter through a combination of narrative description and still images; it is yet another to allow users to witness the shifting soundscape amid a choreographed virtual tour. Third, the open-ended quality of digital scholarship enables the project's research and findings to develop in dynamic ways. Theoretical and methodological insights can be integrated into our curatorial work as they emerge, generating a deeply iterative process constantly oscillating between analysis and representation.

This latter observation, that digital scholarship can operate in a never finalized way and is therefore open to iterative development, is especially vital in the context of responding to peer reviews. As of July 2019, *MB* has received two critically constructive reviews, published in the *Journal of Heritage Tourism* and

American Anthropologist.[24] Both reviews raise useful observations for refining the project's design and content. For example, Engberg helpfully suggests that a more interactive search function would allow project users to explore attractions by multiple variables, "dates for the building of the parks, denominational affiliation, regions, size, number of visitors, etc.".[25] In response, we are building a database of comparable variables to help guide users through the project. And, Mohan and O'Dell-Chaib astutely observe that our multi-modal representation of attractions in the "Tours" portal would be improved by accounting for "the possible pluralities of encounters at these sites".[26] In response, future tours will commit to integrating the voices of multiple visitors, ideally visitors who bring divergent ideological or somatic dispositions to the attraction. This iterative process illustrates the dynamism of digital scholarship projects, as they are able to revise in response to reviews as they are published over time.

To close, I (Vaughn) reflect on my role as an undergraduate research assistant for *Materializing the Bible*. I began working on the project in October 2016, my first semester at Miami University. After reviewing *MB*, we engaged work on digital scholarship and religious tourism, and discussed potential project additions.[27] My first major contribution was to assist arranging archival, textual, and fieldwork material for the "Tours" portal. I then assisted with ethnographic fieldwork at the Garden of Hope, including interviews, participant observation, writing field notes, recording audio-visual data, and audio transcription. This led to the 3-D imaging analysis discussed earlier and, ultimately, working independently with archival material at the University of Dayton's Marian Library and fieldwork at Catholic shrines in Alabama and Ohio.

Throughout my work on *MB*, it has been essential to foreground questions of audience and how different technologies provide opportunities for rendering data. My experience with 3-D imaging was particularly instructive. When we first discussed this possibility, we knew we were interested, but were unsure how to proceed. With support from archaeological colleagues (namely, Dr. Jeb Card and fellow undergraduate Megan Ashbrook), I began the process and refined my understanding of the technology and its application to *MB*. The best

---

24 Aron Engberg, Digital Scholarship Review: "Materializing the Bible," *Journal of Heritage Tourism* 14, no. 1 (2018): 83–84; Urmila Mohan and Courtney O'Dell-Chaib, "Curatorial Authority in Digital Scholarship: A Review of Materializing the Bible," *American Anthropologist* 120, no. 4 (2018): 843–848.
25 Engberg, Digital Scholarship Review: "Materializing the Bible," 84.
26 Mohan and O'Dell-Chaib, "Curatorial Authority in Digital Scholarship," 846.
27 Our texts included Cantwell, and Rashid, *Religion, Media, and the Digital Turn*; Beal, *Roadside Religion*.

way to learn is by doing, which I found to be true repeatedly in this project. The more I practiced a particular research activity, the more confident I became in my abilities as a researcher and scholar. I have relished the opportunity to develop these skills and hone my approach to fieldwork and analysis.

Ultimately, I am grateful for the experiential learning opportunities *MB* has provided. Writing this, nearly three years after first joining the project and now entering my final year as an undergraduate, I am reminded of why this faculty-student collaboration has been impactful for me. The freedom to try and fail, and try again, was instrumental in my development as a student. To be trusted and taken seriously as an undergraduate researcher encouraged me to continually accept new responsibilities and to creatively explore the affordances of digital scholarship.

Acknowledgments

We would like to thank Kristian Petersen and Christopher Cantwell for their editorial guidance. We are also grateful to Steve Cummings for his hospitality at the Garden of Hope. Selected portions of this chapter appear courtesy of the journal *Religion*, revised from "Immersion as shared imperative: entertainment of/in digital scholarship" (48(2): 291–301).

Selected References

Agnew, Vanessa. "History's Affective Turn: Historical Reenactment and Its Work in the Present." *Rethinking History* 11, no. 2 (2007): 299–312.

Agnew, Vanessa "Gooseflesh: Music, Somatosensation, and the Making of Historical Experience." In *The Varieties of Historical Experience*. Stephan Palmie and Charles Stewart, eds. 77–94. London and New York: Routledge, 2019

Beal, Timothy K. *Roadside Religion: In Search of the Sacred, the Strange, and the Substance of Faith*. Boston: Beacon, 2005.

Bishop, Greg, Joe Oesterie, and Mike Marinacci. *Weird California: Your Travel Guide to California's Local Legends and Best Kept Secrets*. New York: Sterling, 2009.

Cantwell, Christopher D. and Hussein Rashid. *Religion, Media, and the Digital Turn: A Report for the Religion and the Public Sphere Program*. New York: Social Science Research Council, 2015.

Engberg, Aron. "Materializing the Bible." *Journal of Heritage Tourism* 14, no. 1 (2018): 83–84.

Engelke, Matthew. "Religion and the Media Turn: A Review Essay." *American Ethnologist* 37, no. 2 (2010): 371–379.

Feldman, Jackie. *A Jewish Guide in the Holy Land: How Christian Pilgrims Made Me Israeli*. Bloomington: Indiana University Press, 2016.

Hood, William. "The *Sacro Monte* of Varallo: Renaissance Art and Popular Religion."
  In *Monasticism and the Arts*, edited by Timothy Gregory Verdon, 291–311. Syracuse: Syracuse University Press, 1984.

Kaell, Hillary. *Walking Where Jesus Walked: American Christians and Holy Land Pilgrimage.* New York: NYU Press, 2014.

Kark, Ruth and Seth J. Frantzman. "The Protestant Garden Tomb in Jerusalem, Englishwomen, and a land transaction in late Ottomon Palestine." *Palestine Exploration Quarterly* 142, no. 3 (2010): 199–216.

Long, Burke O. *Imagining the Holy Land: Maps, Models, and Fantasy Travels.* Bloomington, IN: Indiana University Press, 2003.

Lukens-Bull, Ronald and Mark Fafard. "Next Year in Orlando: (Re)creating Israel in Christian Zionism." *Journal of Religion and Society* 9 (2007): 1–20.

Mohan, Urmila and Courtney O'Dell-Chaib. "Curatorial Authority in Digital Scholarship: A Review of Materializing the Bible." *American Anthropologist* 120, no. 4 (2018): 843–848.

Morgan, David. *The Embodied Eye: Religious Visual Culture and the Social Life of Feeling.* Berkeley: University of California Press, 2012.

Plate, S. Brent, ed. *Key Terms in Material Religion.* London: Bloomsbury, 2015.

Promey, Sally M. ed. *Sensational Religion: Sensory Cultures in Material Practice.* New Haven: Yale University Press, 2014.

Ron, Amos S. and Jackie Feldman. "From Spots to Themed Sites—The Evolution of the Protestant Holy Land." *Journal of Heritage Tourism* 4, no. 3 (2009): 201–216.

Rowan, Y. "Repacking the Pilgrimage: Visiting the Holy Land in Orlando." In *Marketing Heritage: Archaeology and the Consumption of the Past*, edited by Y. Rowan and U. Baram, 249–266. Walnut Creek, CA: Alta Mira Press, 2004.

Stevenson, Jill. *Sensational Devotion: Evangelical Performance in 21st Century America.* Ann Arbor, MI: University of Michigan Press, 2013.

Supp-Montgomerie, Jenna. "Affect and the Study of Religion." *Religion Compass* 9, no. 10 (2015): 335–345.

Wharton, Annabel Jane. *Selling Jerusalem: Relics, Replicas, Theme Parks.* Chicago: University of Chicago Press. 2006.

Emily C. Floyd and Sally M. Promey
Collaboration and Access in the Study of Material and Visual Cultures of Religion

Materialities & Secularization Theory

As the study of material and visual religion has taken shape over roughly the past four decades, the field has faced a unique set of challenges, both ideological and technical. Secularization theory looms large from an ideological perspective. Scholars in multiple disciplines have rehearsed, many times over, the challenge of the secularization paradigm for studies of religion, and for studies of religious materialities in particular. In brief, secularization theory maintained that if religion survived into modernity, it would be an immaterial and interior sort, a set of beliefs rather than behaviors or practices with their material, spatial, and sensory agencies, implements, and accouterments. This set of secularist assumptions constrained the study of material religion by situating it almost exclusively within the pre-modern, the "primitive," the "less advanced." Although secularization theory, in its earlier reigning forms, has now been fully debunked, its complex effects and aftermaths continue to impede the study of material religion today.

The technical challenges to material religion studies are readily apparent. Chief among these is the necessity for access to large numbers of high quality images for research and teaching. Beyond the simple requirement for many high quality images, scholars of material religion also need access to multiple forms and formats beyond the photograph. When thinking about performative practices, for example, sound, motion, and three-dimensional capacities contribute to an appropriately capacious archive.

Earlier visual technologies came with specific kinds of restrictions. Slide libraries archiving images generally catered to discipline-specific audiences, art historians and anthropologists, among them, and the discipline shaped the nature of the slide collection. Slide libraries specific to the discipline of art history, for example, largely consisted of works of fine art. Furthermore, slides were themselves discrete material objects. If someone in a department needed a slide for teaching, and a scholar from elsewhere in the university had removed it for the day, an entire lecture required retooling. This made disciplines highly

---

Note: The authors wish to thank Camille Angelo for her careful and thoughtful manuscript editing.

protective of and territorial about these objects. Slides also required support technologies (projectors, screens, darkened rooms) that were for years the purview of only those departments (art history, archaeology, anthropology) presumed to require them. While scholars in other disciplines (religious studies, history, sociology) focused occasionally on the study of objects, or incorporated object study into scholarship on specific topics like ritual, for example, those disciplines did not develop classroom support technologies to make possible robust visual teaching. The advent of what is now called digital humanities (DH) represented enormous opportunities, fairly quickly inspiring smart classrooms that allowed for image projection in every discipline.

The Center for the Study of Material and Visual Cultures of Religion

The Center for the Study of Material and Visual Cultures of Religion (MAVCOR) took shape in 2008 as one means of addressing these and other challenges to the study of its subjects. At full scale, MAVCOR has three major components. First, the Center is a programmatic hub at Yale University. Here its work includes, for example, the organization of conferences and symposia, the development of seminars in its fields, the curation of occasional exhibitions, and the work of an active interdisciplinary research group of graduate students and faculty. Second, MAVCOR convenes groups of fellows to work together as part of a project cycle. Project cycles have a roughly five-year duration and aim to produce collaborative scholarship around a designated theme as well as to encourage lasting networks of intellectual exchange. The first project cycle devoted its energies to the intersections of religion, materiality, and sensation. The collective produced a volume titled *Sensational Religion: Sensory Cultures in Material Practice* (Yale University Press, 2014). The current project cycle is dedicated to the study of Material Economies of Religion in the Americas (MERA). Invited Fellows represent all ranks in the academy (from graduate students to senior faculty) and from international institutions as well as universities and museums in the United States. Third, and though MAVCOR is indeed a center, it does not simply, or even most fundamentally, shape a physical space at Yale University. Instead, a core dimension of the center's work takes place digitally through MAVCOR's website (https://mavcor.yale.edu), which went live in 2011. Consisting primarily of a born-digital peer-reviewed journal, electronic exhibition spaces, and a material objects archive, MAVCOR's website intended to address two of the critical challenges that have historically faced the study of material religion: the ideological framework of secularization theory that might suggest the decline and disappearance of material dimensions of religion over time, and the technical hurdles that reinforced barriers among disciplines and that encouraged the study of certain forms and kinds of materials over others.

New technological developments and intellectual frameworks and perspectives have allowed MAVCOR, in concert with other key contributions like the scholarly print journal *Material Religion*, to dismantle earlier impediments to the study of material religion. Secularization theory favored the fine arts, for example, and thus dismissed many objects and monuments important to the study of material religion. In the past, then, these objects and monuments were neither robustly collected and preserved nor usefully curated. MAVCOR's born-digital Material Objects Archive is intended to provide a space for objects neglected under earlier paradigms. The reasons for this neglect were multiple. They included the privileging of some religious traditions over others; gendered understandings of production and usage; differential valuation of materials and media of production; and the economic class of makers and/or users. Racialization also played an enormous part in what attracted study and the interpretive and evaluative frameworks applied.

Responding to the call of the editors of this volume to discuss MAVCOR as a case study for DH scholarship, in what follows we, Sally Promey, the Center Director, and Emily C. Floyd, MAVCOR site Editor and Curator, describe the evolution of the Center over time, discuss current MAVCOR projects intended to expand the scope of digital publishing, and address some aspects of the ethics and complications of digitization. We intend to present MAVCOR as one way in which DH might be leveraged to address theoretical and methodological shifts in the study of religion.

Emergence and Development of MAVCOR

Late in 2008, when Promey and the MAVCOR team first started to plan for MAVCOR's digital dimension, we had two primary aims. First, we wanted to provide substantial peer-reviewed scholarship, about a range of material objects, to a wider set of audiences than those usually reached by scholarly print journals. From the start, this set of aspirational audiences included religious practitioners in diverse traditions as well as scholars. Second, we wanted to encourage scholarship on, and to archive and make available, diverse economies of images and objects and architectures, including, but also vastly exceeding, those high cultural productions conventionally studied by art historians. As contributions started to come to us, we realized we could accomplish more than these two goals, prompting us to expand the scope and kinds of material we published even before the site went live in 2011.

MAVCOR came together at a time when DH was not fully developed at Yale University. For example, Yale hired its first DH Librarian in 2013, and opened its DHLab in 2015. Getting started thus required navigating by considerable trial and error to make our website function in a way that fulfilled even our two initial goals. From the outset, building MAVCOR necessitated that we, as academics with little expertise in digital media, learn to communicate clearly with designers and programmers. We now do a fair amount of our own designing and we contract, and collaborate, with a programmer who has a substantial track record of successful scholarly projects. However, at first, we did not know enough about what we were asking from the technical and design experts to achieve the result we wanted. We also felt the limitations of the then current templates for Yale University websites, which we had opted to use at the beginning because this was the way most websites on university servers were imagined at the time. We were, moreover, encouraged in this direction by the IT professionals with whom we first worked. We were unaware, at that point, whether what we really wanted was possible to attain, and if it was possible, whether we could do it for the amount of funding available to us.

When we began, Yale University websites were largely departmental sites with templates designed to provide introduction to the disciplinary departments and information about them. Fortunately, the university's IT specialists agreed that a customized site, produced by Drupal programmers outside the university, would offer us more creative ways of working around these challenges. Despite customization, however, we felt that the initial design still lent a "departmental aesthetic" to the MAVCOR site and constricted the site's functions. These frustrations prompted our first major post-launch redesign, which we debuted in 2015.

In the summer of 2017, we introduced further changes to the site in order to make it more intuitive, informative, and accessible for our users. Two "Think Sessions" framed the 2017 redesign. The first of these sessions took place at the American Academy of Religion (AAR) conference in November 2016 in San Antonio where MAVCOR gathered a group of religious studies scholars known for innovative and energetic ideas about this medium. The second one coincided with the February 2017 annual meeting of the College Art Association (CAA) in New York City where innovators in the field of art history sat down with us to have a similar conversation. These meetings offered valuable insights about new ways users might wish to encounter the site, opportunities to clarify and more directly convey MAVCOR's mission, and accessibility challenges to which we might respond.

One of the primary changes we instituted in 2017 was a redesigned home page, intended to clarify for first-time visitors the variety of possibilities the site provides for organizing, studying, and (re)imagining content (See Figure 1). The new home page features a large hero image comparable to the 2015 design, but we now have the ability to overlay text on the hero image so that visitors receive both visual and textual information about the project. In the header, the site displays a redesigned logo that emphasizes the project's acronym, "MAVCOR" in addition to its extended title "Center for the Study of Material and Visual Cultures of Religion." The new site also includes expanded content below the hero image: two columns of announcements, one sharing recent contributions to *MAVCOR Journal* and one communicating information about upcoming events. The latter column also provides space for us to share information about developments and opportunities of interest to our readership and to highlight our Twitter feed.

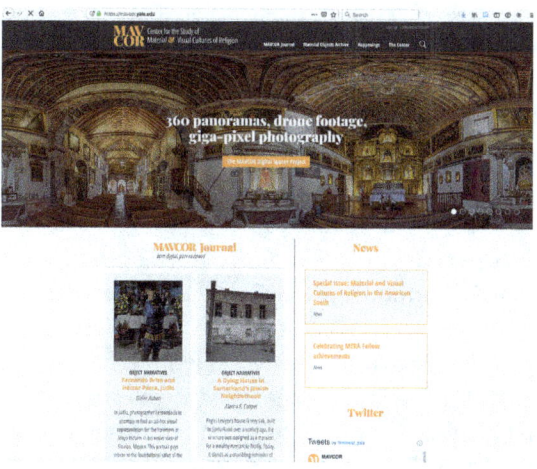

Figure 1: Screenshot of the new MAVCOR homepage.

Among the most substantial feedback we received from the AAR and CAA events was that many site users did not realize that *MAVCOR Journal,* formerly titled *Conversations,* was a peer-reviewed, academic journal with a distinguished scholarly editorial board. The former title, *Conversations,* reflected an important aspect of the original vision for MAVCOR. Fundamental to the initial conception of the site was a commitment to encourage multiple contributors, from a range of disciplines, to write Object Narratives on a single item, from different perspectives, in *conversation* with one another. We continue to find this idea compelling and retain it as one goal for the site. However, the journal title, *Conversations,* lacked precision and failed to specify the nature of the publication, revealing neither its review process, nor its content. The name change, along with the new inclusion on the homepage of the tagline "born-digital, peer-reviewed" more explicitly broadcasts key information about the publication; *MAVCOR Journal* is not an "informal" blog, but a formal scholarly serial, circulated in a nontraditional manner.

In response to the Think Session feedback we received, we have also added other hallmarks of the traditional scholarly serial publication: a journal ISSN number from the United States government; DOI numbers from CrossRef to identify each article; and volume and issue numbers to conform to traditional journal structure.[1] We also made changes to the display of the articles themselves in order to signal the publication's professional character and facilitate citation and user engagement. These include working with the digital preservation service Portico (www.portico.org) to archive articles, making them available in perpetuity even in the unlikely event of the disappearance of MAVCOR from the web; the addition of .pdf downloads so that users can choose to print out articles (and to facilitate Portico's archival process); the inclusion of paragraph numbers to aid in citation; and the addition of a section designated for author biographies. These markers of a traditional, peer-reviewed publication make it possible for librarians to catalog the journal in their databases. Such additions also assist scholars and students to locate MAVCOR content and assure them of the journal's status as a sustainable, peer-reviewed publication.

The recent addition of volumes and issues to *MAVCOR Journal* opens up new opportunities for collaboration and for grouping of articles around specific topics. *MAVCOR Journal* is currently working with guest editor Dana Leibsohn, an art historian specializing in colonial Mexico and early modern exchanges, on a special issue (forthcoming in 2020). The issue is dedicated to the conception, materiality, motion, and spatiality of exchange, centered on how trade created new landscapes and new networks in the ancient and colonial Americas. We are also collaborating with the born-digital *Journal of Southern Religion* to publish a joint issue on material cultures of religion in the American South, with completion expected late 2019. Conversations with MAVCOR's editorial board members, advisors, and contributors have informed and continue to shape the site in meaningful ways.

---

1 A DOI number is a "unique alphanumeric string assigned by a registration agency to identify content and provide a persistent link to its location on the Internet," see "Instructional Aids," American Psychological Accosiation, http://www.apastyle.org/learn/faqs/what-is-doi.aspx.

Academic Conventions and the Digital Humanities

Digital formats afford new freedoms: the capacity to make arguments in novel ways, to form new scholarly partnerships across disciplines and distance, to enhance accessibility to academic and larger publics, to accumulate materials and varieties of contributions over time, to modify and improve formats, and to expand aims in relation to discerned needs and innovations. At the same time, the constraints of tenure requirements and broader scholarly conventions for evaluating academic work, standards often framed by the outward forms and appearances of the scholarly print journal and academic press, frequently do not reward scholars who employ digital media. In imagining MAVCOR as a Yale University site, we hoped the university's reputation would, at least in some small measure, make digital publication more legible as scholarly enterprise. This, we thought, would be especially important with respect to the tenure reviews of junior scholars who publish with us.

MAVCOR endeavors to balance these two sets of interests then: to elicit new ways of viewing, thinking about, and engaging with material and visual cultures of religion, while also addressing institutional requirements for promotion and tenure. We are invested in designing and producing a collaborative, accessible, substantial, interdisciplinary, experimental, and intellectually and methodologically responsible space and experience. In *Religion, Media, and the Digital Turn: A Report for the Religion and the Public Sphere Program Social Science Research Council* (2015), Christopher D. Cantwell and Hussein Rashid note,

> Digital media's inherently participatory nature means that to engage with the medium successfully, one must be aware of its characteristics or rules. Communities of print, visual, or audio culture that migrate online simply by creating 'e-' versions of themselves not only underutilize the medium, but also risk alienating the communities of use that are rapidly emerging around digital content.[2]

At the same time, it is also important to recognize and learn from traditional print approaches to scholarship. The rhetoric of replacement critiqued by Cantwell and Rashid, in which digital projects create little more than online replicas of their analog originals, is inadequate in at least two ways. First, it fails to ac-

---

2 Christopher D. Cantwell and Hussein Rashid, *Religion, Media, and the Digital Turn A Report for the Religion and the Public Sphere Program Social Science Research Council*, (New York: Social Science Research Council, 2015), 7.

knowledge the expansive potential of the digital medium; and, second, it constrains digital possibilities by framing the medium primarily as a new, more durable and accessible, but otherwise largely identical, version of analog.

As increasing numbers of scholars launch web-based projects, content creators, consumers, professional organizations, and tenure committees have wondered how to assess or weigh non-print scholarship. In response, in recent years DH publications and projects have adopted a range of innovative approaches to peer review that present new models for evaluating digital content. Presses and professional organizations have initiated many of these developments. A collaborative imprint of University of Michigan Press, Digital Culture Books, for example, publishes work on new media studies and the DH. As part of this effort, the press seeks to develop "open platforms that make openness part of the scholarly peer review process."[3] Similarly, the Roy Rosenzweig Center for History and New Media at George Mason University designed PressForward, a platform that attempts to harness internet virality as a form of peer review. Both of these presses move away from the anonymity that has historically characterized peer review in an effort to promote transparency and encourage larger communities of experts to assess future publications prior to the emergence of the final versions. Indeed, the volume in which this essay appears takes an "open" approach to peer review as developed by DeGruyter. Professional organizations like the College Art Association, American Historical Association, and, soon, American Academy of Religion have responded to the growth of DH projects by issuing guidelines for evaluating born-digital content on its own terms.

Within this context of peer review experimentation and innovation, MAVCOR approaches print and digital publishing as independent scholarly media but also ideally understands print and digital forms and formats to be mutually reinforcing, each with respect to the other. We recognize the potential benefits of employing some of the outward forms of print publishing and acknowledge the advantages of some print conventions for vetting of academic content. MAVCOR's Editorial Board and double-blind review process, for example, closely resemble print models, offering means by which scholars gain feedback, produce knowledge, and enhance collaborative dimensions of research, teaching, and mentoring. Peer-review not only ensures a high standard of scholarship, but also serves as an expression of scholarly community, providing a venue for idea exchange. Within MAVCOR's interdisciplinary context, this feedback is especially useful. When scholars of religion write on materiality, for example, they benefit from

---

[3] "About Us," Digital Culture Books, http://www.digitalculture.org/about/.

an art historian's perspective, just as art historians with scant training in religious studies frameworks may benefit from the insights of a reader in that discipline. In following certain established norms of print publication, MAVCOR acknowledges the current benefits and strengths of these norms and asks how best to adopt or adapt them for publishing within DH media. As colleagues in digital publishing invent new best practices and standards, we will watch closely to consider whether these developments in review processes may allow academic publication in this new arena to expand and enhance what print already does well.

One of the ways MAVCOR has sought to innovate on the traditional model of journal publishing is in the diversity of article types we publish. Indeed, the born-digital nature of the project has allowed a multiplicity of content types. This array of formats would be more challenging, if not impossible, to include in a print publication. MAVCOR's article types encourage creative ways of thinking about the roles of material objects in religious practice. In addition to *Essays*, modeled after conventional journal articles, we also publish *Object Narratives* (roughly 1,500-word contributions dedicated to exploring a single object), *Medium Studies* (texts of any length that focus attention on materials, media, and techniques), *Mediations* (theoretical musings of varying lengths on subjects related to MAVCOR's areas of inquiry), *Interviews* (transcripts of conversations with relevant thinkers or artists optionally accompanied by contextual information and author commentary), *Constellations* (focused exhibitions of four to twenty objects brought into conversation with one another, with the option of including descriptive text), and *Collections* (large groups of related objects curated by a single individual, potentially divided into smaller sections, and accompanied by discursive text). Some of these formats, including *Mediations* and *Collections*, emerged in response to user feedback.

The digital platform has enabled MAVCOR to provide users with new ways of interacting with and making arguments about materiality. Unlike print media, which is largely static, MAVCOR can publish audio and visual material, as well as text. The journal welcomes video-first contributions. We have also begun collaborating with Internet Archive (https://archive.org) in order to digitize books, including photo albums, prayer guides, and other devotional materials, held in the personal collections of our contributors, so that contributors can produce articles that comprise books, as well as pictures, performances, buildings, and sculptures. Internet Archive's unique book viewer platform allows visitors the experience of virtually turning pages. Although viewers cannot achieve the actual sensory experience of physically holding a book and manually manipulating its pages, this technology allows site visitors to see the layout of pages in relationship to one another, rather than in the isolation in which they are generally displayed in print publications.

We began publishing *Collections* in 2016 in response to a common scholarly problem: many researchers accumulate large databases of images, audio files, and video over the course of their investigations. The vast majority of these are never shared with the scholarly community or general reading public. Academic monographs, the typical venue for publication of long-term projects, set limits on the numbers of images they can include, even if greater numbers might be relevant to the argument at hand. MAVCOR's Collections provide a space for curated display of the full set of these accumulated materials accompanied by introductory text. As with other categories of contribution, the scholar/author retains copyright to these materials, or works with MAVCOR to secure permissions from the relevant venues, as appropriate.

In October 2017, MAVCOR published its first *Collection* realized in concert with the print publication of a scholarly monograph, Rachel McBride Lindsey's *A Communion of Shadows: Religion and Photography in Nineteenth-Century America* (University of North Carolina Press). The print book includes a link to Lindsey's MAVCOR Collection, which in turn links to the press's publicity page for her book. While MAVCOR has from the start planned to archive such assemblies of images and research project materials (audio and video), it is largely thanks to the initiative of University of North Carolina Press Executive Editor, Elaine Maisner, that we now imagine Collections as a means of establishing formal and mutually advantageous relations with scholarly print publication. This first collaboration with UNC press has been enthusiastically received and other authors have approached us about undertaking similar projects to accompany their books. In 2019 we published a second Collection in collaboration with the University of North Carolina Press, coinciding with the publication of Alex Seggerman's monograph *Modernism on the Nile*, on Egyptian art movements, Islam, and the construction of Middle Eastern modernity.

Many DH projects represent the robust and innovative visions of single scholars or small groups of scholars, tightly focused and dedicated to exploring a specific question or area of study. These projects make valuable and exciting innovations and contributions to scholarship. MAVCOR consults and learns from them, but imagines a different contributor model for itself. MAVCOR's platform offers a unified forum in which multiple scholars' work can come together around a single, interdisciplinary set of mutually engaged subjects and in which scholars are invited to join the DH community without necessarily investing in building a website or developing these infrastructural tools on their own.

Accessibility, Accumulation, and Flexibility in the Digital Humanities

While MAVCOR is invested in substantial and original scholarship, it is not *solely* a community of scholars. As an open-access publication, we regularly receive emails from individuals outside the academy, inquiring about the dates of religious festivals described in the journal, wishing to contribute a religious object from their own collection to the Material Objects Archive, or simply hoping to identify an object located in a relative's attic. The open access nature of MAVCOR is one of its greatest strengths—furnishing a space on the Internet where anyone (in university, public school, museum, civic organization, religious community, e. g.) can find thoroughly researched and thoughtful information about material religion. Our commitment to a broad readership has influenced our approach to editing and publishing content. Since we aim to attract the larger public, as well as scholarly readers from a wide range of academic backgrounds and interests, it is important that all contributions define terms, offer historical context, and explain their theoretical concepts. MAVCOR promotes scholarship that is simultaneously substantive *and* accessible.

In order to ensure MAVCOR's content is both comprehensive and accessible, the Center depends upon success in attaining financial support as well as on a small and highly trained staff (several graduate associates in addition to the editor/curator and director). A continuing barrier to website development for most DH projects is the need to raise substantial funds to build and maintain these sites. MAVCOR has been fortunate to attract generous support from, for example, the Henry Luce Foundation, Yale Institute of Sacred Music, and Yale University. Although our financial stability requires ongoing preparation of new grant materials, MAVCOR, thanks to this institutional generosity, is able to ensure continuity in this work. This continuity is an undeniable asset in shaping MAVCOR and making it possible to showcase the work of multiple scholars on a shared platform.

Another key feature and distinct advantage of digital technologies is that they allow modification over time. From the beginning, we have valued and counted upon these prospects for accumulation and malleability. We can continue building the website and multiplying contributors and contributions for as long as there is interest and meaningful rationale for this labor. These two characteristics (accumulation and amenability to refinement and change) work well in tandem. In April 2018, MAVCOR extended this collaborative and open format by inviting users to create accounts on the site through which they are able to curate their own *Galleries* of site content. *Galleries* can include material objects as well as articles from the journal. Furthermore, users are able to add their own textual glosses and commentary. We envision these *Galleries* as pedagogical, research, and exploratory tools. Professors might create a *Gallery* of articles and images and then invite their students to read the articles in conversation with the selected images. A course might assign students to shape *Galleries* in response to a specific prompt. A researcher might employ *Galleries* to gather together material related to a subject of interest. A religious practitioner might compile objects and articles connected to devotions, holidays, or specific iconographies. With the addition of *Galleries*, MAVCOR seeks to engage a broader collaborative community.

The MAVCOR Digital Spaces Project: A Case Study

In summer 2017, MAVCOR launched the Digital Spaces Project. These high-resolution giga-pixel, 360º panorama, and drone photographs record religious spaces and form part of the Material Objects Archive (See Figure 2). For the first round of photography, in March 2017, MAVCOR collaborated with *cusqueño* photographer Raúl Montero Quispe and the Society of Jesus in Cuzco to document seven churches in the Cuzco area, all administered by the Jesuits: La Compañía de Jesús de Cuzco, San Pedro de Andahuaylillas, San Juan Bautista de Huaro, Nuestra Señora de la Candelaria de Canincunca, San Pablo de Ocongate, San Juan Bautista de Ccatcca, and San Francisco de Asís de Marcapata. This kind of imagery records both architectural interiors and exteriors (using drone photography), as well as the location of works within these spaces and their current states of restoration. The photographs also indicate aspects of current religious practice, documenting flowers and candles left before images as well as other elements of devotional ephemera.

The MAVCOR Digital Spaces Project's work in Peru raises ethical and methodological considerations frequently faced by other DH projects. Petit, Yang, and Huang raise similar concerns in their essay for this volume, where they grapple with questions raised by documenting unofficial religious sites in China. In inaugurating our project, we first needed to study, understand, and come to some decisions about the complicated historical relationship between cultural heritage photography and the theft of national patrimony in Latin America, and especially in Peru.

Countries, organizations, and individuals in this region have substantial reason for anxiety about the attention that photography draws to their already en-

Figure 2: 360º panorama at Huaro.

dangered and looted patrimony. In contrast to churches in Europe and the United States where tourists, locals, and pilgrims are generally welcome to photograph at will, churches in South America regularly feature prominent signs prohibiting all photography. These signs represent fears about possible theft, although occasionally caretakers cite conservation concerns as well (i.e. the potential damage caused by flash photography to delicate artworks). These anxieties about theft are grounded in painful experiences: *Memoria Robada* (Stolen Memory), a transnational investigation into trafficking of Latin American cultural heritage carried out by investigative reporting team Ojo Público (*Public Eye*), declares:

> From 2008 to 2016 alone, the main auction houses of Europe and the United States sold more than 7,000 objects of Peruvian archaeological patrimony ... The volume of Latin American works sold to collectors in the major world capitals is even greater than the 4,907 cultural objects that Interpol is currently investigating as stolen from all of South America, Central America, and Mexico. This is what one could call a disappeared museum.[4]

---

[4] "Las historias ocultas del saqueo cultural de América," *Memoria Robada*, https://memoriarobada.ojo-publico.com/investigaciones/memoria-robada-las-historias-ocultas-del-saqueo-cultural-de-america-latina/. All translations from Spanish by Emily C. Floyd. In addition to their reporting, *Memoria Robada* also includes a database of stolen objects, of repatriated objects, and of auction houses known to have sold stolen objects. "Disappeared museum" is a reference to Héctor Feliciano's *The Lost Museum: The Nazi Conspiracy to Steal the World's Greatest Works of Art* (New York: Basic Books, 1997).

Peru's Ministerio de Cultura (Ministry of Culture, formerly the Instituto Nacional de Cultura) keeps an extensive digital archive of all the paintings, sculpture, and objects stolen, most of these from church and monastic collections.[5] Museum, church, and culture officials in Peru frequently describe thieves using scholarly publications as "catalogs" to select objects to steal.

Although some culture officials in Peru cite anecdotal support of scholarly publications serving as catalogs for potential thieves, there is also substantial evidence that photographic documentation of cultural heritage can provide an important means of defense of these materials. In *Memoria Robada,* Ojo Público presents the example of an eighteenth-century Quechua manuscript stolen from the Biblioteca Nacional del Perú and eventually sold to the library of Dumbarton Oaks in Washington, D.C. In 2012 Dumbarton Oaks summer fellow Isabel Yaya McKenzie informed French Quechua expert César Itier of the manuscript. He recognized it and was able to prove, on the basis of scans he had had made years earlier, that it was the manuscript he had consulted at the Peruvian national library. Itier's scans convinced Dumbarton Oaks they had purchased stolen goods. The research library immediately initiated efforts to repatriate the manuscript, though the original theft has yet to be solved. Similarly, "No robes el pasado," (*Don't steal the past*), an educational guide for young people designed by the Peruvian Ministerio de Cultura in collaboration with UNESCO, tells the story of two colonial paintings returned to churches in rural Bolivia on the basis of photographs and technical documentation.[6] Recently, Ojo Público's *Memoria Robada* project published an account of a seventeenth-century painting of the Virgin of Guadalupe stolen from a Cuzco-area church.[7] The painting ultimately ended up in the collection of the Catholic Diocese of Orange County, California, where the Ojo Público team was able to identify it, thanks to a photographic registry compiled in 1983 by the lawyer Frederic J. Truslow. The painting returned to Peru in 2018.

Parish priests who attempt to report a stolen painting, at times without clear recollection of its specific iconography, will be challenged to prove the work rightfully belongs in their holdings. *Memoria Robada* recounts numerous cases of this nature, describing, for example, 214 objects stolen from Costa Rica that

---

[5] See Consulta de Robos de Bienes Culturales, http://www.robodebienesculturales.cultura.pe/MINC-BMU/pages/principal/principalWeb.jsfx.

[6] UNESCO Office, Lima, *No Robes el Pasado* (Lima, Peru: UNESDOC Digital Library, 2013), 30–31. Available online at: http://unesdoc.unesco.org/images/0022/002269/226971S.pdf.

[7] David Hidalgo, "El robo de la Virgen de Guadalupe: la pintura del Cuzco que reapareció en California," *Memoria Robada* (2018). https://memoriarobada.ojo-publico.com/investigaciones/el-robo-de-la-virgen-de-guadalupe-la-pintura-del-cusco-que-reaparecio-en-california/.

the "Museo Nacional de Costa Rica was unable to reclaim as it was unable to supply documents demonstrating their origin."[8] Proper documentation of cultural heritage contributes to the safe maintenance of that heritage. MAVCOR hopes that its documentary collaboration with the Jesuit churches in Cuzco will assist these seven churches, some of them rural and relatively little-known, to establish and defend ownership of their holdings so that objects can be reclaimed should theft occur. The MAVCOR Digital Spaces Project photographs, and other projects of this sort, may even help to discourage some potential thieves by increasing public awareness of the churches' collections and thus reducing the market for stolen works.

In addition to combating theft, photography can also assist in the conservation and study of spaces compromised by other cultural processes. In some instances, dramatic restorations of these spaces obscure the historic appearance of buildings, painting over or filling in missing details to produce radically revised visual and spatial programs. Even the most careful restoration comes with advantages and disadvantages. Restorations may be celebrated by some of the populations who live with these spaces and appreciate the experience of renewed colors and the completion of fragmentary or damaged murals. For scholars aiming to place these churches within a historical context, however, excellent documentation of the churches' pre-restoration appearance can be immensely important.

The churches photographed by MAVCOR represent a range of interventions by restorers, from the highly scientific, recent work at Andahuaylillas to the heavily over-painted early-twenty-first-century work at Ocongate, to the completely untouched state of Marcapata.[9] The Digital Spaces Project preserves a record of the appearance of the Jesuit churches in Cuzco at a specific moment in their use, documenting the appearance of Marcapata prior to restoration, and the aspect of the other churches in their current states. Even the most skilled and meticulous restoration efforts disguise as well as reveal. They may uncover hidden layers of earlier mural painting, but in doing so they inevitably remove later additions. At Ocongate it is difficult to discern the building's pre-restoration state due to the thick layer of modern paint added by the restorers. Andahuaylillas

---

8 *Memoria Robada*, https://memoriarobada.ojo-publico.com.
9 "Cusco: Restaurarán el Templo de San Francisco de Asís de Marcapata," Perú.com, February 17, 2013, http://peru.com/viajes/noticia-de-viajes/cusco-restauraran-templo-san-francisco-asis-marcapata-noticia-122046. When MAVCOR met with Meritxell Oms Arias, Director of the Asociación SEMPA which administrates the Ruta del Barroco Andino and the six churches including Marcapata, Oms asserted that although plans for restoration had been proposed no work had begun at the time of our photography.

and Ccatcca, in contrast, demonstrate two different approaches to careful restoration/conservation. At Ccatcca the restorers left large expanses of wall blank, occasionally filling in some details with precise, if subtle parallel lines of paint, which allow the viewer to identify previously absent sections. In general, however, the restorers at Ccatcca avoided such efforts to fill lacunae and left the surviving murals to float within wide white expanses of empty wall. At Andahuaylillas the restorers similarly employed parallel lines of paint to indicate areas of restoration, but a quick survey of the church reveals a greater number of such areas than at Ccatcca. Certainly more mural painting had survived intact at Andahuaylillas up to the date of the restoration, but Ccatcca and Andahuaylilas nonetheless represent slightly different approaches to the task. Huaro, restored from 2004-ca. 2007 by the Instituto Nacional de Cultura and the World Monuments Fund, represents another example of careful restoration.[10]

Unrestored Marcapata is a rural church some six hours' drive from Cuzco, over the 5000-meter Ausangate pass and plunging into the *ceja de selva* or "eyebrow of the jungle," where the Andes meet the Amazon rainforest. When Meritxell Oms Arias, Director of the Society of Jesus's Ruta del Barroco Andino (Route of the Andean Baroque) proposed Marcapata as a potential site to include within the digitization project, she referenced its untouched state and the plan to restore it within the next years as particularly compelling reasons to make the long, perilous drive to visit the church. Marcapata's unique thatch roof, well documented in the 360º drone photos that we also took at each site, formed an additional draw (See Figure 3). Photography at Marcapata represented a unique opportunity for MAVCOR to document the site prior to restoration, recording both the remarkably well-preserved nature of the murals and the single area of water damage, where water broke through the roof over the painted image of Saint James the Greater, washing away portions of the mural (See Figure 4). These 360º photographs will allow future generations to assess the extent of later restorations as even careful restorations like those at Andahuaylillas or Huaro can disguise aspects of the lived history of the church that might otherwise be of interest to scholars.

---

[10] See "Restauración del Templo de San Juan Bautista de Huaro: En la ruta del barroco andino," *Conservación de la Historia: Restauración y puesta en valor del patrimonio cultural*, no. 26 (February–March 2007): 5–9.

Figure 3: Drone photography at Marcapata.

Location and Scale

A primary rationale for MAVCOR's photographic efforts is the work's utility to scholars, students, and professors in researching, studying, and teaching these objects and places. As Averett and Counts also explore in their essay for this volume, traditional photography offers only a single view of a space and is limiting in what it can tell the viewer about the location of objects and people within that space. In contrast to traditional photography, 360º panoramas offer a sense of scale, location, and relationship to other works in a specific space. A major and singular challenge of illustrating works of art is conveying their scale. As anyone can attest whose knowledge of an object comes primarily from illustrations of that object in a book, encountering the original can present something of a shock. This is because the object in question is very likely to be substantially smaller or larger than the book's photography was taken to imply. Such texts often show tiny objects at much larger than actual scale in order to offer greater detail to the reader, while drastically condensing larger objects to fit on the page. The 15th edition of Gardner's classic art history textbook, *Art Through the Ages*, offers one approach to this problem by including rubrics alongside each image, suggesting the photograph's relationship to the original by detailing how many millimeters, centimeters, or meters high the illustration would be were it to scale. These measurements are a valuable innovation, but they still cannot offer the intuitive, visceral sense of scale that best comes from comparison to an object of known size or to the body itself. 360º panoramas offer an improved, if still imperfect, solution to this problem. In the Jesuit church panoramas, for example, pews scaled to fit the human body help provide a sense of the dimensions of the rest of the space.

Another crucial element often missing in single-point photographic illustrations is the location of the object in question and its positioning relative to other objects within a space. 360º panoramas allow the viewer to see the object's positioning on a wall, its location relative to other objects, or its situation within a broader religious context. Instead of being told that a mural stands at the entrance to a church, viewers can witness the positioning of this work for themselves, and consider how exactly it then engages with the lateral walls, envisioning how a visitor to the church might encounter this iconographic program. The ability to view 360º panoramas with virtual reality glasses contributes to the immersive experience of viewing such images.

MAVCOR is far from the first project to recognize the value of virtual representations of architecture to the study of historic sites, or even to the study of material religion. Indeed, one of the earliest religion and DH projects was the Virtual Qumran project, which sought to avoid contentious debates about provenance or ownership of the Qumran site and the Dead Sea Scrolls by focusing on the information gleaned from archaeological research at the site itself.[11] As part of this endeavor, the project built a 3D virtual model of Qumran's architecture. In developing the MAVCOR Digital Spaces project we were inspired by the Virtual Tour of the Sistine Chapel developed by the Villanova team lead by Paul Wilson, a communication professor, Chad Fahs, Assistant Digital Media Coordinator for the communication department, and Frank Klassner, a member of the computing sciences.[12] In addition to the Sistine Chapel, this group is also working to produce tours of St. John Lateran, St. Paul's Outside the Walls, St. Mary Major, the Pauline Chapel and St. Peter's Basilica. Between 2015 and 2018, Benjamin E. Zeller at Lake Forest College in Chicago worked with undergraduate student research assistants to develop 360º tours of several historic religious spaces in Chicago.[13] Zeller's tours combine 360º photography with historic photographs and other information about the sites. MAVCOR looks to the models offered by these earlier projects, but moves beyond a geographically defined focus area, instead choosing sites to photograph in collaboration with individual scholars and with preference for lesser known sites, or at least less well-documented sites.

---

11 *The Orion Virtual Qumran Tour*, http://virtualqumran.huji.ac.il/.
12 *Sistine Chapel Virtual Tour*, http://www.vatican.va/various/cappelle/index_sistina_en.htm.
13 "Sacred Spaces in 360," *Digital Chicago*, http://digitalchicagohistory.org/exhibits/show/digital-chicago-churches/sacred-spaces-introduction.

Promoters of 360° panoramic photography often suggest that panoramas are "just like being in the original space." Such an argument may imply that panoramas are a substitute equal to or even better than a visit to the original space. Viewers of panoramas, however, should not be fooled by this rhetoric: panoramas create a distinct viewing experience, profoundly different from the experience of a visitor to the original site. This experience offers advantages unavailable to the in-person visitor, but also distorts the experience of the site in several ways. MAVCOR's 360° panoramas are produced at relatively high resolution, compiling as many as 250 high definitive range (HDR) photographs in order to produce a single panorama. HDR photographs are composed of multiple exposures taken at different shutter speeds in each camera position. These exposures are then stitched together in post processing to create a final image in which detail is visible in both dark shadow and bright light. The high resolution of MAVCOR's panoramas allows MAVCOR to limit the level of distortion in the images while the HDR photographs compensate for the camera's more limited ability to distinguish between light and dark in contrast to the human eye. The result is that sometimes *more* is visible than would be discernable to a visitor in the actual space, as when the viewer of the photographic works is able to zoom in to details on the ceiling or into areas of darkness revealed by the HDR photography. At the same time, the viewer is also capable of zooming out so far that the space of the church takes on unreal and potentially misleading proportions, offering a radically new and distorted way of viewing the space that would be impossible without the 360° capability.

The same could perhaps be said of MAVCOR's giga-pixel photography. In order to produce giga-pixel photographs, MAVCOR used a telephoto lens to photograph small portions of a larger image at very high resolution. Then, as with the 360° panoramas, MAVCOR stitched together that mosaic of images in post-processing to produce a final, massive whole. In the Jesuit churches we used giga-pixel photography on large murals. The final images allow the viewer to zoom in close enough to see individual brushstrokes and the texture of the wall behind them. Such images offer an unusually detailed kind of close looking. Scholars might use them to consider techniques of production or to teach about those techniques to their students, or simply to focus in on a discrete segment of a larger composition. These super high-resolution images also allow the viewer to analyze restored areas and to assess what segments remain of the original composition. They thus serve to further document restoration efforts. Photography of this nature has been quite popular in museum contexts in which it complements the work of restorers, curators, and art historians more broadly. The Getty Foundation employed giga-pixel photographs in their *Closer to Van Eyck* project, which focuses on the artist Jan van Eyck's famous Ghent Altarpiece.[14] The project offers not only extreme close-ups and high-resolution photographs of the entire altarpiece, but also data drawn from the conservation of the work, including dendrochronology, conservator's reports, and x-ray and infrared photography.

MAVCOR's photography in Cuzco churches represents an initial foray, a first glimpse of a project we intend to expand significantly over time. One of the MAVCOR site's primary functions is to provide shared space for material religion scholars to come together to do DH work without each individual scholar needing to fund their own isolated project. Future Digital Spaces Project photography will take place in collaboration with specific scholars, the photographs intended to complement and further their scholarship. This is not unlike MAVCOR's approach to Collections, which the site has published in concert with print publications. For new contributions to the Digital Spaces Project, then, MAVCOR will work with scholars to take photographs and record audio and video material, as well as to aid them in constructing tours or other kinds of interpretative tools using the new images, videos, and sound recordings. In November 2017, we began this first round of scholar-centered photography, partnering with medieval art historians Meg Bernstein and James Alexander Cameron to photograph four neighboring churches in the Lincolnshire Fens in eastern England: Bicker (St. Swithin), Algarkirk (St. Peter and St. Paul), Sutterton (St. Mary), and Kirton-in-Holland (St. Peter and St. Paul). The 360º panoramas and drone photographs taken in concert with Bernstein and Cameron can now be seen on the MAVCOR site. In March 2019 we partnered with art historians Amara Solari and Linda Williams to photograph a series of colonial Maya churches in the northern Yucatán Peninsula. In the near future, religious studies scholar Kambiz GhaneaBassiri will advise and guide a similar effort to photograph a series of important mosques in the United States and to offer scholarship about them, along with this new photography, on the MAVCOR site. In December 2018 we began this collaboration with GhaneaBassiri, photographing at the Islamic Center of Washington DC (See Figure 5).

Conclusion

MAVCOR Journal, the MAVCOR Digital Spaces Project, and the website as a whole are part of the larger project that is MAVCOR. This larger project has many mov-

---

14 *Closer to Van Ecyk*, http://legacy.closertovaneyck.be/#home.

Figure 4: Marcapata 360º panorama.

Figure 5: The prayer hall at the Islamic Center of Washington DC.

ing parts, generated, shaped, and overseen by the Center for the Study of Material and Visual Cultures of Religion at Yale University. It should by now be clear that the MAVCOR website places a premium on collaborative scholarship of many sorts. It is a cumulative interdisciplinary site, rather than a comparative one. It nonetheless also imagines itself as a possible space where new kinds of comparison might emerge across a span of traditions, geographies, times, media, and disciplines. MAVCOR does not endorse a particular comparative methodology (or even comparative methodologies in general). Rather, in addition to many other aspirations, it hopes to elicit serious reflection from site contributors and visitors, on the usefulness of comparison as one strategy for considering varieties of things that assume a new adjacency simply by virtue of their presence on the site. We do not imagine any aspect of this project to be static, though we are fully committed to responsibly maintaining the work entrusted to us by contributors, each of whom retains copyright to their own texts and/or materials.

We conclude with an expansion of our introductory discussion, and with what may seem like an extravagant claim for DH: the advent of electronic technologies has been crucial to the wider emergence and development of material religion as a sub-specialization within religious studies. In their essays in this volume, Bielo and Vaughn and Averett and Counts present ways in which the affordances of digital technologies offer new analytical and representational possibilities and have transformed the capture, analysis, curation, and publication of research data. We go beyond this to argue that, without the advent of DH, the robust study of material and visual cultures of religion, and thus MAVCOR itself, would have been unlikely to have assumed its present shape as a formally recognized and widely practiced specialization. The academic study of objects, images, monuments, and spaces requires the possibility of displaying such materials in the classroom. This necessitates both projection devices and rooms in which projected images can be seen. In the first years of the 1990s, such display was extremely difficult to achieve outside the discipline of art history. Art history, alone among disciplinary identities, had long demanded interior architectural and spatial structures that facilitated image projection. Each art history classroom included machinery for dual slide projection and each room could be darkened so that everyone could see the projected images. If anthropologists or historians in other disciplines (like religious studies) wished to display images, they had to carry their own heavy equipment into the classroom, and they needed to arrange for a classroom space that had adequate surface for projection and that either had no windows or had windows with curtains or blinds adequate to blocking out light in order to facilitate visibility of projected works.

By the early 2000s, and into the early 2010s, significant changes to educational infrastructure placed so-called smart classrooms in numerous university departments in the United States, including religious studies.[15] In the present moment, on many campuses, art history departments continue to demand superior image projection capabilities—but virtually nowhere in the United States

---

15 The first such smart classrooms were generally provided to the scientific disciplines rather than the humanities.

would one need to carry their own heavy and expensive equipment with them in order to use the current technologies for image projection in the university classroom. This is not to suggest that religious studies scholars universally incorporate material sources in their teaching and research, but rather to trace the rise of a new sub-specialization, called material religion, within the field. Although there are other factors at work here, one reason that material religion has caught on faster in departments of religious studies than has the study of religion in modernity in departments of art history has everything to do with electronic technology and DH. This raises larger questions about the roles of available technologies in shaping disciplinarities and interdisciplinarities in scholarship and classrooms. MAVCOR aims to create spaces for interdisciplinary conversations and offers new resources for collaboration and access. In this work, we imagine more permeable (inter)disciplinary domains, open to a wider range of actively engaged contributors and participants.

Selected References

Cantwell, Christopher D. and Hussein Rashid. *Religion, Media, and the Digital Turn: A Report for the Religion and the Public Sphere Program Social Science Research Council.* Social Science Research Council, 2015.

Feliciano, Héctor *The Lost Museum: The Nazi Conspiracy to Steal the World's Greatest Works of Art.* New York: Basic Books, 1997.

Hidalgo, David. "El robo de la Virgen de Guadalupe: la pintura del Cuzco que reapareció en California." *Memoria Robada* (2018). https://memoriarobada.ojo-publico.com/investigaciones/el-robo-de-la-virgen-de-guadalupe-la-pintura-del-cusco-que-reaparecio-en-california/.

UNESCO, Lima Office. *No Robes el Pasado.* Lima, Peru: UNESDOC Digital Library, 2013.

Erin Walcek Averett and Derek B. Counts
Scaling Religious Practice from Landscape to Artifact: Digital Approaches to Ancient Cyprus

Introduction

Although now hidden by the sprawl of modern cities and towns and the aggressive reclamation of land by farmers and shepherds, sanctuaries were once a persistent feature across the ancient landscapes of the island of Cyprus. During the first millennium BCE sanctuaries occupied prominent locations in and around coastal and inland urban centers, as well as in secondary towns, especially those in the agriculturally rich plains of the Mesaoria in southcentral Cyprus. Yet, the importance of Cypriot sanctuaries extends far beyond their physical footprints, which are often simple, open-air precincts delineated by stone and mudbrick walls. Cypriot cult spaces articulated a variety of complex relationships from the entanglement of foreign and local religious iconography to the negotiation of socio-economic identities, and perhaps even the boundaries of political authority. The paucity of textual evidence, however, requires an understanding of Cypriot religion almost entirely derived from material remains. Common archaeological features of Iron Age sanctuaries include boundary walls, intramural buildings, altars and hearths, ceramic assemblages (including vessels and lamps), faunal remains (from animal sacrifice and feasting), coins, glass, and,

Note: We would like to thank Michael Toumazou, Director of the Athienou Archaeological Project, for his continued support of our 3D project. We also wish to acknowledge the work of our collaborator, Kevin Garstki, who among other intellectual contributions has assumed a primary role in the creation and curation of our 3D models. We also owe a debt of gratitude to Marina Solomidou-Ieronymidou, the Director of the Department of Antiquities of Cyprus, for permission to scan artifacts, and to Anna Satraki (archaeological officer at the Larnaka District Archaeological Museum and Noni Papasianti (curator of the Kallinikeion Municipal Museum of Athienou) for graciously accommodating our team during our summer scanning sessions. We wish to thank Reinhard Sneff for permission to use the images of his model. This ongoing project would not have been possible without generous funding and support from the University of Wisconsin-Milwaukee (Research Growth Initiative grant and a FRACAS award), and Creighton University (George F. Haddix Faculty Research Grant, College of Arts and Sciences Faculty Summer Research Grant, a Kripke Center for Religion and Society grant, and a Scheerer grant from the Department of Fine and Performing Arts), and Davidson College and the Athienou Archaeological Project.

https://doi.org/10.1515/9783110573022-009

especially, limestone and terracotta sculpture of varying sizes displaying a range of both human and divine iconography. Nevertheless, the intricacies of Cypriot Iron Age belief systems and the nature of Cypriot cults, including even the theonyms used to evoke the divinities worshipped in the island's sanctuaries, remain elusive.

Because of their rich and distinctive physical remains, which are often signaled above ground by fragmentary sculpture, Cypriot sanctuaries have been a focal point of research since they were first discovered, recorded, and plundered in the nineteenth century. Still, not surprisingly, the shortcomings of early antiquarian exploration,[1] coupled with twentieth-century looting, urban development, and inaccessibility (in the case of sites located in the Turkish Occupied area), continue to derail a complete study of Cypriot ritual spaces. Moreover, the sheer quantity of material excavated from these sanctuaries—disbursed far and wide across the world's museums, as well as throughout the many regional museums on the island—is overwhelming and not easily organized or published using traditional methods of archaeological recording and analysis.

The recent adoption of a suite of digital tools, however, is transforming the way scholars capture, analyze, archive, curate, and share archaeological data.[2] Likewise, this shift is also pushing the boundaries of traditional publication in archaeology as scholars find new and more dynamic online platforms and, si-

---

[1] Veronica Tatton-Brown, ed., *Cyprus in the 19th Century AD. Fact, Fantasy and Fiction. Papers of the 22nd British Museum Classical Colloquium December 1998* (Oxford: Oxbow Books, 2001).
[2] For recent overviews of the impact of "digital archaeology," see e. g. Kevin Garstki, ed., *Critical Archaeolgy in the Digital Age* (Albany, NY: SUNY Press, forthcoming); Erin Walcek Averett, Jody M. Gordon, and Derek B. Counts, eds., *Mobilizing the Past for a Digital Future: The Potential of Digital Archaeology* (Grand Forks, ND: The Digital Press at the University of North Dakota, 2016); William Caraher, "Understanding Digital Archaeology," *Archaeology of the Mediterranean World*, 17 July 2015, https://mediterraneanworld.wordpress.com/2015/07/17/understanding-digital-archaeology/; Maurizio Forte, "Introduction to Cyber-Archaeology," in *Cyber-Archaeology. British Archaeological Reports International Series* 2177, ed. Maurizio Forte (Oxford: Archaeopress, 2010): 9–14; Maurizio Forte, "Cyber Archaeology: A Post-virtual Perspective," in *Between Humanities and the Digital*, ed. Patrik Svensson and David T. Goldberg (Cambridge, MA: The MIT Press, 2015): 295–309; Jeremy Huggett, "Archaeology and the New Technological Fetishism," *Archeologia e Calcolatori* 15 (2004): 81–92; Jeremy Huggett, "Challenging Digital Archaeology," *Open Archaeology* 1 (2015): 79–85; Jeremy Huggett, "A Manifesto for an Introspective Digital Archaeology," *Open Archaeology* 1 (2015): 86–95; Thomas Levy, ed. "Cyber-Archaeology," Special issue, *Near Eastern Archaeology* 77.3 (2014); Christopher Roosevelt, Peter Cobb, Emanuel Moss, Brandon Olson, and Sinan Ünlüsoy, "Excavation is ~~Destruction~~ Digitization: Advances in Archaeological Practice," *Journal of Field Archaeology* 40 (2015): 325–46; Ezra Zubrow, "Digital Archaeology: A Historical Context," in *Digital Archaeology: Bridging Method and Theory*, ed. Thomas L. Evans (New York: Routledge, 2006): 10–32.

multaneously, harness digital publication as a way to provide open access to research and stimulate public dialogues.³

After an overview of the ways in which Cypriot Iron Age religion has been captured through traditional archaeological inquiry, this chapter explores how digital tools have begun to reshape our approach to religious practice on the island. Our examination begins at a wider angle, highlighting how digital technologies enhance our ability to view sites, including religious sites, within a broader island-wide context. We end with a narrower, micro-level of analysis, where digital tools are providing nuanced views of individual practice, ritual behavior, and votive objects. Most importantly, we hope to show not only how digital tools have aided and enhanced our interpretations of Cypriot material culture, but also how such tools have allowed us to formulate new research questions that will provide valuable contexts for application by researchers confronting similar data sets in different archaeological or cultural contexts.

Iron Age Cypriot Religion

There are no detailed, ancient literary descriptions of Cypriot sanctuaries. For example, the famed second-century CE travel writer, Pausanias, did not visit Cyprus and therefore we lack the sort of peripatetic musings that revealed the twilight of religious life at many sanctuaries in mainland Greece. Moreover, the epigraphic record is scarce compared to mainland Greek sanctuaries. Local inscriptions from Cypriot sanctuaries, while useful, are often fragmentary, short, and rarely reference ritual practices or identify specific deities (especially before the fifth century BCE), tending instead to read as generic formulae of pious worshippers. Scholars of Cypriot religion thus have little with which to augment the archaeological record. Luckily, since the nineteenth century, sanctuary remains and an astonishing array of votive objects discovered through antiquarian exploration and, later, systematic excavations have shaped an "archaeology of religion" for the island.⁴

---

3 Eric Kansa, Sarah Kansa, and Ethan Watrall, eds., *Archaeology 2.0: New Tools for Communication and Collaboration* (Los Angeles: Cotsen Institute of Archaeology, 2012); Mark Lake, "Open Archaeology," *World Archaeology* 44, no.4 (2012): 471–78; William Caraher, "Announcing the Digital Edition of Pyla-Koutsopetria I: A Free Download," *Archaeology of the Mediterranean World*, 15 Feb. 2017, https://mediterraneanworld.wordpress.com/2017/02/15/announcing-the-digital-edition-of-pyla-koutsopetria-1-a-free-download/.
4 For the beginning of an interest in the material remains of religion with the excavations of Ohnefalsch-Richter (Max Ohnefalsch-Richter, *Kypros, the Bible, and Homer. Oriental Civilization,*

Cypriot sanctuaries consisted primarily of stone and mudbrick enclosure walls defining sacred areas of varying sizes, complemented by interior structures as well as features associated with a variety of pyrotechnic (e.g., altars and hearths) and other activities. Located in and around urban centers, but also in more isolated sites away from towns, these sacred precincts were important spaces for the display of votive art and the performance of Cypriot religion. Indeed, the arrangement of space within the sacred precincts, coupled with votive objects and other cult paraphernalia found within these areas, reveals a range of activities that occurred inside: banqueting, music and dancing, processions, and the offering of prayers and gifts. Votive sculpture in limestone and terracotta displayed in sanctuaries often acted as proxies for worshippers, providing constant prayer to the god or, more simply, represented a gift given with the intent of receiving good fortune in return—such reciprocity was at the heart of votive religious practice in the ancient Mediterranean and Near East.[5] These votive sculptures, therefore, represent a primary vehicle for interaction with the divine, acting as a physical attendant to ephemeral prayers. Ranging in size from a few centimeters to two meters, male and female depictions of human worshippers display a wide-range of garments, headgear, and jewelry; they carry an equally wide variety of hand-held attributes (e.g., vessels, birds, animals, branches, weapons, etc.). In addition, divine and otherworldly epiphanies are found in images of male and female deities, as well as uncanny representations of monsters and fantastical creatures. Worshippers sacrificed animals (e.g., sheep, goats, and cows), but also prepared feasts of food and drink, accompanied by burnt incense, music, and dancing. A limestone relief discovered in a sanctuary in the area around ancient Golgoi, near the modern town of Athienou, pictorially narrates, in striking episodic detail, the vast range of sights, smells, sounds, and sensations that one might feel in the presence of the god: worshipping and offering gifts to the seated god, dancing, and feasting (Figure 1).

Art and Religion in Ancient Times. Elucidated by the Author's Own Researches and Excavations During Twelve Years Work in Cyprus, 2 vols. [London: Asher & Co, 1893]), see Anja Ulbrich, "An Archaeology of Cult? Cypriot Sanctuaries in 19th century Archaeology," in *Cyprus in the 19th Century AD. Fact, Fancy and Fiction. Papers of the 22nd British Museum Classical Colloquium December 1998*, ed. Veronica Tatton-Brown (Oxford: Oxbow, 2001): 93–106.

5 Erik Sjöqvist, "Die Kultgeschichte eines cyprischen Temenos," *Archiv für Religionswissenschaft* 30 (1933): 308–59; Joan Connelly, "Standing Before One's God: Votive Sculpture and the Cypriot Religious Tradition," *Biblical Archaeology* 52 (1989): 210–18; Christopher Gill, Norman Postlethwaite, and Richard Seaford, eds., *Reciprocity in Ancient Greece* (Oxford: Oxford University Press, 1998); Folkert van Straten, "Votives and Votaries in Greek Sanctuaries," in *Oxford Readings in Greek Religion*, ed. Richard Buxton (Oxford: Oxford University Press, 2000): 191–226.

Figure 1: Limestone votive relief depicting worshippers at a sanctuary, Cypro-Classical period, 4th century BCE. Sanctuary at Golgoi-Ayios Photios, Cyprus. (Inv. no. 74.51.2338. The Metropolitan Museum of Art, New York, The Cesnola Collection, Purchased by subscription, 1874–76.

Based on several decades of research on the religious, political, socio-economic, and artistic importance of Cypriot sanctuaries, we now understand these sacred spaces—both large and small, rural and urban—as key places for ancient Cypriots to encounter the divine, express social status and wealth, control resources, mediate local and foreign cultural influences, and negotiate political power.

Visualizing Cypriot Sanctuaries

For the most part, documentation of Cypriot sanctuaries has been restricted to site-specific publications of plans and artifact descriptions with illustrations and photographs, usually in black and white and with one view. While the relationship between features and artifacts is often provided, the nature of the documentation, coupled with traditional publication strategies, has limited the ability of researchers to bring together multiple data sets (e.g., geospatial, artifact, metadata, external/linked data, etc.) or ask more nuanced questions about the way sites develop, function, and change over time.

There are compelling exceptions in the secondary literature on Cypriot sanctuaries, which include some imaginative visual (and even one literary) narratives of cult activity in Cyprus during the Iron Age. These narratives counter the difficulties in reconstructing the Cypriot religious environment based solely on object lists and site descriptions and further illustrate the severe limitations of traditional publication formats to visualize archaeological contexts. As early as the nineteenth century, the amateur archaeologist Max Ohnefalsch-Richter produced a groundbreaking monograph that catalogued, analyzed, and compared 72 sanctuaries on the island (including their locations, architecture, and finds). He also attempted to reconstruct what he imagined a sanctuary at the ancient site of Idalion looked like in its "twilight," accompanied by a narrative description (Figure 2).

Figure 2: Max Ohnefalsch-Richter's artistic reconstruction of sanctuary of "Aphrodite" at Idalion. (After Ohnefalsch-Richter 1893, vol. 2: pl. LVI).

Set in a wooded, idyllic landscape in the midst of stone precinct foundations, scattered fragments of sculptures and a few fully-preserved standing votaries imaginatively illustrate the sacred precinct after its destruction as viewed through the eyes of the excavator.[6] In addition to votive sculptures, the sacred space is dominated by a pyramidal-shaped baetyl, while votive masks hang on

---

6 Anja Ulbrich, "An Archaeology of Cult? Cypriot Sanctuaries in 19th century Archaeology," in *Cyprus in the 19th Century AD. Fact, Fancy and Fiction. Papers of the 22nd British Museum Classical Colloquium December 1998*, ed. Veronica Tatton-Brown (Oxford: Oxbow, 2001): 98–99.

trees around the precinct (recalling Ohnefalsch-Richter's own interpretation of their use at another site at Idalion, but creatively inserted here). A group of contemporary Cypriots in the background adds a living element to the otherwise cluttered, yet neglected, remains of the sanctuary's afterlife. Ohnefalsch-Richter's juxtaposition of the ancient with the contemporary brings ancient Cypriot worship to life by connecting the ancient stone proxies directly to the local nineteenth-century population.[7] Three men in traditional Cypriot *vrakas* (trousers)—one with a pickaxe—recall local workmen who would have excavated with Ohnefalsch-Richter; joining them is a lone female in white, whom Ohnefalsch-Richter links to the ancient statues in the foreground by noting their similarity in dress.[8]

Twentieth-century scholars were still struggling to offer a proper visualization of the complexity of objects, space, and context in Cypriot sanctuaries. The 1929 scientific excavation of the rural sanctuary of Ayia Irini along the north coast of Cyprus by the pioneering Swedish Cyprus Expedition (SCE) revealed an astonishing *in situ* discovery of more than two thousand terracotta statues (ranging from figurines to life-size statues), arranged in concentric semi-circles around a limestone altar, upon which sat a spherical, aniconic stone (Figure 3).[9] The excavation photographs provided a hitherto unparalleled and unadulterated snapshot of a sanctuary in action and the impact of the SCE's discovery at Ayia Irini on modern perceptions of Cypriot ritual space cannot be overestimated. While the excavations at Ayia Irini complemented earlier, antiquarian accounts of sanctuaries lined with votaries, the distinct and orderly arrangement of sculptures around the altar—larger figures at the back and smaller figurines in the front—inspired both the Cyprus Museum (Nicosia) and the Medelhavsmuseet (Stockholm, Sweden) to reproduce the discovery in their respective galleries with statues arranged in a semi-circular pattern according to scale, maintaining a recreation of how the statues were originally found.[10] Reflecting

---

[7] On the antiquarian tendency to discover unbroken links from antiquity to the present, see Anastasia Serghidou "Imaginary Cyprus. Revisiting the Past and Redefining the Ancient Landscape," in *Cyprus in the 19th Century AD. Fact, Fancy and Fiction. Papers of the 22nd British Museum Classical Colloquium December 1998*, ed. Veronica Tatton-Brown (Oxford: Oxbow, 2001): 23–24.

[8] Max Ohnefalsch-Richter, *Kypros, the Bible, and Homer. Oriental Civilization, Art and Religion in Ancient Times. Elucidated by the Author's Own Researches and Excavations During Twelve Years Work in Cyprus*, vol. 1 (London: Asher & Co, 1893): 399.

[9] Einar Gjerstad, John Lindros, Erik Sjöqvist, and Alfred Westholm, *The Swedish Cyprus Expedition II. Finds and Results of the Excavations in Cyprus 1927–31* (Stockholm: The Swedish Cyprus Expedition, 1935): 777–91.

[10] Kristian Göransson, "The Swedish Cyprus Expedition, The Cyprus Collections in Stockholm and the Swedish Excacvations After the SCE," *Cahiers du Centre d'Études Chypriotees* 42 (2012):

the post-excavation agreement to divide the finds equally between the two countries, each museum displayed roughly half of the total Ayia Irini assemblage (ca. 1,000 figures); in Sweden, the focal point remained the enigmatic spherical stone, which was retained by the Swedes after the division of artifacts. Interestingly, in 2009, when the Medelhavsmuseet reinstalled its Cypriot galleries, a mirror was added behind the statues, which offers visitors a creative visualization of the full assemblage.[11]

Figure 3: Swedish Cyprus Expedition excavation photograph of terracotta votive statues and figurines found in-situ placed in a semicircle around the altar. (Wikimedia Commons).

Nevertheless, traditional print publications still found it challenging to articulate votive statuary in sanctuaries using 2D drawings and photographs. In his publication of the sculptures from the Sanctuary of Apollo at Idalion, Reinhard Senff offered his own reconstruction of a Cypriot sanctuary. Drawing on the

---

413–16; Porphyrios Dikaios, *A Guide to the Cyprus Museum* (Nicosia: The Cyprus Government Printing Office, 1947): 60.
11 Sanne Houby-Nielsen, "Annual Report of the Medelhavsmuseet 2005–2008," *Medelhavsmuseet. Focus on the Medterranean* 4 (2009): 123–44.

plans of the earlier excavation of the sanctuary by Robert Hamilton Lang,[12] Senff constructed a small model and placed a selection of the sculptures (represented by paper cut-outs on stands) within the sacred precinct; this hypothetical visualization allows one to appreciate the distribution of statues within the sacred space, as well as the likelihood of both open and partially-covered "stoas," or gallery-like spaces (Figure 4). More importantly, Senff "populated' the space of the sanctuary, activating the lifeless sculptures from their passive museum display and static catalogue entries. Senff's simple, yet admirable, attempt to visualize the Idalion sanctuary in 3D underscores once again the limitations of working with significant, but essentially static, data sets and paper publication formats.

Figure 4: Model reconstruction of the Sanctuary of Apollo at Idalion. (© Reinhard Senff).

In her preface to the publication of the Archaic Precinct at the Sanctuary of Apollo Hylates at Kourion, Diana Buitron-Oliver also attempted to imagine, this time through written narrative, not only the Cypriot religious landscape, but also the highly structured, yet personal, acts of religious practice:

---

[12] Robert Lang and R. Stuart Poole, "Narrative of Excavations in a Temple at Dali (Idalium) in Cyprus," *Transactions of the Royal Society of Literature* 11 (1878): 30–54.

Now, in the distance, the man could see smoke rising and knew he was close to the sanctuary. Relief that his long journey was nearly over became mixed with apprehension of having to face the god and his priests. He had made the journey to ask for the god's help, for prosperity for his herds and crops, and also to give thanks for his good fortune so far, for his growing family and his expanding possessions. Approaching the outskirts of the sanctuary, he noticed the massively built stonewall encircling the sacred area. As he approached, the man passed potters and coroplasts displaying pots and figures of all sizes ... The coroplasts displayed statuettes of men wearing conical helmets, or human or bull masks, and other figures carrying small animals or musical instruments ... In back of the displays were larger and more expensive, hollow statues of figures wearing long tunics, some them armed with swords ... Local farmers had brought another popular offering: lambs and kids, new-born or very young, whose piteous bleating could be heard amidst the clamor of the artisans and groups of food sellers. The man had brought with him an object he wished to present to the god: an arrow with a bronze point, a token of thanks for safety in battle ... Soon the pilgrims were admitted into the sanctuary by a priest wearing a mask and dressed in long, dark robes ... As the pilgrims approached, the priests moved and swayed to the slow, solemn beat of drums, pipes, and cymbals played by other priests and attendants ... the smell of incense and burning flesh permeated the air and the rhythmic music was hypnotic, inducing a sort of awed trance in the pilgrims. The priests called the pilgrims forward one by one, questioning each man on what he hoped for from the god and what he had to offer. According to the response, the priest directed the pilgrim to a specific area in the sanctuary. The man with the arrow and young kid was directed to the south end of a narrow platform that ran through the circular altar. There he surrendered the kid which was quickly slaughtered and cut into portions ... The man also placed his arrow on the altar amid many other similar gifts. As he looked around, he saw mounds of offerings, vases and terracotta figures like the one he himself had purchased, neatly arranged in groups of like objects, around the circular altar, along the length of the platform, and on the smaller altars of the sanctuary. Satisfied and pleased that the goal of this long trip was accomplished, the man was escorted to the sanctuary gate ...[13]

Like the visual representations of Ohnefalsch-Richter and Senff, Buitron-Oliver's literary trope includes details that cater specifically to the archaeological discoveries of that site, yet nevertheless attempt to populate and enliven the sanctuaries that dominated the physical landscape of Cyprus in antiquity. These attempts offer important, even if hypothetical, insights into Cypriot cult and the experience and functioning of sanctuaries, as well as create a context for the ritual objects found within them. New digital technologies, however, now offer new ways for scholars to visualize religious landscapes, sanctuaries, and the objects used within them.

---

13 Diana Buitron-Oliver, *The Sanctuary of Apollo Hylates at Kourion: Excavations in the Archaic Precinct*. SIMA 109 (Jonsered: Paul Åström, 1996): xix–xx.

Digital Approaches to Cypriot Religion

New digital tools and approaches have enhanced our understanding of Cypriot sanctuaries. In some ways the emergence of something termed "digital" archaeology as methodologically or analytically novel is misleading.[14] From its beginnings, archaeology has always adopted/adapted new technology to record, measure, organize, and analyze data. Unlike many other disciplines that fall into the broader category of "digital humanities (DH)," archaeology occupies a rather unique place (or, better, a bridge?) at the intersection between the humanities, social sciences, and the natural/material sciences.[15] Such connections and collaborations have introduced a wide range of digital tools into archaeological practice at different levels of analysis from regional surveys that look at landscape and ancient ecology to site- and object-based queries that search for patterns of production and cultural interaction. Like their impact in other corners of DH, the scalability of digital applications in archaeological research has provided a robust means to examine both large and small data sets.

With the increasing adoption of new digital technologies in Cypriot archaeology, especially field-based, mobile computing, we are also witnessing the emergence of 'born-digital' data (i.e., data derived directly from digital tools rather than existing data *digitized*), which has prompted important shifts in the interpretation and dissemination of research. And while these newer digital approaches have yet to produce a paradigm-shifting moment in the broader discipline,[16] there is little doubt that the data they have produced has allowed us to ask new research questions (or approach old questions differently) and, more importantly, has transformed the nature, reproducibility, scalability, and accessibility of archaeological data itself. In fact, the true promise of DH is the ease with which one is able to reproduce and analyze data within different interpretive frameworks and at different levels of resolution, but also to compress scales (landscape, site, object) into a single, integrated picture.

---

14 Jeremy Huggett, "Challenging Digital Archaeology," *Open Archaeology* 1 (2015): 79–85.
15 Jeremy Huggett, "A Manifesto for an Introspective Digital Archaeology," *Open Archaeology* 1 (2015): 86–95.
16 Jody Gordon, Erin Walcek Averett, and Derek B. Counts, "Mobile Computing in Archaeology: Exploring and Interpreting Current Practices," in *Mobilizing the Past for a Digital Future: the Potential of Digital Archaeology*, ed. Erin Walcek Averett, Jody Gordon, and Derek B. Counts (Grand Forks, ND: Digital Press at the University of North Dakota, 2016): 1–32; *contra* Christopher Roosevelt, Peter Cobb, Emanuel Moss, Brandon Olson, and Sinan Ünlüsoy, "Excavation is Destruction Digitization: Advances in Archaeological Practice," *Journal of Field Archaeology* 40 (2015): 325–46.

The impact of the gradual but steady deployment of digital technologies in archaeological research on Cyprus can be witnessed in several projects. While by no means exhaustive, the following examples illustrate the use of digital approaches at differing scales of resolution to answer different questions, but also help us consider the potential for digital archaeology to reshape the study of Cypriot sanctuaries. The results of survey projects like the Troodos Archaeological and Environmental Survey Project (TAESP) provided a comprehensive "big picture" view of how landscapes (including religious landscapes) develop, change, and adapt over time. TAESP implemented an array of digital technologies to build a robust archaeological and geomorphological data set that addressed human occupation in the survey area (ca. 160 km^2) from the Neolithic to modern period; the results included a comprehensive landscape data set (e.g., land use patterns, communication networks, soils and sediments, vegetation, water, and resource exploitation) combined with interpretive approaches grounded in landscape theory. While the accumulation of large and diverse data sets has always been a feature (and, perhaps a bug!) of excavation and survey, TAESP was one of the first projects in Cyprus constructed with a solid digital backbone that combined analyses of large data sets (GIS) with interpretive approaches.[17] Notably, TAESP was also the first project on the island to harness the potential of digital data, as well as digital platforms for dissemination, to publish their results in an online, open-access article that linked data, discussion, and interpretation to an interactive GIS and integrated data archive.[18] As important as these results have been for the archaeology of Cyprus, TAESP's digital component and its accessibility via an online archive represent an early "fulfillment" of the on-going promise of the DH to create, curate, and disseminate information in an open platform.

Two ongoing projects in Cyprus are employing digital technologies to understand better the relationship between sacred sites and their environmental and urban contexts. Maria Iacovou and colleagues implemented the Palaepaphos Digital Atlas in 2002–2003, with a goal of creating a digital "umbrella," which

---

[17] Michael Given, Bernard Knapp, Jay Noller, Luke Sollars, and Vasiliki Kassianidou, *Landscape and Interaction: The Troodos Archaeological and Environmental Survey Project, Cyprus*, vol. 1, *Methodology, Analysis and Interpretation* (London: Council for British Research in the Levant, 2013); Michael Given, Bernard Knapp, Jay Noller, Luke Sollars, and Vasiliki Kassianidou, *Landscape and Interaction: The Troodos Archaeological and Environmental Survey Project, Cyprus*, vol. 2, *The TAESP Landscape* (London: Council for British Research in the Levant, 2013).

[18] Michael Given, Hugh Corley, and Luke Sollars, "Joining the Dots: Continuous Survey, Routine Practice and the Interpretation of a Cypriot Landscape (with interactive GIS and integrated data archive)," *Internet Archaeology* 20 (2007): https://doi.org/10.11141/ia.20.4.

also functions as a heritage management tool for the ancient kingdom of Palaepaphos that features a GIS linked to an entity-related geo-database combined with a multi-sensor geophysical survey. The Digital Atlas, coupled with the creation of a digital land relief map, enabled a more precise understanding of the diachronic land use of the area surrounding the Sanctuary of Aphrodite in Palaepaphos through sophisticated spatial and temporal landscape analysis. Using this high-resolution data, Iacovou and her team discovered that many of the archaeologically-rich areas surrounding the sanctuary were not designated as protected parcels by the Department of Antiquities and thus could potentially be destroyed by new urban development. This ultimately led to the Palaepaphos Urban Landscape Project (PULP) in 2006, whose aim was to harness cutting-edge digital methods to reassess the urban zones of Palaepahos and explore how the Aphrodite sanctuary related spatially and functionally to the ancient city using multiple scales of analysis (micro, medium, and macro). PULP made use of a multidimensional digital platform with GIS that organized archaeological and cartographic information in a relational database in addition to advanced documentation and imaging technologies, including Global Satellite units, UAVs, and geospatial analyses. Through harnessing the power and potential of digital tools and data, PULP has dramatically altered our understanding of the relationship between the famed goddess sanctuary and the urban core of Palaepaphos.[19]

Another ongoing, region-based project with a particular focus on sacred spaces is the Unlocking Sacred Landscapes Network, which is a collaborative project that aims to bring together different approaches to study the sacred landscapes of the Mediterranean from all time periods. Notably, many scholars working within this network employ a range of digital approaches, from 3D modelling to GIS analysis.[20] Under the broader aegis of the Unlocking Sacred Landscapes Network, the Settled and Sacred Landscapes of Cyprus project is focused on the Xeros River Valley and Cyprus more broadly by implementing digital technologies and integrated data sets to enrich our understanding of religious landscapes.[21] This initiative uses GIS to explore the diachronic use and the relation-

---

19 For a recent overview, with references to earlier publications, see Maria Iacovou, "Palaepaphos: Unlocking the Landscape Context of the Sanctuary of thee Cypriot Goddess," *Open Archaeology* 5 (2019): 204–34.
20 Giorgos Papantoniou, Apostolos Sarris, Christine E. Morris, Athanasios Vionis, "Digital Humanities and Ritual Space: A Reappraisal," *Open Archaeology* (2019) 598–614.
21 Giorgos Papantoniou and Athanasios Vionis, "Landscape Archaeology and Sacred Space in the Eastern Mediterranean: A Glimpse from Cyprus," *Land* 6 (2017): 40; Giorgos Papantoniou,

ship between sacred landscapes and political structures from the Iron Age through Hellenistic age and into the Roman and Early Christian periods. Such large-scale projects provide the social, cultural, and environmental contexts for deeper understanding of the role of religious spaces through time and point the way forward for more comprehensive syntheses of the island's religious landscapes.

The transition by Cypriot scholars to innovative, integrated, and open publications of archaeological data has been slow, with some noteworthy exceptions. For example, William Caraher, David Pettegrew, and Scott Moore followed TAESP's lead in making their data open and available via an online digital repository[22] for the Pyla-Koutsopetria Archaeological Project (PKAP) and even "retrofitted" their first excavation volume to provide an open-access, linked digital version.[23] PKAP's integration of digital data and publication foreshadows our own attempts, described below, to leverage the agile and dynamic nature of digital data to change how we study and present Cypriot religion.

Going from a regional to a site scale, digital tools have been applied to specific sites in Cyprus to provide creative ways of visualizing and recreating the use and construction of buildings and spaces, including sacred spaces. Paralleling the approaches above, which utilize digital platforms to organize and query large data sets, Nicholas Blackwell and James Johnson isolated depositional patterns for specific object types discovered at the sanctuary of Athienou-*Malloura* and analyzed them using GIS.[24] As Blackwell and Johnson note, at a basic level GIS functions as a database management system to plot and examine spatial relationships among data points; nevertheless, its powerful analytical functions allow for the integration of multiple layers of data that has the potential to answer a series of problem-based research questions about sanctuary use across

---

Apostolos Sarris, Christine E. Morris, Athanasios Vionis, "Digital Humanities and Ritual Space: A Reappraisal," *Open Archaeology* (2019) 598–614.

22 William Caraher, R. Scott Moore, and David Pettegrew, "Pyla-Koutsopetria Archaeological Project I: Pedestrian Survey," Open Context, 2013: DOI: https://doi.org/10.6078/M7B56GNS.

23 William Caraher, "Announcing the Digital Edition of Pyla-Koutsopetria I: A Free Download," *Archaeology of the Mediterranean World*, 15 Feb. 2017, https://mediterraneanworld.wordpress.com/2017/02/15/announcing-the-digital-edition-of-pyla-koutsopetria-1-a-free-download/; William Caraher, Scott Moore, and David Pettegrew, *Pyla-Koutsopetria I: Archaeological Survey of an Ancient Coastal Town*. Linked Version (Boston: American Schools of Oriental Research, 2017) http://dx.doi.org/10.17613/M6GB7K.

24 Nicholas Blackwell and James Johnson, "Exploring Sacred Space: GIS Applications for Analyzing the Athienou-*Malloura* Sanctuary," in *Crossroads and Boundaries. The Archaeology of Past and Present in the Malloura Valley, Cyprus*, eds. Michael Toumazou, P. Nick Kardulias, and Derek B. Counts (Boston: American Schools of Oriental Research, 2011): 291–302.

both space and time. In the case of the Malloura sanctuary, a series of hypotheses were "tested" to illustrate the potential of digital tools to address longstanding questions about the relationship between artifact findspots and Cypriot ritual behavior within sanctuaries (Figure 5). The authors acknowledged that some questions were more difficult to address (e.g., social status or the origin of worshippers [local vs. visitor]); nevertheless, the merging of artifact and geospatial data clearly identified distinct clusters of symbolically-charged objects (e.g., divine images, ritual objects), as well as clusters of distinct ritual activity associated with the practice of Cypriot religion.[25]

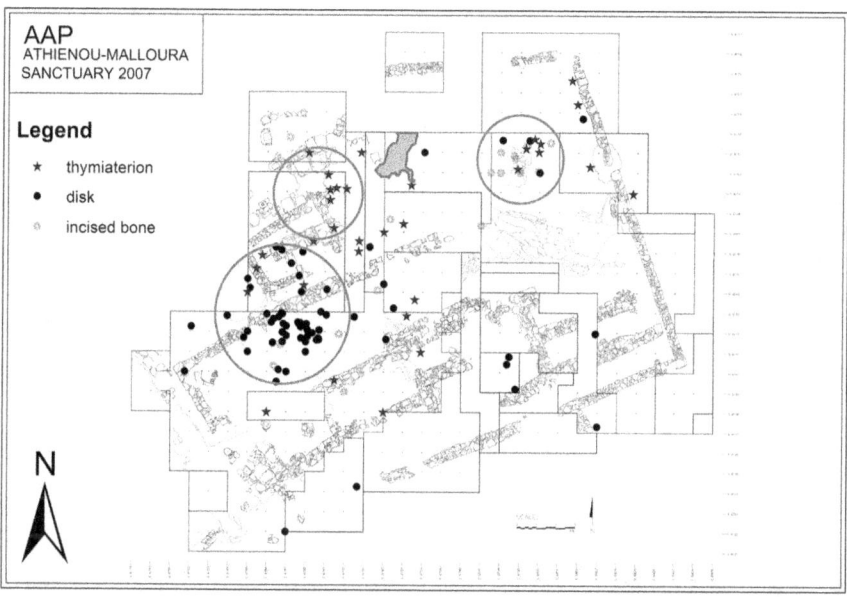

Figure 5: GIS map of votive distribution in the Athienou-*Malloura* Sanctuary. (After Blackwell and Johnson 2011).

Another example, the "Modeling the Past" project, created virtual walkthroughs of buildings and associated objects excavated by the Princeton Archaeological Expedition to Polis at the site of Polis Chrysochous on the north-

---

25 Nicholas Blackwell and James Johnson, "Exploring Sacred Space: GIS Applications for Analyzing the Athienou-*Malloura* Sanctuary," in *Crossroads and Boundaries, The Archaeology of Past and Present in the Malloura Valley, Cyprus*, eds. Michael Toumazou, P. Nick Kardulias, and Derek B. Counts (Boston: American Schools of Oriental Research, 2011): 300.

west coast of Cyprus, with a specific use for both research and education.[26] Two important Iron Age sanctuaries in Polis at the locales of Peristeries[27] and Maratheri[28] are brought to life through 3D visualization technology, which provides an enhanced understanding of visitor movement and experience in the sanctuary and allows the excavators not only to propose, but also to reconstruct virtually the spaces and a selection of associated artifacts. The project originated from a multi-disciplinary course at Princeton for an exhibition on the site, "City of Gold," but the long-term goal includes using the recreations for a permanent exhibition of objects in Cyprus and the development of an interactive web application.[29]

Digital tools now allow for architectural and artifactual recreations of fragmentary material previously executed by artists or through physical models in more interactive and dynamic platforms, which provide a more immersive experience of sanctuaries (for an innovative and compelling look at the application of emergent AR [augmented reality] and VR [virtual reality] technologies to stage and map in virtual space "counterfactual histories" based on archival data, see Kaplan and Schiff, this volume).[30] This approach has the potential to be expanded to allow users to experience multiple scales where in theory they start with a large regional map and have the ability to zoom into a specific sanctuary, and from there they can experience the architecture and setting of the sacred area including models of artifacts found there.

---

[26] Joanna Smith and Szymon Rusinkiewicz, "Modeling the Past for Scholars and the Public," *International Journal of Heritage in the Digital Era* 2:1 (2013): 167–94; see also "Polis, Cyprus—B.D7—Peristeries Sanctuary," Princeton University, http://polis-cyprus.princeton.edu/BD7/index.html#.

[27] "Polis, Cyprus—B.D7—Peristeries Sanctuary," Princeton University, http://polis-cyprus.princeton.edu/BD7/index.html.

[28] "Polis, Cyprus—A.H9—Maratheri Temple," Princeton University, http://polis-cyprus.princeton.edu/AH9/.

[29] For a more recent example of the use of new digital tools to map and visualize a sanctuary outside Cyprus, see Filippo Carraro, Alessandra Marinello, Daniele Morabito, and Jacopoo Bonetto, "New Perspectives on the Sanctuary of Aesculapius in Nora (Sardinia): From Photogrammetry to Visualizing and Querying Tools," *Open Archaeology* 5 (2019): 263–73.

[30] For a recent exploration of the use of digital models to create sensory experiences, see Costas Papadopoulos, Yannis Hamilakis, Nina Kyparissi-Apostoolika, and Marta Díaz-Guardamino, "Digital Sensoriality: The Neolithic Figurines from Koutroulou Magoula, Greece," *Cambridge Archaeological Journal* (2019): 1–28.

The Athienou Archaeological Project's 3D Initiative

Both the Malloura GIS and Polis 3D projects discussed above intentionally overlay site and object in their digital approaches to the material remains of Cypriot religion. In fact, at the level of individual artifact, digital visualization tools can be used to improve documentation, research, and publication of objects; likewise, these data can be easily re-integrated into larger site and landscape projects via digital platforms. There are some imaging projects already underway that are creating 3D repositories of individual artifacts, notably figurines and statuary from Nea Paphos, Salamis, and the region of Nicosia.[31] Since 2014, the authors have co-directed an innovative 3D documentation and publication initiative (under the umbrella of Davidson College's Athienou Archaeological Project, AAP, mentioned above), which fits well into this larger trend of exploring new technologies to document, understand, and analyze the material remains of the Cypriot past, including ancient religion (several papers in the volume focus on how digital tools are helping map *material* manifestations of religion in the past and in the present, e.g., Bielo and Vaughn; Pettit and Yang; Bingenheimer). AAP has been documenting the cultural landscape of the Malloura Valley in the hinterland of Cyprus through pedestrian survey and systematic excavation since 1990 (Figure 6). The Malloura Valley was occupied for nearly 3,000 years, beginning in the early first millennium BCE; our investigations have unearthed domestic, religious, and funerary contexts, with an impressive assemblage of material remains.[32] In 2014, AAP started experimenting with new visualization technolo-

---

[31] Giorgos Papantoniou, Fernando Loizides, Andreas Lanitis, and Dimitrios Michaelides, "Digitization, Restoration and Visualization of Terracotta Figurines from the 'House of Orpheus,' Nea Paphos, Cyprus," in *Progress in Cultural Heritage Preservation 4th International Conference, EuroMed 2012, Limassol, Cyprus, October 29–November 3, 2012. Proceedings*, Lecture Notes on Computer Sciences 7616, eds. Marinos Ioannides, Dieter Fritsch, Johanna Leissner, Rob Davies, Fabio Remondino, and Rossa Caffo (Heidelberg: Springer, 2012): 543–50; Gabriele Koiner, Nicole Reitinger, and Paul Bayer, "3D-Scans – erstmalig angewandt an Skulpturen im Cyprus Museum in Nikosia," Conference Talk, Archaeologie im 21. Jahrhundert-Fragen und Methoden einer modernen Geisteswissenschaft, 22 Juni 2018. Hauskolloquium am Institut fur Archaeologie, 2018; Vassos Karageorghis and Thomas Kiely, *Salamis*-Toumba. *An Iron Age Sanctuary in Cyprus Rediscovered. (Excavations of the Cyprus Exploration Fund, 1890)* (Nicosia: The Cyprus Institute, 2019): 128–29.

[32] Michael Toumazou, P. Nick Kardulias, and Derek B. Counts, eds., *Crossroads and Boundaries: The Archaeology of Past and Present in the Malloura Valley, Cyprus. Annual of the American Schools of Oriental Research 65* (Boston: American Schools of Oriental Research, 2011); Michael

gies to document the ancient past at Malloura at multiple scales from landscapes to features and artifacts. Although multiple sites have been excavated, the focus of excavations for the past 20 years has been an impressive rural sanctuary, which has yielded thousands of objects from its long use beginning in the eighth century BCE and continuing into the fourth century CE. The artifact assemblage from the sanctuary includes ceramic vessels, coins, animal bones, and other cult objects, as well as close to 4,000 pieces of figural limestone and terracotta sculpture, which are critical for reconstructing Cypriot cult practices and beliefs given the dearth of inscriptions or textual references (Figure 7).

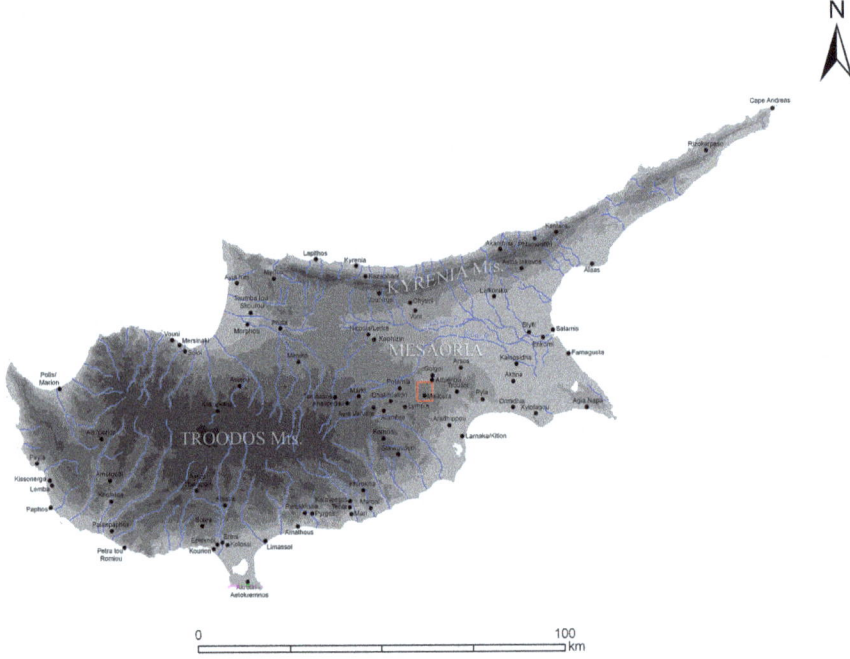

Figure 6: Map of Cyprus with AAP project area in red rectangular. (Map by D. Massey).

The sculptures, the material remains of dedications of local and perhaps even visiting worshippers, depict human, divine, and animal figures ranging in size from several centimeters to over-life-size. Given the quantity and impor-

---

Toumazou, Derek B. Counts, Erin Walcek Averett, Jody Gordon, and P. Nick Kardulias, "Shedding Light on the Cypriot Rural Countryside: Investigations of the Athienou Archaeological Project in the Malloura Valley, Cyprus, 2011–2013," *Journal of Field Archaeology* 40.2 (2015): 204–20.

tance of these limestone and terracotta votive offerings, we initiated our digital project to create 3D models with three primary goals in mind:
1. to document artifacts three-dimensionally to supplement traditional 2D photographs and illustrations;
2. to enable remote study of artifacts and to test (via digital tools) whether or not fragments of sculpture could potentially join together;
3. to allow for dynamic, open, linked, and interactive methods of disseminating information, from academic publication to public education and outreach.

Figure 7: Terracotta votive figurine of a chariot model (AAP-AM-1218+1459+2007, Kallinikeion Municipal Museum, Athienou, © AAP) and limestone statuette of Zeus-Ammon (AAP-AM-184, Larnaca District Archaeological Museum, © AAP).

Our work was conducted with the permission of the Department of Antiquities of Cyprus, who supports our ultimate aspiration of opening up the privileged experience of working with ancient artifacts to the broader scholarly and public communities (in working closely with the town of Athienou and the Department of Antiquities of Cyprus, we do not encounter the anxieties and issues regarding digital colonialism, access, and looting that Floyd and Promey, this volume, and others discuss).

We identified structured light scanning as the best technology to meet our research goals. Structured light technology entails the projection of a series of parallel light stripes onto an object; based on the displacement of the stripes, as viewed through the camera, the system can identify and retrieve the 3D coordinates on the surface of any object in view. This technique is a type of range-based modeling, which collects the surface geometry of the artifact as an absolute measurement (i.e., a feature measuring 7.5 cm on the artifact will be 7.5 cm on the digital model). 3D acquisition via structured light stands in contrast to image-based modeling systems such as photogrammetry, in which the scale of the resulting model is relative. In our case, structured light technology offered the right combination of high metric accuracy and photo-realistic texture at an efficient speed of production, with minimal post-processing. After an informative, but challenging, pilot season where we used a DIY approach in collaboration with colleagues at the University of Kentucky,[33] we researched and ultimately selected GoMeasure3D's HDI R1X structured light scanner (https://gomeasure3d.com/hdi/advance/) as an optimal system. The complete scanner set-up (which included an automated turntable to increase efficiency, hard cases for international transport, and a high-performance laptop) cost approximately $ 17,000 (supported by an internal grant program at the University of Wisconsin-Milwaukee). This system matched our particular research and publication goals, which required a fast scanner and user-friendly, efficient post-processing software that would create high-resolution 3D models. Primary scanning with the HDI R1X began in 2015 at the Larnaka Archaeological District Museum and at the Kallinikeion Municipal Museum in Athienou. At its most efficient, the scanner generates 1.1 million points per scan, with a normal accuracy range of 65 to 125 microns. One specific benefit of this system is the ability to change the field of view, allowing us to capture a wide range of artifact sizes —from roughly 5 cm to 65 cm. The output is a model with high resolution— around 1.5 million faces for a medium size object—and accurate photo texture. Once set up and properly calibrated, a complete scan, including post-processing, can be achieved in less than 3 minutes.

---

33 Derek B. Counts, Erin Walcek Averett, and Kevin Garstki, "A Fragmented Past: (Re)constructing Antiquity through 3D Artefact Modeling and Customised Structured Light Scanning at Athienou-*Malloura*, Cyprus," *Antiquity* 90 (2016): 206–18.

The AAP 3D Models

Why 3D? 3D models provide numerous benefits for studying votive offerings that are simply not provided by traditional illustrations and figures. For example, since 3D models produced from structured light technology retain their absolute geometry, they can be digitally measured using tools available in open-access visualization software, such as MeshLab, Blender, Unity, or Sketchup. 3D software also allows one to adjust lighting or turn off the photo-texture to accentuate variations in surface details like inscriptions, tool marks, or fingerprints that might otherwise be invisible to a camera or even the naked eye when studying objects in person (Figure 8).

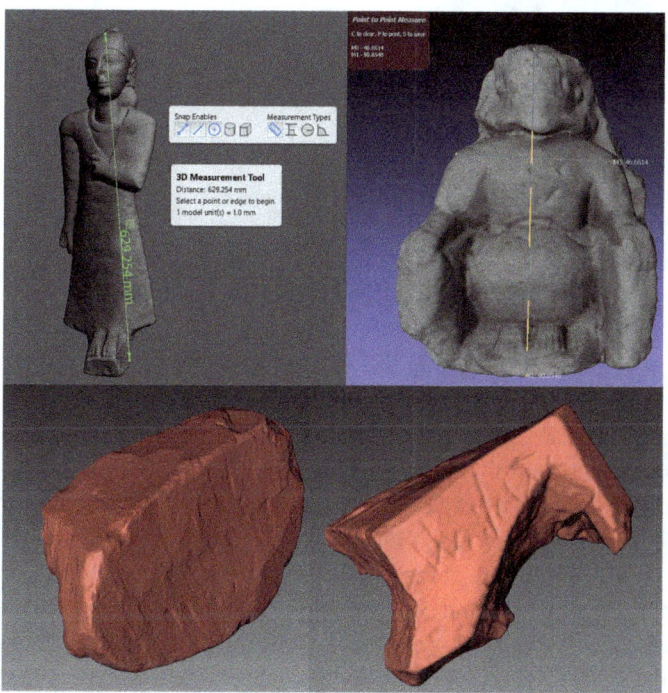

Figure 8: Image showing different viewing options for models: digital measuring, full rotation, texture off to display tools marks and inscriptions. (Image by K. Garstki).

In addition, 3D models allow viewers to experience objects three dimensionally through rotation and zoom at any scale, providing unlimited remote access for more dynamic investigation. The ability to study, measure, and manipulate objects more closely replicates the physical experience of studying the artifacts firsthand than working with 2D photographs or illustrations, which provide only selected views.[34] Finally, digital models allow users to join—virtually—broken pieces of sculpture as single pieces (Figure 9). All of these advantages have immense implications for secondary research applications as well as for public outreach and education.

Figure 9: Two fragments from a life-size limestone male statue, preserving the sandaled feet on a base, digitally joined (AAP-AM-3900, Larnaca District Archaeological Museum, © AAP).

From Born-Digital 3D Data to Digital Publication

One of our primary goals from the beginning of this project was to incorporate 3D data into our final publication volumes for the site of Athienou-*Malloura*, and to do this in an open, transparent, and dynamic format that does not detach artifacts from their archaeological contexts. While the potential of 3D models to enhance the documentation and analysis of archaeological material is tremendous, the difficulties in publishing these in a holistic and reproducible way remain challenging. Due to the nature of the 3D data, traditional print publications are simply not an option. Instead, our 3D scanning project has culminated in a digital, open-access interactive catalogue of a representative selection of limestone and terracotta sculptures recovered from the Malloura sanctuary together with a series of essays detailing our methods in addition to providing analysis of this body of material: *Visualizing Votive Practice: Exploring Limestone and Terra-*

---

[34] On the use of digital models to replicate the physical experience of handling artifacts, see Costas Papadopoulos, Yannis Hamilakis, Nina Kyparissi-Apostoolika, and Marta Díaz-Guardamino, "Digital Sensoriality: The Neolithic Figurines from Koutroulou Magoula, Greece," *Cambridge Archaeological Journal* (2019): 1–28. DOI: https://doi.org/10.1017/S0959774319000271.

cotta Sculpture from Athienou-Malloura Through 3D Models.³⁵ While traditional print catalogues include photographs or illustrations with each entry, the *Visualizing Votive Practice* digital catalogue instead includes an embedded visualization interface (using Adobe 3D PDF), which allows the user direct interaction with the 3D artifact model. This publication thus retains a traditional framework for publishing the Malloura sculptures but, significantly, is born digital and published open-access through a Creative Common license (see Floyd and Promey, this volume, who also discuss the benefits of combining aspects of analogue and digital publication formats). Information on the archaeological context, formal description, and analysis accompanies high-resolution, photorealistic 3D digital models, providing new information and a dynamic visualization platform for researchers not available through 2D images. Comparative material, when available via open-access licenses (e. g., the Cypriot collection of the Metropolitan Museum of Art, now in the public domain), and even secondary literature (e. g., antiquarian volumes available via Internet Archive) is directly linked within the object commentaries, giving the reader instant access.

Published by The Digital Press at The University of North Dakota (https://thedigitalpress.org/VVP/) the core platform of our open-access digital publication is Adobe Acrobat, which allows for the creation of robust 3D PDFs. Adobe's PDF format is easily accessible through the full suite of licensed Adobe products or the free-to-download Adobe Reader DC® and its 3D visualization tools. In tandem with this digital book, however, higher-resolution models and their associated metadata are published online by Open Context (https://opencontext.org/projects/116-visualizing-votive-practice-exploring-limestone-and-te). This provides a stable database repository for the published 3D data by providing Digital Object Identifiers (DOIs) for each artifact and its associated metadata. Significantly, the linked data published through Open Context in database form is queryable, enabling researchers to aggregate different information. Finally, a third access point to the material is provided by the presentation of high-resolution models on SketchFab, a popular platform for sharing, buying, and selling 3D models online; many museums and archaeological projects have published their models to this site. The *Visualize Votive Practice* models are presented on the Athienou Archaeological Project's SketchFab collection, with key features annotations and condensed contextual information (https://sketchfab.com/AthienouArchaeologicalProject/collections/vis

---

35 Derek B. Counts, Erin Walcek Averett, Kevin Gartski, and Michael Toumazou, *Visualizing Votive Practice: Exploring Limestone and Terracotta Sculpture from Athienou-Malloura Through 3D Models* (Grand Forks, ND: Digital Press at The University of North Dakota, 2020).

ualizing-votive-practice). The primary value in this venue is a platform built for a more general audience that will benefit from public-facing scholarship, a superior visualization of the models, providing the most photorealistic viewing platform, and from a practical viewpoint the ability to view all models together in thumbnail view.

Our experimental approach to collaborative, open-access publication engages the ongoing discussions of best practices in publishing 3D data and new approaches to disseminating archaeological data and (we hope) sets an important precedent for future archaeological publication. More importantly, we envision the volume and digital publication as providing a substantial and dynamic resource for the study of Cypriot religion, ritual, and sanctuaries (for more discussion on the advantages of digital publication models see, for example, Bielo and Vaughn and Promey and Floyd in this volume).

Future Directions

The urge to recreate the past through visual representation is not new. After all, we experience the world around us in 3D and so it is only natural to replicate this experience to understand the past. Visualizing fragmentary artifacts and buildings and imagining past landscapes and sites has always been part of the creative process of archaeology—a process that interprets, but also humanizes events that took place long ago.[36] After all, archaeology is as much about reconstructing the parts we don't know as it is about presenting the parts we do. Archaeologists have traditionally utilized whatever tools were available to achieve this end: narrative or even poetic description, photography, illustration, artistic reconstructions, physical models, and now digital tools. The "turn to the digital" thus in some ways does not represent a radically new approach, but is simply the newest tool to aid in envisaging the past (see Floyd and Promey, this volume, with a similar view against a radical break with past practices in digital publication and on the significance of digital tools to render three-dimensional artifacts more dynamically, realistically, and accurately).

---

36 Joanne Pillsbury, ed., *Past Presented. Archaeological Illustration and the Ancient Americas* (Washington, D.C.: Dumbarton Oaks, 2012); Michael Shanks and Connie Svabo, "Archaeology and Photography: A Pragmatology," in *Reclaiming Archaeology: Beyond the Tropes of Modernity*. Archaeological Orientations, ed. Alfredo González-Ruibal (New York: Routledge, 2013): 89–102.

At the same time, however, the tools we use impact and shape the knowledge we create.[37] Digital technologies offer new ways to see, record, interpret, and present the past. The use of new tools allows us to ask new research questions that enhance, not replace, traditional scholarship on Cypriot votive religion. As we have demonstrated, our 3D visualization project was the direct result of a need for more dynamic ways of recording and studying artifacts. Attempts to create 3D renderings of Cypriot art began in the late 1990s using QuickTime VR, developed for Apple, which allowed for three-dimensional, multi-angle viewing of artifacts; in particular, the Cobb Institute of Archaeology provided web-based views of artifacts from the Pierides Museum in Larnaka. AAP was also an early adopter of 3D imaging in the early 2000s, showcasing a series of artifacts for web-viewing. Still, such early attempts to capture the materiality of objects were in some ways superficial, their primary purpose was to display objects on websites without much concern for contextual information, metadata, accurate geometry, or visualization tools that allow for more nuanced examination, let alone proper data archiving. This mirrors the larger divide that persists today in the applications of digital tools to humanistic inquiries. Noting the schism between scholars who use digital technologies in studying traditional humanities subjects and those who use the methods of the contemporary humanities in studying digital objects, Kathleen Fitzpatrick states, "Those differences often produce significant tension, particularly between those who suggest DH should always be about *making* (whether making archives, tools, or new digital methods) and those who argue that it must expand to include *interpreting*."[38] Our 3D project ambitiously melds these two goals in creating an open, stable, and linked archive of votive statuary from our site in addition to harnessing this data to yield new insights and interpretations not only for our sanctuary, but for Cypriot religion more broadly.

As this contribution has highlighted, the future of scholarly study of Cypriot religion is bright. A new generation of scholars has already begun to explore the true potential of digital approaches to enhance the study of the Cypriot past

---

37 Colleen Morgan and Holly Wright, "Pencils and Pixels: Drawing and Digital Media in Archaeological Field Recording," *Journal of Field Archaeology* (2018): 1–16. DOI: 10.1080/00934690.2018.1428488; Matt Edgeworth, "Multiple Origins, Development, and Potential of Ethnographies of Archaeology," in *Ethnographies of Archaeological Practice. Cultural Encounters, Material Transformations*, ed. Matt Edgeworth (Lanham: AltaMira Press, 2006): 1–19; Bruno Latour and Steve Woolgar, *Laboratory Life: The Construction of Scientific Facts*. Second edition (Princeton: Princeton University Press, 1986).
38 Kathleen Fitzpatrick, "The Humanities, Done Digitally," in *Debates in the Digital Humanities*, ed. Matthew Gold (Minneapolis: University of Minnesota Press, 2012): 12–15.

broadly and the religious landscape more specifically. At the same time archaeologists are looking to new, open-access platforms for publication, institutions are increasingly turning to open science models to allow access to objects/data as well as research.[39] On a large scale, digital tools allow for the collection of extensive data sets, but more importantly they also allow for linked, open data that can be analyzed, compared, and interpreted in myriad ways. On the one hand, more regional-scale projects across the island will enhance our picture of the Cypriot religious landscape and allow more comparative analysis across regions and sites; and on the other hand, projects like ours that focus on specific sanctuaries or specific classes of religious materials are beginning to provide higher-resolution, granular details of ritual and cult activity. These two approaches (at a macro to micro scale) are transforming our understanding of Cypriot Iron Age religion.

Selected References

Averett, Erin Walcek, Jody M. Gordon, and Derek B. Counts, eds. *Mobilizing the Past for a Digital Future: The Potential of Digital Archaeology*. Grand Forks, ND: Digital Press at the University of North Dakota, 2016.

Blackwell, Nicholas G., and James A. Johnson. "Exploring Sacred Space: GIS Applications for Analyzing the Athienou-*Malloura* Sanctuary." In *Crossroads and Boundaries. The Archaeology of Past and Present in the Malloura Valley, Cyprus*, edited by Michael K. Toumazou, P. Nick Kardulias, and Derek B. Counts, 291–302. Boston: American Schools of Oriental Research, 2011.

Buitron-Oliver, Diana. *The Sanctuary of Apollo Hylates at Kourion: Excavations in the Archaic Precinct*. SIMA 109. Jonsered: Paul Åström, 1996.

Caraher, William. "Understanding Digital Archaeology." *Archaeology of the Mediterranean World*, https://mediterraneanworld.wordpress.com/2015/07/17/understanding-digital-archaeology/, 2015.

Caraher, William. "Announcing the Digital Edition of Pyla-Koutsopetria I: A Free Download." *Archaeology of the Mediterranean World*. https://mediterraneanworld.wordpress.com/2017/02/15/announcing-the-digital-edition-of-pyla-koutsopetria-1-a-free-download/, 2017.

Caraher, William, R. Scott Moore, and David Pettegrew. "Pyla-Koutsopetria Archaeological Project I: Pedestrian Survey." Open Context (2013). DOI: https://doi.org/10.6078/M7B56GNS.

Caraher, William, R. Scott Moore, and David K. Pettegrew. *Pyla-Koutsopetria I: Archaeological Survey of an Ancient Coastal Town*. Linked Version. Boston: American Schools of Oriental Research, 2017. http://dx.doi.org/10.17613/M6GB7K

---

[39] e.g., Derek B. Counts, Review of Antoine Hermary and Joan Mertens, *The Cesnola Collection of Cypriot Art: Stone Sculpture* 1st rev. ed. (New York, 2015). *Bulletin of the American Schools of Oriental Research* 378 (2017): 242–44.

Carraro, Filippo, Alessandra Marinello, Daniele Morabito, and Jacopoo Bonetto. "New Perspectives on the Sanctuary of Aesculapius in Nora (Sardinia): From Photogrammetry to Visualizing and Querying Tools." *Open Archaeology* 5 (2019): 263–73.

Connelly, Joan B. "Standing Before One's God: Votive Sculpture and the Cypriot Religious Tradition." *Biblical Archaeology* 52 (1989): 210–18.

Counts, Derek B. "Review of Antoine Hermary and Joan Mertens, *The Cesnola Collection of Cypriot Art: Stone Sculpture* 1st rev. ed. (New York, 2015)." *Bulletin of the American Schools of Oriental Research* 378 (2017): 242–44.

Counts, Derek B., Erin W. Averett, and Kevin J. Garstki. "A Fragmented Past: (Re)constructing Antiquity through 3D Artefact Modeling and Customised Structured Light Scanning at Athienou-*Malloura*, Cyprus." *Antiquity* 90 (2016): 206–18.

Counts, Derek B., Erin Walcek Averett, Kevin J. Garstki, and Michael K. Toumazou. Visualizing Votive Practice: Exploring Limestone and Terracotta Sculpture from Athienou-Malloura through 3D Models. Grand Forks, ND: Digital Press at the University of North Dakota, 2020.

Dikaios, Porphyrios. *A Guide to the Cyprus Museum*. Nicosia: The Cyprus Government Printing Office, 1947.

Edgeworth, Matt. "Multiple Origins, Development, and Potential of Ethnographies of Archaeology." In *Ethnographies of Archaeological Practice. Cultural Encounters, Material Transformations*, edited by Matt Edgeworth, 1–19. Lanham: AltaMira Press, 2006.

Fitzpatrick, Kathleen. "The Humanities, Done Digitally." In *Debates in the Digital Humanities*, edited by Matthew Gold, 12–15. Minneapolis: University of Minnesota Press, 2012.

Forte, Maurizio. "Introduction to Cyber-Archaeology." In *Cyber-Archaeology. British Archaeological Reports International Series* 2177, edited by Maurizio Forte, 9–14. Oxford: Archaeopress, 2010.

Forte, Maurizio. "Cyber Archaeology: A Post-virtual Perspective." In *Between Humanities and the Digital*, edited by Patrik Svensson and David T. Goldberg, 295–309. Cambridge, MA: MIT Press, 2015.

Garstki, Kevin, ed. *Critical Archaeology in the Digital* Age. Los Angeles, CA: Cotsen Institute of Archaeology, forthcoming.

Gill, Christopher, Norman Postlethwaite, and Richard Seaford, eds. *Reciprocity in Ancient Greece*. Oxford: Oxford University Press, 1998.

Given, Michael, Hugh Corley, and Luke Sollars. "Joining the Dots: Continuous Survey, Routine Practice and the Interpretation of a Cypriot Landscape (with interactive GIS and integrated data archive)." *Internet Archaeology* 20 (2007): https://doi.org/10.11141/ia.20.4.

Given, Michael, A. Bernard Knapp, Jay Noller, Luke Sollars, and Vasiliki Kassianidou. *Landscape and Interaction: The Troodos Archaeological and Environmental Survey Project, Cyprus*. Vol I, *Methodology, Analysis and Interpretation*. London: Council for British Research in the Levant, 2013.

Given, Michael, A. Bernard Knapp, Jay Noller, Luke Sollars, and Vasiliki Kassianidou. *Landscape and Interaction: The Troodos Archaeological and Environmental Survey Project, Cyprus*. Vol 2, *The TAESP Landscape*. London: Council for British Research in the Levant, 2013.

Gjerstad, Einar, John Lindros, Erik Sjöqvist, and Alfred Westholm. *The Swedish Cyprus Expedition*. Vol. II, *Finds and Results of the Excavations in Cyprus 1927–31*. Stockholm: The Swedish Cyprus Expedition, 1935.

Göransson, Kristian. "The Swedish Cyprus Expedition, The Cyprus Collections in Stockholm and the Swedish Excacvations After the SCE." *Cahiers du Centre d'Études Chypriotees* 42 (2012): 413–16.

Gordon, Jody M., Erin W. Averett, Derek B. Counts, Kyo Koo, and Michael K. Toumazou. "Mobile Computing in Archaeology: Exploring and Interpreting Current Practices." In *Mobilizing the Past for a Digital Future: The Potential of Digital Archaeology*, edited by Erin W. Averett, Jody M. Gordon, and Derek B. Counts, 1–32. Grand Forks, ND: Digital Press at the University of North Dakota, 2016.

Houby-Nielsen, Sanne. "Annual Report of the Medelhavsmuseet 2005–2008." *Medelhavsmuseet. Focus on the Medterranean* 4 (2009): 123–44.

Huggett, Jeremy. "Archaeology and the New Technological Fetishism." *Archeologia e Calcolatori* 15 (2004): 81–92.

Huggett, Jeremy. "Challenging Digital Archaeology." *Open Archaeology* 1 (2015): 79–85.

Huggett, Jeremy. "A Manifesto for an Introspective Digital Archaeology." *Open Archaeology* 1 (2015): 86–95.

Iacovou, Maria. "Palaepaphos: Unlocking the Landscape Context of the Sanctuary of the Cypriot Goddess." *Open Archaeology* 5 (2019): 204–34.

Kansa, Eric, Sarah Kansa, and Ethan Watrall, eds. *Archaeology 2.0: New Tools for Communication and Collaboration*. Los Angeles: Cotsen Institute of Archaeology, 2012.

Karageorghis, Vassos, and Thomas Kiely. *Salamis*-Toumba. *An Iron Age Sanctuary in Cyprus Rediscovered* (Excavations of the Cyprus Exploration Fund, 1890). Nicosia: The Cyprus Institute, 2019.

Koiner, Gabriele, Nicole Reitinger and Paul Bayer. "3D-Scans – erstmalig angewandt an Skulpturen im Cyprus Museum in Nikosia." Conference talk, Archäologie im 21. Jahrhundert-Fragen und Methoden einer modernen Geisteswissenschaft, 22 Juni 2018. Hauskolloquium am Institut für Archäologie. 2018.

Lake, Mark. "Open Archaeology." *World Archaeology* 44, no.4 (2012): 471–78. DOI: 10.1080/00438243.2012.748521

Lang, Robert H., and R. Stuart Poole. "Narrative of Excavations in a Temple at Dali (Idalium) in Cyprus." *Transactions of the Royal Society of Literature* 11 (1878): 30–54.

Latour, Bruno, and Steve Woolgar. *Laboratory Life: The Construction of Scientific Facts*. Second edition. Princeton: Princeton University Press, 1986.

Levy, Thomas, ed. "Cyber-Archaeology." Special issue, *Near Eastern Archaeology* 77.3 (2014).

Morgan, Colleen, and Holly Wright. "Pencils and Pixels: Drawing and Digital Media in Archaeological Field Recording." *Journal of Field Archaeology* (2018): 1–16. DOI: 10.1080/00934690.2018.1428488.

Ohnefalsch-Richter, Max. *Kypros, the Bible, and Homer. Oriental Civilization, Art and Religion in Ancient Times. Elucidated by the Author's Own Researches and Excavations During Twelve Years Work in Cyprus*. 2 vols. London: Asher & Co, 1893.

Papadopoulos, Costas, Yannis Hamilakis, Nina Kyparissi-Apostoolika, and Marta Díaz-Guardamino. "Digital Sensoriality: The Neolithic Figurines from Koutroulou Magoula, Greece." *Cambridge Archaeological Journal* (2019): 1–28. DOI: https://doi.org/10.1017/S0959774319000271.

Papantoniou, Giorgos, and Athanasios Vionis. "Landscape Archaeology and Sacred Space in the Eastern Mediterranean: A Glimpse from Cyprus." *Land* 6 (2017). https://doi.org/10.3390/land6020040.

Papantoniou, Giorgos, Fernando Loizides, Andreas Lanitis, and Michaelides Dimitrios. "Digitization, Restoration and Visualization of Terracotta Figurines from the 'House of Orpheus,' Nea Paphos, Cyprus." In *Progress in Cultural Heritage Preservation 4th International Conference, EuroMed 2012, Limassol, Cyprus, October 29–November 3, 2012. Proceedings, Lecture Notes on Computer Sciences 7616*, edited by Marinos Ioannides, Dieter Fritsch, Johanna Leissner, Rob Davies, Fabio Remondino, and Rossa Caffo, 543–50. Heidelberg: Springer, 2012.

Papantoniou, Giorgos, Sarris Apostolos, Christine E. Morris, and Vionis K. Athanasios. "Digital Humanities and Ritual Space: A Reappraisal." *Open Archaeoology* (2019): 598–614.

Pillsbury, Joanne, ed. *Past Presented. Archaeological Illustration and the Ancient Americas*, edited by J. Pillsbury. Washington, D.C.: Dumbarton Oaks, 2012.

Roosevelt, Christopher, Peter Cobb, Emanuel Moss, Brandon Olson, and Sinan Ünlüsoy. "Excavation is ~~Destruction~~ Digitization: Advances in Archaeological Practice." *Journal of Field Archaeology* 40 (2015): 325–46.

Senff, Reinhard. *Das Apollonheiligtum von Idalion. Architektur und Statuenausstattung eines zyprischen Heiligtums.* Jonsered: Paul Åströms, 1993.

Serghidou, Anastasia. "Imaginary Cyprus. Revisiting the Past and Redefining the Ancient Landscape." In *Cyprus in the 19th Century AD. Fact, Fancy and Fiction. Papers of the 22nd British Museum Classical Colloquium December 1998*, edited by Veronica Tatton-Brown, 21–31. Oxford: Oxbow, 2001.

Shanks, Michael, and Connie Svabo. "Archaeology and Photography: A Pragmatology." In *Reclaiming Archaeology: Beyond the Tropes of Modernity*. Archaeological Orientations, edited by A. González-Ruibal, 89–102. Milton Park, Albingdon: Routledge, 2013.

Sjöqvist, Erik. "Die Kultgeschichte eines cyprischen Temenos." *Archiv für Religionswissenschaft* 30 (1933): 308–59.

Smith, Joanna S., and Szymon M. Rusinkiewicz. "Modeling the Past for Scholars and the Public." *International Journal of Heritage in the Digital Era* 2, no.1 (2013): 167–94.

Tatton-Brown, Veronica, ed. *Cyprus in the 19th Century AD. Fact, Fantasy and Fiction. Papers of the 22nd British Museum Classical Colloquium December 1998*. Oxford: Oxbow Books, 2001.

Toumazou, Michael K., P. Nick Kardulias, and Derek B. Counts, eds. *Crossroads and Boundaries: The Archaeology of Past and Present in the Malloura Valley, Cyprus*. Annual of the American Schools of Oriental Research 65. Boston: American Schools of Oriental Research, 2011.

Toumazou, Michael K., Derek B. Counts, Erin W. Averett, Jody M. Gordon, and P. Nick Kardulias. "Shedding Light on the Cypriot Rural Countryside: Investigations of the Athienou Archaeological Project in the Malloura Valley, Cyprus, 2011–2013." *Journal of Field Archaeology* 40.2 (2015): 204–20.

Ulbrich, Anja. "An Archaeology of Cult? Cypriot Sanctuaries in 19th century Archaeology." In *Cyprus in the 19th Century AD. Fact, Fancy and Fiction. Papers of the 22nd British Museum Classical Colloquium December 1998*, edited by Veronica Tatton-Brown, 93–106. Oxford: Oxbow, 2001.

van Straten, Folkert T. "Votives and Votaries in Greek Sanctuaries." In *Oxford Readings in Greek Religion*, edited by Richard Buxton, 191–226. Oxford: Oxford University Press, 2000.

Zubrow, Ezra B. W. "Digital Archaeology: A Historical Context." In *Digital Archaeology: Bridging Method and Theory,* edited by Thomas L. Evans, 10–32. New York: Routledge, 2006.

Part III: **Places**

Abhishek Amar
Sacred Centers in India: Archiving Temples and Images of a Hindu City

Introduction

The '*Sacred Centers in India*' project is a digital archive of Hindu Gaya and Buddhist Bodhgaya, which began in April 2013 at the Digital Humanities Initiative (DHi) at Hamilton College.[1] Through this interdisciplinary and collaborative archive, the project seeks to examine the complex historical development of these two sites. The first phase of the project, 2013–2018, focused exclusively on the development of a multi-layered archive of material remains housed in the twenty Hindu shrines and temples of Gaya. Equally important during this phase was to develop a 3-D and a VR model for pedagogical purposes. The second phase will focus at completing the archive by including data from thirty-five additional shrines and temples. DHi, funded initially through the Andrew W. Mellon Foundation grant in 2010, is a research and teaching collaboration that uses new media and computing technologies to promote humanities-based research, scholarship and teaching across the liberal arts. So far, it has received two grants from the Mellon foundation, which has helped in developing a sustainable digital infrastructure at Hamilton.

Being the preeminent site for the performance of the ancestral rites, the city of Gaya in the Bihar province of India has been accorded a central place in the oldest descriptions of pilgrimage in the Hindu scriptures.[2] Innumerable pilgrims from different parts of India have visited the city for millennia, and shaped its contours through their ritual practices and patronage. This multi-layered history of Gaya has not been critically examined. Previous studies have accorded primacy to the texts and scriptures, specifically *Gaya-Māhātmya*, and treated material remains as supplementary. Often these works have relied exclusively on texts to develop a framework, within which they have placed the inscriptions, sculp-

---

[1] I would like to acknowledge my thanks to Janet Simons and the DHi team for their help and support in developing this project. My UG research students, especially Lauren Scutt, were instrumental in the first phase of this project.
[2] Matthew Sayers, "Feeding the Dead at Gaya: From the Buddha's Enlightenment to a Modern Pilgrimage," speech at the South Asian Studies Colloquium at The University of Pennsylvania, Philadelphia, PA, 2012; Lalita Prasad Vidyarthi, *The Sacred Complex in Hindu Gaya* (London: Asia Publishing House, 1961).

tures, and architectural remains. This treatment of material remains ignores their ability to convey an alternative and a much more complex history of the city. This lacuna led me to conduct an archaeological survey of the temples and shrines of Gaya in 2011, which resulted in the documentation of fifty-five temples and shrines, each of which contained sculptures and images. I began to explore new ways of organizing the documented materials to pursue my research questions, which made me contact the then director of DHi, Janet Simons. She has been instrumental in conceiving and developing this archive project since then. In short, this project has adopted an interdisciplinary approach to organize and digitally curate a collected dataset and pursue my research about the multi-layered past of the Hindu city of Gaya. In the last six years, we have developed a web archive that includes a Virtual Reality component, a 3-D model of the Vishnupada temple, a digital database, and GIS information of twenty shrines and temples of Gaya (http://sci.dhinitiative.org/).

Background and History of Gaya

Gaya has been a sacred center from at least the time of the Buddha, in the middle of the first millennium before the Common Era. The Pali canon, especially *Sutta Piṭaka*, describes Gaya as a place renowned for ascetic activities and purificatory baths.[3] By the beginning of the Common Era, this site began to emerge as a place for the performance of ancestral rituals, *Śrāddha*.[4] Over the next several centuries this place of ancestral ritual evolved into a place centered around the Hindu god Vishnu, whose footprint became the focal point for ancestral offerings. In this period, the history of Gaya truly begins to evolve; it is expressed most often in legendary accounts of the myth of its origin and a detailed description of the pilgrimage. These accounts become more formalized in the early medieval period and come to be called the *Gaya-Māhātmya*, literally "the greatness of Gaya." Material remains indicate that during this burgeoning of textual historiography, an increasing number of images and new shrines emerged in this city; these religious artifacts were the focus of a growing cult dedicated to Vishnu.[5] Because of its strategic location, along routes between eastern and central

---

[3] Matthew Sayers, "Gaya-BodhGaya: The Origin of a Pilgrimage Complex," *Religions of South Asia* 4.1 (2010): 9–25.

[4] Sayers, "Feeding the Dead at Gaya"; Claude Jacques, *Gaya Māhātmya* (Pondichéry: Institut Français d'Indologie, 1962); Vidyarthi, *The Sacred Complex in Hindu Gaya*.

[5] Abhishek S. Amar, "Buddhist Responses to Brāhmaṇa Challenges in Medieval India: Bodhgayā and Gayā." *Journal of the Royal Asiatic Society* 22 (2012): 155–85.

India, the city underwent massive reconstruction in the eighteenth and nineteenth centuries. Gaya found itself at the nexus of a major road and rail network in the late nineteenth century, and this central position resulted in a significant increase in the number of pilgrims.[6] The previously textual or theoretical pan-Indian appeal became a logistically possible reality. Today pilgrims come from all over India for the pilgrimage. This is a pilgrimage that has been repeated over the last two millennia and resulted in the addition of multiple layers of sacrality to Gaya. It is this layered history that my project seeks to uncover. In doing so, the project will explore questions about the demarcation and negotiation of space and ritual, geographic and historical dimensions of pilgrimage, the ideological and material construction of sacred centers and Hinduism broadly, and the relationships between texts and material culture.

Archaeological Survey and Data Collection

During my doctoral research on the history of Bodhgaya, I conducted a preliminary survey of the temples and shrines of Gaya in 2005–2006. This survey exposed me to the vast collection that had neither been documented nor carefully examined in the past. In summer 2011, I carried out a survey of Gaya and documented fifty-five *vedis* (shrines for performance of ancestor-rituals) and temples jointly with Dr. Matthew Sayers. This survey was undertaken in cooperation with the K. P. Jayaswal Research Institute, Patna (KPJRI), hereafter previously, the KPJRI was authorized by the state government in 2006 to record and document all the historical sites and objects in the state of Bihar as a part of recent efforts to preserve the archaeological heritage of the state. An important goal for this documentation project was to create a documentary record to prevent any further destruction and theft of the valuable material remains of Indian history. Smuggling and theft has been a major problem for the state of Bihar, especially of the early medieval Buddhist and Hindu sculptures. Even though several villages of Gaya were surveyed previously in this project, the city of Gaya had yet to be surveyed. Therefore, it was ideal to collaborate with the institute for conducting this survey. For documentation of these sites, a detailed form was designed by Dr. B. K. Choudhary, director of the KPJRI, in conjunction with Amar (http://sci.dhinitiative.org/). The documentation included a careful recording of each object—image, architectural element, shrine, etc.—its context, associated

---

6 L.S.S. O'Malley, "Gaya Çrāddha and Gayawāls," *Journal of the Asiatic Society of Bengal* 72.3 (1903): 1–11.

legends and stories, ritual usage, and ownership, as well as relevant GIS data. In addition to the completed form, each object was photographed digitally.

During this survey, we also observed and participated in the ritual activity performed at different shrines and interviewed several ritual experts, priests, and local intellectuals from among the Gayawala Brahmana, the local lineage of priests who have historically held the rite to ritual performance at Gaya, and other stakeholders including pilgrimage-operators, managers, and tour-guides in or around Gaya. Several of these rituals and interviews were recorded and added to the wealth of data. Several of those interviewed included older Gayawala Brahmana priests, who were considered local scholars in their community. We asked questions about the changes in rituals over time, the new generation and their interest in the priesthood, the movement of images and objects, the economic aspect of pilgrimage, and the role of ritual experts at the site.

Early Meetings

After the survey in June 2011, I realized the enormity of the task of organizing and processing this data in order to delve into some of the above raised questions. I worked with some of the data for academic research but mostly did nothing to organize this material until March 2013. Then, I contacted Janet Simons from DHi at Hamilton College to explore the possibilities of developing a digital database and different ways in which I could share this material. The early discussions, often filled with technological terms and questions about the methods and organizational schema in the first year, were somewhat difficult. These discussions made me aware of the digital modes of research and how I could engage with digital tools to not only organize my materials but also develop an archive that could be shared through an online platform. Organizing a database required collating the collected data from the forms and generating tables of sites and their materials, which I had used for my dissertation research previously and more recently, for publishing an archaeological Gazetteer. Site or regional Gazetteers have been a useful mechanism to share a large amount of data but they are often inadequate in presenting artifacts, their specific contexts, and how they are being engaged with on a regular basis. Scholars often have to develop a static standardized format to present the data and make difficult decisions about including illustrations and visual materials. It also limits one's ability to present multiple datasets and demonstrate interconnections, as pointed by Averett and Counts in this volume. In fact, digital documentation of the material remains makes it much easier to develop a digital database that provides a better understanding of the material as well its specific context of usage and daily engagements. This is a contrast from museum displays, which often lack specific contexts and limit our ability to engage with the objects in a productive manner.

In order to effectively engage with digital tools, I conducted a brief review of the several digital projects and focused specifically on premodern site-specific projects with a GIS component. Projects such as *'Digital Karnak'*, *'MayaCityBuilder'* and *'Pure Land Inside the Mogao Grottoes of Dunhuang'* provided useful insights for developing this project. The *'Digital Karnak'* and *'MayaCityBuilder'* are excellent models in terms of spatio-temporal organization of structures within their geographical context. I also drew from the *'story map ideas of ESRI'* that encouraged me to collect stories for several images and objects, and subsequently tag them with their stories. Particularly helpful was the *'Project and Resource listings of Geo-Humanities Special Interest group'*, which provides information about projects with a GIS (geographical information system) component for plotting sites on Google Earth maps. For 3-D and VR models, I looked at the *'The Oplontis Project'*, *'The Pompei Quadriporticus Project'*, and *'Pure Land Inside the Mogao Grottoes of Dunhuang'*. These projects document multiple layers of their respective sites/structures within their 3-D and VR models, and provide a nuanced understanding of the shifting contexts. I also spoke to directors of some of these projects, when they came to Hamilton for talks and workshops.

An awareness of these projects made it possible for me to think about the development, digital curation, and different modes of presentation of my web archive. We decided to share the database in multiple formats. One format was to list and display the forms, which had all the data with specific descriptions. Another format was to organize the object along with its specific context and geographical information to demonstrate the interconnections (Fig. 1).

Developing and presenting multiple formats helped capture diachronic rearrangements and reconstructions within Gaya and examine the generation of new myths as a result of the ongoing practice of pilgrimage. This also emphasized the interdisciplinary nature of the project, which drew not merely from historically grounded sources but also from the current context. In addition to outlining the historical development of the Gaya in its geographical and cultural context, the project also developed materially grounded stories of temples, shrines, and sculptures that allows visitors to experience multiple layers of this sacred city. Digital tools such as GIS made it possible to explore the links between different shrines and also analyze the pattern of their growth and development overtime through the patronage of pilgrims.

In addition, one of the prerequisites for the Mellon-funded DHi was to find pedagogical importance of the project for our undergraduate students at Hamilton. This was clearly articulated as the fifth goal in the first phase grant application (http://www.dhinitiative.org/about/goals-phase1). I worked with some of my

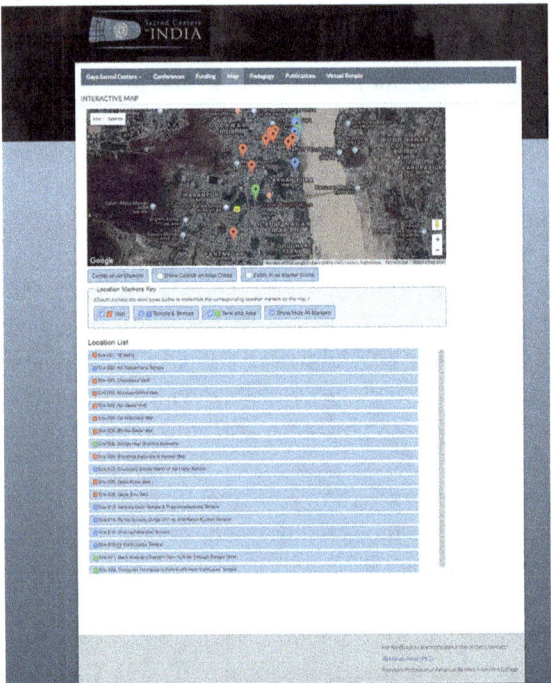

Figure 1: GIS based Presentation.

students in summer 2013 to develop digital forms and geographic alignment of the shrines within Gaya. Along with the DHi team, I also decided that the students of the study abroad program to India in fall 2013 would use the field processes and methods for digital collection development. We decided to develop a Google based collaboration site for the data collection since students of the program were to visit Gaya and Bodhgaya. We also planned to migrate the collected data into a longer-term digital repository collection for continued scholarship. This was an ambitious and somewhat unrealistic goal, which was never accomplished.

With the archive and its broader goals in mind, I have participated in innumerable meetings to discuss and develop work plans over the last six years with the members of the DHi Collection Development team (http://www.dhinitiative.org/community/collectiondev). The team was led by Janet Simons, who as a co-director, managed technology development implementation and sustainability of the web portal and digital archive, research projects, and coordinated the activities of the DHi Collection Development team. The team included lead designer and software engineer, Gregory Lord; library information systems specialist,

Peter MacDonald; library metadata specialist, Lisa McFall; a Unix/High Performance Computing Network Administrator, Steve Young; and multiple undergraduate students working as DHi Interns in digitization, multimedia communication, research and development. Discussions with the collection team were instrumental in developing a step-by-step method for the first phase of the project, which focused on twenty sites. I will list and discuss the steps in the next section.

Methodology

Organizing Forms

The first and foremost goal was to create a standardized form and format for the collected data, which needed to be organized in a format in accordance to library of Congress standards. This included:
1. Creating a standardized form and format for listing all data including photographs
2. Cleaning up forms and relabeling of the photographs in a standardized manner
3. Organizing photographs of images, architectural materials, inscriptions, fragments, shrines, and temples in a sequential order to clarify context, their specific positions, and relationships with other remains
4. Organizing sites into a specific site-unit on the basis of geographical proximity. This was the most difficult task and required some spatial thinking. There are several shrines that are out in the open and did not have an immediate ritual context. An excellent example is a group of early medieval images, which are plastered on top of a small bridge over a drain. I decided to group this 'bridge-site' with the nearest *Vedi* based on geographical proximity to the nearest important shrine.

I worked with five undergraduate students, who were interested in pursuing DH. They were chosen through a competitive selection process, which included an application and interview. The DHi gave these students summer fellowships. During the term of the fellowship, these students were trained in specific areas to undertake the above listed tasks. In this early phase, I spent considerable time working closely with these students to complete these tasks. In fact, it took almost a year to accomplish these tasks for the twenty sites in the first phase. While working on these, I also noticed gaps in the data, particularly about the specific contexts of images and sculptures and the lack of GIS data for mapping. At this time, I decided to use GIS pro on an IPad to collect and create a new database of maps. I returned to the site to collect the missing-data, which was crucial for creating and maintaining the standardized format.

Metadata

The next step was to create a metadata schema based on the documentation form and research questions. Creating strong metadata-schema is the foundation of any archive and requires one to think carefully about different curatorial strategies. Metadata, in a simple way, is a spreadsheet of all of your data. It records how information is related as well as the file structure, itself. Successful metadata will store all information in a single space to create a cohesive, easily navigated organizational structure. Multiple meetings with the metadata expert Dr. Lisa McFall were crucial in developing a schema that would facilitate the exploration of questions of space, geographical locations, relationships between water bodies, rivers, hills and religious structures, distribution-patterns of shrines, temples, and images, and a wide range of material remains. I was also interested in examining the rematerialization of past historical objects and the newer stories around them, which have led to the creation of ritual spaces and new layers of sacrality on preexisting landscape. Developing a metadata schema was, therefore, crucial.

This metadata schema was developed based on the database of twenty sites. In doing so, we also had to be mindful of adding other surveyed sites to the project later. This may include a different sculpture, architectural piece, or a shrine that was not documented at previously discussed sites. For instance, there was an image of a Buddhist goddess '*Aparajita trampling Hindu deity Ganesha*' at the Chitragupta temple. This was the only image of its kind, which needed to be accounted for. Similarly, some shrines have Buddhist votive stupa, which are being worshipped as Hindu objects. These artifacts made us anticipate further additions to the project without disrupting its uniformity and organization. Keeping this in mind, we decided to create a '*Buddha-Buddhist*' category to ensure that every object affiliated with Buddhism is adequately represented in the metadata. Additionally, it was important to anticipate the 'knowledgebase' and familiar categories of potential audiences, who might use this archive. This also explains why we created categories such as '*Buddha-Buddhists*' and '*Other Hindu gods-goddess*' in addition to '*Shiva*', '*Vishnu*' and '*Devi-goddess*' as major metadata categories. These categories facilitated access of organized and curated information to a broad range of audience including the experts and novices. Bielo and Vaughn point towards similar curatorial strategies in this volume when they discuss specific attractions that are materialized on the Bible at particular places. Each of these places have distinctive history and afford an embodied experience, like the shrines and temples of Gaya. The process of curation must be intuitive and informed by the awareness of potential audiences.

Lauren Scutt (a Hamilton graduate of 2017) worked extensively (and excellently) on the metadata schema and populated the spreadsheet with minute details. Like the file structure on the server, there are different levels of metadata. The most basic levels correspond the documentation sheets with folders of images. More detailed levels link each image to its form, file, and supplementary information. Just to provide a sense of complexity, the spreadsheet held over thirty columns of information. A typical metadata schema consists of predetermined parameters for each data point (in this case each image). Examples of parameters in the spreadsheet are *General Remarks* or *Description of Artifact*. This ensures that each image corresponds with the correct information in the archive. More importantly, each image was tagged with keywords. Tagging is an important process because it allows the user to search for images directly with specific key terms. To facilitate tagging, we created certain "special phrases" such as "Goddess Images" or "Recent Shrines." Therefore, even if the images are recorded on different forms, all of these files will become available to the user. Placing the images into assigned categories created consistency, which also simplifies searching the archive (Fig. 2).

Another layer of complexity was to ensure that the metadata and forms speak to each other in a consistent and coherent manner. Because they are based upon one another, both files needed to be constantly updated and reworked. Once a structural problem surfaced in one, it needed to be addressed in the other. Spelling and phrasing needed to remain consistent, as well. We decided to do away with diacritical marks and translate Hindi words generally. It seems obvious that spelling needs to be accurate and consistent while working on any academic project. Creating consistency across forms and metadata is crucial for users. If a database is tediously organized, users may notice patterns, which will help them determine better keywords to access information more effectively.

Another major challenge during this process was the conversion of forms to metadata. At times, it was necessary to change sentences to phrases and vice-versa. It was difficult to maintain the crispness and verity of the facts. One example of this was the identifiers used in the parameters: *Present condition, Conservation assessment,* and *Protection status.* Many of the adjectives recorded during the field research were vague. They did not give a good sense of the site. For example, for just one parameter we used *alright, average,* and *decent,* which all mean about the same thing. It is difficult to compare sites when the adjectives

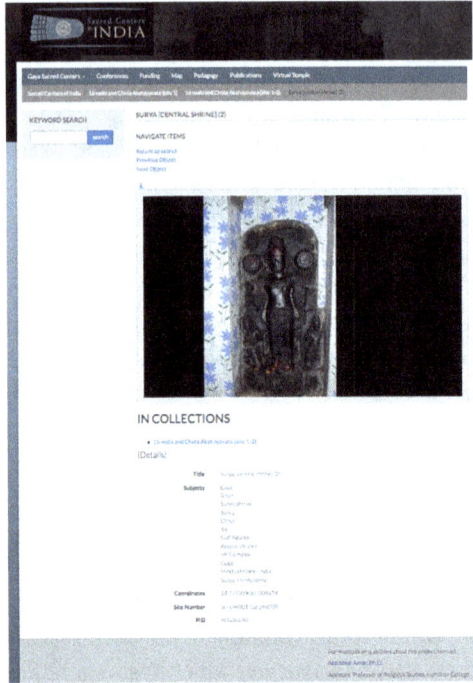

Figure 2: Surya Image with Metadata.

are not codified. Therefore, we developed a fixed list of identifiers to minimize any confusion for users. These discrepancies only became apparent in the process of entering information into the metadata spreadsheet. The juxtaposition of all of the terms allowed us to clearly view the shortcomings of the structure of the forms. In this way, one can notice the interrelationship between forms and metadata, and how they continually inform each other.

Digital Collection and Presentation

Hamilton's DHi has existing infrastructure and processes for archiving, development of data, and long-term preservation in Fedora Commons (http://www.fedora-commons.org/). Fedora was chosen for its scalability and extreme flexibility in the manner in which objects can be accessed. Fedora also has built-in flexibility for the creation and maintenance over time of relationships between objects and across digital collections. DHi uses Islandora (http://islandora.ca/about) and other open source collaborative tools to interface with collections in

Fedora Commons. Islandora can be used to create customized themes for faculty collections and projects. The DHi collection development team works with Islandora and Fedora Commons consultants at Common Media (http://commonmedia.com/) and with the Liberal Arts Islandora Collaboration Group members (founding members include Grinnell, Hamilton, Lafayette, Vassar, and Williams) to create digital scholarship infrastructure, which is open source. DHi also provides Drupal multisite web presences that connect to digital collections allowing ongoing living digital scholarly activity. All of these infrastructural preferences and decisions have been made to ensure long-term sustainability of this project (and other projects) on the web.

The next step, after the organization of metadata, was to organize the digital collection on an internal server with metadata and prepare for next steps to input the data into the archive. Additionally, we also decided to develop postcards to brand and advertise the project within Hamilton. The postcard included developing a project logo and a brief description of the project. To organize and present the database, we also began to discuss different formats including the content and possible ways in which a visitor would navigate this website. We decided to include site descriptions, documentation forms, photographs, a 3D model of the Vishnupada temple, GIS information, and a VR model of the temple for pedagogical purpose. The DHi team also suggested that we create some tag words (categories) to foster a discovery mechanism and enhanced accessibility for a common browser. My research students were particularly helpful in choosing these tag words, which they thought were repeated often and could lead one to the archive from a web browser search engine. Tag words provided another entry point for a browser to locate and engage with this project. Another entry point is through the interactive map application as every site/shrine/temple is plotted on a map and a simple click on a marker leads one to the detailed information on the site along with the documentation form and associated pictures. All of these show our efforts to provide multiple entry points to this database.

A 3-D model of the Vishnupada temple was developed in Blender- an open source software- by the lead design of DHi team Greg Lord (Fig. 3). This model was developed on the basis of sketches/diagrams of the temple, and specific photographs. The sketch was based on accurate measurement of every part of the temple, for which I used a tape measure and a laser measure in a separate trip in December 2014. I also hired a professional photographer for generating higher resolution pictures, which could be used for the 3-D textures of the temple. By the end of spring 2015, the 3-D model was created. This model was constructed to an accurate scale, and provides an interactive display of both the interior and exterior of the temple and shrine. In the meantime, the metadata was also incorporated within the archive.

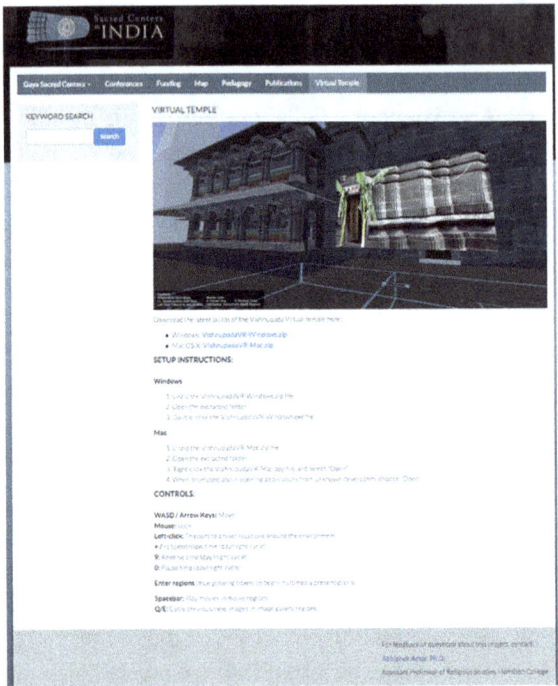

Figure 3: 3-D Model/Virtual temple.

Given the pedagogical goals of the project, we then decided to develop a VR model of the Vishnupada temple and its surrounding context. A contextual understanding of this temple is crucial to explain the layered development of Gaya and its shrines. The completed 3-D temple was made interactive by importing it into the popular Unity game engine, allowing for both real-time interaction with, and movement through the model. The foundations of what later became an interactive virtual reality experience were put in place here, with a series of control scripts that allow a user to move freely through the temple and its surrounding context. Now, the 3-D Vishnupada Temple in Unity is enabled with full VR support through the powerful SteamVR system, allowing realtime exploration in a virtual reality space, using the HTC Vive VR hardware. This version of the virtual experience affords free movement, head tracking, tours through preset locations, realtime day/night lighting including control over time passage, interactive zooming, and audio support for realistic sound within the experience. This version also can incorporate multimedia from the project archive, allowing for the display of image galleries and videos of the Sacred Centers in India collection directly within the VR experience, linking virtual locations to real-world research materials. Many videos of rituals, within the interior and exterior parts of temple were recorded in 2014 fieldwork, which now can be accessed through this VR.

In contrast to Kaplan and Shiff's discussion of speculative modeling of Jewish homelands in this volume, this project has produced a recreative VR to facilitate a realistic and experiential understanding of the Vishnupada temple complex. The VR model makes a geographically distant location digitally accessible to a much broader audience, many of which may never visit or see the temple complex. The temple management committee also has a strict policy of not allowing non-Hindus inside the complex, which has limited the access to the temple complex and its countless sculptures to non-Hindu scholars. The digital reconstruction has also made it possible to examine the spatial arrangement of shrines and gradual development of the complex, which may not be apparent to a visitor. Concurrently, 3-D model of the temple and it surroundings can lead one to examine the layers of development. In my seminar classes, I have utilized the 3-D model to discuss the issue of gradual development and the VR to walk through the different parts of the temple complex. The walk provides students an understanding of spatial arrangement, placement of images and sculptures, and their associated rituals. These experiences have inspired my students to produce creative assignment projects for the class, many of which are available in the pedagogy section of the archive (Fig. 4).

In short, I have collaborated and relied extensively on the DHi collection team's expertise to develop this archive. The last five years have been a steep learning curve for me to explore multiple technological tools and understand the challenges of developing this web archive. I must admit that many conversations about the tools and presentation formats seemed difficult and incomprehensible at times. However, discussions with the DHi team and their patience in explaining every minute technological detail was immensely productive in making an informed decision about the archive. This also illustrates the collaborative nature of the DH projects.

Future Goals

At this stage, there are three important goals for this project. The first goal is to transition to phase two and complete the database for the shrines and temples of Gaya, which includes processing and inputting data of an additional thirty temples and shrines. The metadata for this phase has been completed. The second goal is to develop an archive of materials that have been found from Bodhgaya. In addition to existing temples, shrines and images at Bodhgaya, many objects and remains were moved to different museums within India and the UK. There

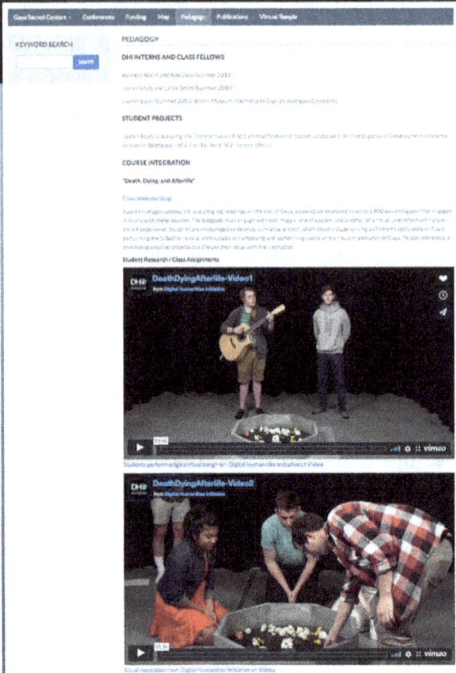

Figure 4: Pedagogy and Student Projects.

also exist photographs from nineteenth century excavation and restoration of Bodhgaya, which I have collected through my museum and archival research. The third goal is to develop a virtual museum of images and sculptures that I have found from the villages of Gaya and the Nalanda district. Many of these villages have a long-distinguished past, which I have examined and studied in detail. By developing a database of these objects, I hope to protect, preserve, and generate awareness about them.

Conclusion

The development of a digital archive requires intense collaboration. Multiple steps of the pilot phase have demonstrated that a digital scholarship project requires a strong grasp on both the material and technology behind the project. However, it is not mandatory for one to develop an expertise in digital before embarking on an ambitious project. Like other forms of scholarship, it can also be gradual and developed over time. Given the challenges of engaging productively with digital forms of scholarship, it is important to work collaboratively with tech experts to explore the possibilities of what can be accomplished. This archive has been helpful in rethinking the method that I have been trained in, which often focused on the qualitative questions. However, a much more careful examination and organization of the database has also forced me to develop quantitative questions related to the historical production of sculptures and architectural materials. In addition, the archive has been helpful in protecting, preserving, and researching heritage. In fact, this database can be used productively in the fields of heritage studies, or tourism, which are on the priority list of the Government of Bihar. More recently, I have successfully integrated this project into my pedagogy. As shown within the pedagogy segment of this project, students from my seminar course titled 'Death, Dying and Afterlife' visited the DHi lab to experience the VR model, which inspired them to develop creative projects including writing a short story book on Gaya, creating a replica of the Vishnu's footprint, enacting rituals, and writing blogs on the themes of space and ritual through an analysis of images and shrines of the database.

Selected References

Amar, Abhishek S. "Buddhist Responses to Brāhmaṇa Challenges in Medieval India: Bodhgayā and Gayā." *Journal of the Royal Asiatic Society* 22 (2012): 155–85.
Jacques, Claude. "*Gaya Māhātmya*—Introductions etc. (Cont.)" Giorgio Bonazzoli, trans. *Purāṇam* 22.1 (1980): 33–70.
Jacques, Claude. "*Gaya Māhātmya*—Introductions etc." Giorgio Bonazzoli, trans. *Purāṇam* 21.2 (1979): 1–32.
Jacques, Claude. *Gaya Māhātmya*. Pondichéry: Institut Français d'Indologie, 1962.
O'Malley, L.S.S. "Gaya Çrāddha and Gayawāls." *Journal of the Asiatic Society of Bengal* 72.3 (1903): 1–11.
Reed, Ashley. "Digital Humanities and the Study and Teaching of North American Religions." *Religion Compass* 10 (2016): 307–316.
Sayers, Matthew. "Feeding the Dead at Gaya: From the Buddha's Enlightenment to a Modern Pilgrimage." Speech presented at the South Asian Studies Colloquium at The University of Pennsylvania, Philadelphia, PA, 2012.
Sayers, Matthew. "Gaya-BodhGaya: The Origin of a Pilgrimage Complex." *Religions of South Asia* 4.1 (2010): 9–25.
Vidyarthi, Lalita Prasad. 1961. *The Sacred Complex in Hindu Gaya*. London: Asia Publishing House.

J. E. E. Pettit, Fenggang Yang, Yuqian Huang
Developing a Database of Religions in Contemporary China

Over the past decade, information about religions in China has been piecemeal and has differed greatly in government news and social media. The past few years have seen a rise in the destruction and closing of hundreds of churches, suppression of Tibetan monasteries, and the re-reeducation of Muslim minorities.[1] Yet, it has been difficult to gauge how representative these events at national- or regional-level are. Exactly how many churches, mosques, and temples exist in China, and when were these communities founded? Are religious communities increasing in the face of government suppression, or do these news stories indicate a decline or disappearance of religion from the landscape of contemporary China? Are these policies affecting all of China's religions in the same way, or are certain religions developing differently among China's distinct cultural regions?

Our research team at the Center on Religion and Chinese Society (CRCS) at Purdue University developed digital tools as a means to count the religious communities in China. It seemed that digital mapping tools such as Google Maps could enable our team to gain a basic distribution of the sites throughout China, and could determine the changing demographics and its corresponding changes in the religious composition of a country. But while the mapping technologies might open up new datasets on the religions of the United States and other developed nations, the data we could collect on China was limited and difficult to easily form a database of sites. China's government not only holds a tight grip over the data it collects, but technological giants (e. g., Baidu, Weibo) have provided users with comparatively little access to what Chinese people are thinking and doing with regards to religion.

As we started surveying the available geographic data on China's religious communities, we quickly realized that our desire to map out the location of mosques, churches, and temples could have potentially adverse effects not only on our research team, but also the communities we studied. During our survey, changes in official policy towards religious institutions made discussing reli-

---

[1] For more on these events, see Fenggang Yang, *Atlas of Religion in China: Social and Geographical Contexts* (Leiden: Brill, 2018); Ian Johnson's *The Souls of China: The Return of Religion After Mao* (New York: Pantheon, 2017), as well as his writings news articles such as "This Chinese Christian Was Charged With Trying to Subvert the State," *The New York Times*, March 25, 2019.

gions, either in the news or through social media, a potentially dangerous affair. The repressions of religious institutions have been localized in certain cities or provinces, but there is concern that government suppression of religion makes it precarious for groups to make a strong public presence online.

In 2015, researchers at CRCS embarked on a four-year project to document Chinese religious institutions and develop digital tools to achieve new levels of understanding of where religious establishments are in the country and what parts of the country are witnessing growth or decline of religion. We received a generous grant from the John Templeton Foundation that allowed us to bring together a research team of graduate students and postdoctoral fellows. At first, we thought we might simply fund researchers to visit China and document sites with a GIS-enabled cameras and other equipment. We quickly discovered, however, that researchers who had conducted similar geographic studies of China put themselves in peril. Three British geology students from the Imperial College London, for example, had been fined for making similar geographic surveys in northwestern China.[2] Similar groups of Japanese and Korean scholars incurred heavy penalties for research of geographic places, and one University of Chicago professor was even jailed for conducting research on the geographic positions of oil wells in west China.[3] Despite our great interest to know more about the spatial distribution of religions in China, we were unwilling to place graduate students, affiliated researchers or ourselves in harm's way. Given our need to document Chinese religions from afar by using digital resources, we needed collaborators who could help us build tools to collect big data about religion. At present, we have culled information of over 150,000 points via an economic census, "points-of-interest" from online mapping services such as Amap, and the State Administration of Religious Affairs (*guojia zongjiao shiwuju* 国家宗教事务局, hereafter SARA). There would be no way that we could gather this kind of data about religious sites if we were going to collect them one-by-one. We needed to develop a way that we could process big data automatically and clean problematic points.

Our desire to map the religious sites of China was more than simply overcoming the mechanical issues of producing accurate local-level data on Chinese religions. We wanted to be sure that our project was not a "commodity to be controlled, but rather a tool that could be a social good to be shared and reused" by

---

[2] "China Fines UK Students for 'Illegal Map-making,'" *The China Post* (Jan. 6, 2009).
[3] Dingding Xin, "Unlawful Surveys to be Dealt Severely," *China Daily* (March 7, 2007). For more on the dangers of collecting information of religious sites, see Fenggang Yang, "The Failure of the Campaign to Demolish Church Crosses in Zhejiang Province, 2013–2016: A Temporal and Spatial Analysis," *Review of Religion and Chinese Society* 5.1 (2018): 5–25.

making our dataset public.⁴ Our dataset needed to be grounded in computational technologies as well as a foundation from which critical analysis could grow.⁵ As Bingenheimer and Elwert demonstrate in this volume, accessibility to these kinds of datasets is key in moving the study of religious institutions forward into the future. Yet, releasing information on religious institutions also has the potential to place its community under unwelcome scrutiny from the government. We conclude that datasets on the geographic position of Chinese religious institutions needs to exercise some caution. A researcher would not want to inadvertently place a community at risk. In our dataset, we conclude that there is little risk since the Chinese government produced the information. These considerations about the ethical implications can hopefully assist future generations of scholars who face the tension between public accessibility and personal security.

China's 2004 Economic Census and Religious Sites

In the People's Republic of China, five kinds of religions are officially recognized and legally approved by government authorities: Daoism, Buddhism, Catholicism, Protestantism, and Islam. The Chinese government has established a complicated institution to regulate religion. The major control apparatus for religious affairs is the United Front Department (*tongzhanbu* 统战部) of the Chinese Communist Party (CCP), who makes religious policies and rallying religious leaders around the CCP. The daily administration of religious affairs lies in the Religious Affairs Bureau (RAB), including a national level bureau called SARA. Its duties include processing requests for approving the opening of temples, churches, and mosques; approving special religious gatherings and activities; and approving the appointment of leaders of religious associations. Meanwhile, the Ministry of Public Security (*gong'anbu* 公安部, i.e. police) deals with illegal religious activities, and watches some religious groups and active leaders, and the Ministry of State Security (*guo'an bu* 国安部) monitors religious activities involved with foreigners.

The presence of these regulatory bodies in China has created a complicated religious field composed of legal and illegal groups, as well as those in-between. In the last several decades, the number of religious believers in China has con-

---

4 Lisa Spiro, "'This Is Why We Fight:' Defining the Values of the Digital Humanities," In *Debates in the Digital Humanities*, ed. Matthew Gold. (University of Minnesota Press, 2012), 22.
5 James Smithies, *The Digital Humanities and the Digital Modern* (London: Palgrave, 2017), 3.

tinuously increased. Moreover, religious believers have engaged in strenuous contentions against the party-state to reclaim religious sites, restore and construct buildings on traditional religious sites, and sometimes occupy state or collectively owned spaces and construct new religious buildings. As early as 2009, Dr. Shuming Bao from the China Data Center approached us with a dataset extracted from a 2004 Chinese economic census. The file includes 72,887 religious sites from all of China's 31 provinces, provincial-level regions, and municipalities. From the outset, we knew this list certainly did not represent the locations of all religious sites across the country. First, this economic census only listed a church, temple or mosque that was officially registered with the government. But there are many religious communities in China that could not or choose not to register with the government authorities. In parts of China that we have visited there are certainly more unregistered groups than officially sanctioned ones. Second, given the economic nature of the census, there were a great number of places whose annual income did not register with the authorities as noteworthy for inclusion in this census. And for some northwestern cities of China such as Lanzhou, for instance, the census focuses primarily on mosques that double as economic enterprises. When we visited the city in 2016, we discovered nearly three dozen churches not mentioned in the census, and many of them had existed by 2004 when an economic census was conducted.

While the 2004 economic census might not give us a comprehensive look at all religious sites in China, we were convinced that this data set would provide scholars with the best available dataset to analyze the spatial distribution of Chinese sites, as well as the names, locations, relative size, and reported annual income for religious communities throughout China.[6] In an effort to map out these sites, we first used the addresses for each site and used a python code to reference the address in Google and Baidu map APIs. Overall, this automatic process was able to correctly identify the addresses for two-thirds of the sites. A close inspection, however, revealed unforeseen issues. Both Google and Baidu maps were using the administrative categories of 2010, but the economic census used an older address system adopted in 2000. This caused over 8,000 addresses to appear as a discrepancy, i.e., a town belonged to one prefecture in 2000, but was identified in a different prefecture.

As we delved deeper into the addresses and names of the religious sites, we noticed a much more perplexing problem. We found numerous examples where

---

[6] The census has uneven measurements from China's different provinces. For more on these complexities, see Carsten A. Holz, "China's 2004 Economic Census and 2006 Benchmark Revision of GDP Statistics: More Questions than Answers?," *The China Quarterly* 193 (March 2008): 154–156.

the name of a site indicated that it was located in one village, but the village name in the address was listed as a different location. In site #19238 (a church), for example, the name of the religious establishment is listed as a church in Zhangzhuang Village 张庄村in Zengfumiao Town 增府庙乡 (Changge City长葛市), whereas the address of the same site insisted as located in Niutang Village 牛堂村of Changge City 长葛市. There were hundreds of inconsistencies in the data that needed to be examined on a case-by-case basis. Finally, the Chinese government has made many revisions of the administrative boundaries of counties throughout China. Many of the sites included in the Census might have belonged to one administrative division when the data was collected, but was changed in subsequent years. Sometimes the address was misspelled or it turned out that there were two villages with the same name in the same region.

Aside from the inconsistencies in the data, we also faced logistical issues in coordinating a team of researchers to clean the data. First, we wanted at least two people on our research team to examine the problematic points as a way to safeguard misunderstandings of the discrepancies in the data. Given the high volume of errors (roughly 14,000), we needed a system for two editors to evaluate the problematic sites, and a third person to judge if the conclusions reached were the same or different. Furthermore, there would be anywhere between 5–10 different editors working on the dataset at any given time. This required a database that would avoid duplication and help automate the process of queuing points for review. Finally, we wanted the final dataset that would released publically to include simplified and traditional characters, as well as English translations. This would increase the accessibility beyond scholars in China, and hopefully make the data intelligible not only for scholars but also journalists, policymakers, and the general public.

The Translator Tool

To solve these issues, we developed a two-step process to prepare the data for public release. The first step involved translating and preparing the data for deep editing; the second was an editor tool that would help us determine the most accurate location possible. For both web applications, we developed them by hosting the data on a server with two parts, a development side where we could test our application and a production side that the research team would use. Each contained three components. First, we stored the source codes of the projects in HTML, which helped to store the templates for the web interface. These parts of the database helped to store the templates for the web interface. If the data for our project were stored in this layer, we might have issues from third parties who wanted to change our data without authorization. To mitigate this problem, we created a DATA layer that stored the PHP files for database connection. After this, we could manipulate the credentials by remote connection via different web editors or IDEs, e.g. Netbeans, PhpStorm. MySQL Database contained the meta data related to this project.

The top bar of the translator tool allowed the user to navigate between various points. Each time she started the application, it would automatically start with the last point edited. She could navigate to other points in the dataset by either toggling to the "next point" (*hou yi ge* 后一个) or "previous point" (*qian yi ge* 前一个). There was also a space where she could enter the unique ID of the point and go directly to that point (*tiaozhuan* 跳转). The button on the right of the top bar would open a new window to input the common words (see below).

The grid below the top bar included information that the translator used to identify problematic entries in the dataset. The string of Chinese characters on the left was the address that was automatically generated in Google Maps in 2010. The top-right string of characters was the Chinese name from the economic census and the bottom right was the address from the census. To the left of these strings of characters was a button that could be selected when the automatically generated address was "in conflict" (*bu yizhi* 不一致) with one another. This information was used to select the points for the editor tool (see below).

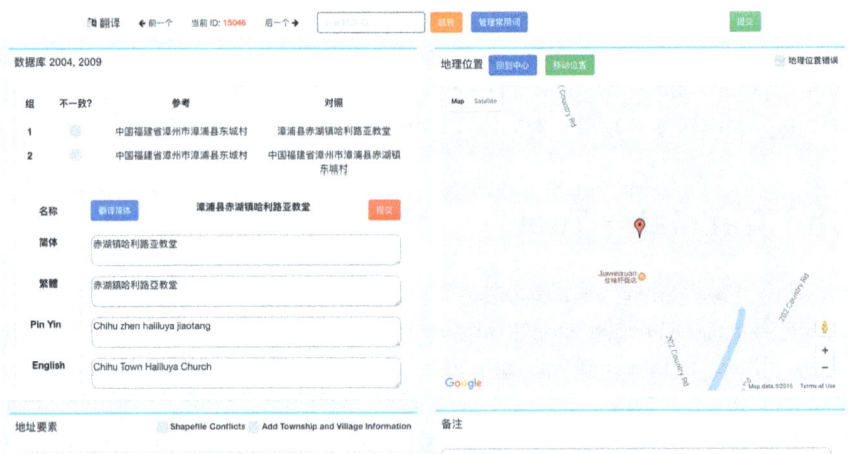

Figure 1: The user interface of the Translator Tool.

Beneath the grid to compare the address and site name was a tool to automatically generate the name of the establishment in simplified and traditional

Chinese, its pinyin romanization, and an English translation. The application would recognize geographic names and remove any national, provincial, prefectural names so that only the place name of a local village or town would appear in the name. In Figure 2, the tool recognized two layers in the original site name, a county (Zhangpu County 漳浦县) and town (Chihu Town 赤湖镇), and removed the first so only one place name would appear. After the name had been properly parsed to its local level, the translator would press the "translate" (*fanyi* 翻译) button, which would generate the corresponding traditional and pinyin equivalent. The translation of traditional Chinese was based on Microsoft Translator Text API (https://www.microsoft.com/enus/translator/translatorapi.aspx), while the translation of Pinyin is based on a third-party library of Pinyin translation (https://github.com/hotoo/pinyin).

The English translation for each site was generated by referencing a list of 550 common words that our team identified while working through the data. We identified nearly 600 different kinds of common names used for religious establishments. For example, there are over 50 different words in the Census used to designate Church (e.g., *jiaohui* 教会, *jiaohuidian*教会点, *jidutang* 基督堂, *jidujiao huidian*基督教会点). These various words would be translated as "Church" and the place name or church would be rendered in pinyin. If we discovered a new word for a temple, church, or mosque, she could select the "common word" button at the top of the page. This would open a new window with a separate database of common words. We implemented this feature with DataTables (https://datatables.net), a powerful table plugin tool for data visualization. The translator tool allowed the translator to identify problematic addresses and names, as well as translate each point into four different names (simplified, traditional, pinyin, and English) in roughly 30 seconds. This automated process enabled each point to be processed in 30–60 seconds as opposed to 4–5 minutes required to input all the Chinese and English names by hand.

The Editor Tool

The second phase of cleaning this data was an Editor Tool (*zongjiao xinxi bianji gongju* 宗教信息编辑工具) that allowed two editors to study problematic points and make the corrections based on existing records, especially the exact location of the religious buildings, as well as the extra village and township information. The landing page for this tool was a national map of China that displayed the working statuses of all Chinese provinces. Only the provinces featured in orange contained data ready for the editors to use and could be clicked. The table is conducted via DataTable. The vector map was conducted via jVectorMap (http://jvectormap.com). This is an open-source platform to display and select shapefiles for countries, provinces, and counties.

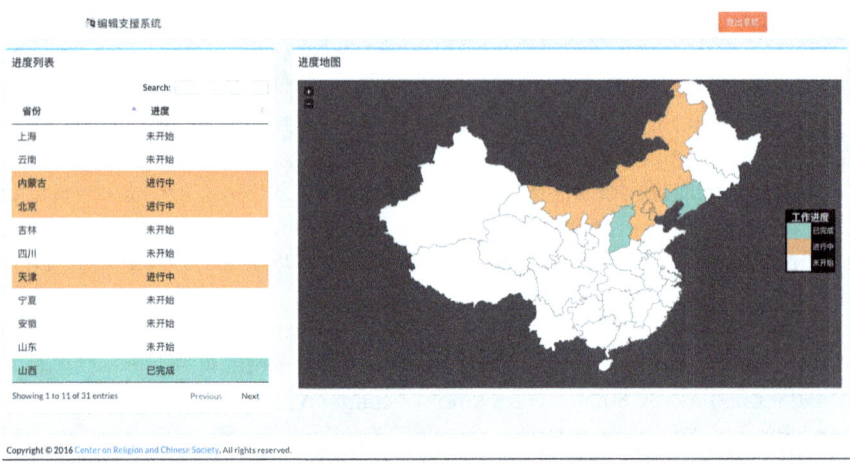

Figure 2: The landing page of the Editor Tool.

After clicking the item in table or map, a new tab appeared with the main Editor Tool interface. The top left corner of this page displayed the site name, address from the Census, and the automatically generated address. The editor would study these discrepancies and judge where the point should most likely be placed. After researching each point, the editor entered the correct data for the town or village in the top right hand of the page. Based on their findings, the editor could also move the point data on the make by pressing the "shift position" (*yidong weizhi* 移动位置) button above the map. There was also a search feature on the map where the editor could search for other points in Google or Baidu maps. Each time the editor moved points, the points were stored in three different coordinate systems of Street Map and Satellite Map, and Baidu Map, so that future users could view the point using different kinds of online maps.

Unlike the simple SQL queries in the previous application, this one proved more difficult since at least two different editors would verify the location for each point. Furthermore, only the problematic items would appear in the queue for editors to work. While the SQL query is similar to the one above, we had to add extra columns to record the work of different editors.

At the time of writing, we reached the end cleaning the data as much as possible. For the census indicated a specific address, we were able to find the corresponding site. Where the census only indicates a village or township name,

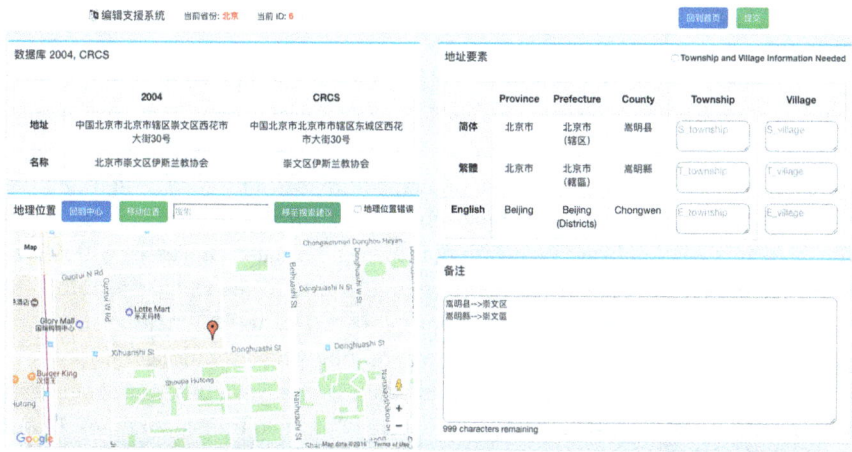

Figure 3: Editor Tool Interface.

we were able to find coordinates for the administrative center of that location. We estimate that at least 90 percent of the religious sites in the dataset have points accurate within 5 km of the actual site. Since the maps in this atlas do not go beyond the county-level, we estimate that the accuracy of the geographic points projected in this atlas is much higher. We have made the coordinates for these points, as well as the names and addresses. This information is now available online at Online Spiritual Atlas of China (https://www.globaleast.org). The current web interface is available and the team is currently working on "publishing" the dataset at Purdue University Research Repository. See https://purr.purdue.edu/publications/3210/2.

Thoughts on Developing Databases on Contemporary Religion

While the exact workflow of our project might not be applicable to scholars outside the study of China, we think that the Translator and Editor Tools might help scholars of religions plan and execute the collaborative processing of large, complicated datasets of geographic data. In particular, these kinds of applications are useful as they allow multiple users to simultaneously work on different aspects of cleaning and analysis of the data. This is important because the processing of this data would be far too much for one person to do, especially without some kind of way generating automatically the addresses and translations for the sites.

In developing the collaborative workspaces, we find four aspects of this project that are relevant for similar studies. First, our use of python code to automate this kind of work is essential to save the time of a research team. If we were to manually look up each point in Google maps or to translate each term in our database by hand, it would take many more months (or years) to deal with large datasets. It is key to always be aware that these automated processes will not always be correct. In our case, we had to scan the data to see where the automatically generated addresses or translations failed, but we estimate that we increased our output ten-fold through incorporating such techniques.

Second, for projects that seek to translate data between different languages, it is helpful to develop an accretive dictionary of ways to translate names. In our dataset, we found nearly 600 common terms for religious establishments. It was beneficial to compile these names as we were translating the data, and it saved a lot of research time when the web application could automatically translate these terms after they had been entered into our common word bank. This was further beneficial as we had multiple translators and editors working with the data. This multi-user dictionary that was specific to our project helped different people adhere to one standard.

A third feature of our two-part research plan was that we were able to delegate the tasks of cleaning, translating, and collaborating geographic data to different team members. The different applications ensured that team members could work simultaneously on separate aspects of the project. This also encouraged team members to develop specialized tasks within their specific project duties. Our experience in working with the religious sites from the economic census reinforced the importance of a healthy dialogue between scholars of religions and technical specialists. During this project, sociologists of religion and civil engineers specializing in geospatial technologies came together to develop new tools. Our ability to process the geospatial information and religious terminology would have been impossible without a substantial knowledge on both of these fields. We found that the creation of collaborative web projects exemplifies the need for cross-disciplinary knowledge that bridges humanities with digital knowledge. Such a project, we find, is best found in teams where specialists from different domains work together.

After cleaning this data, our team produced an atlas of Chinese Religions that is available as a book, *Atlas of Religion in China: Social and Geographical Contexts*. We have also produced six articles that use this data, and we have made the entire 2004 Economic Census data available for public download. We hope that this set of data will assist future generations of scholars explore patterns in the general distribution and size of religious institutions in the early 21st century. We see that the public access to such datasets of China is beneficial to scholars, journalists, policymakers, and the public. Such datasets will provide a window into the religious climate in China and enable us to gain a clearer picture of which areas religion thrives and where it has very limited presence.

Throughout the process, we recognized the potential implications it might have for communities with tense relations to the government. Certainly, if our database was giving the geographic coordinates of illegal or semi-legal institutions, our dataset might be a kind of roadmap to expose a group and put them in harm's way. The dataset we described in this article is derived from the 2004 Economic Census and thus represents legal entities under PRC law. Obviously, such a dataset only gives data on the officially recognized churches, mosques, and temples. It does not give the full picture of all religions in China since these points would not include any points for illegal or semi-legal groups. But this official layer of data on religions makes for an unparalleled insight into the distribution and frequency of religious sites across China. We expect that this will continue to serve as a foundation for analysis in years to come.

Selected References

Holz, Carsten A. "China's 2004 Economic Census and 2006 Benchmark Revision of GDP Statistics: More Questions than Answers?" *The China Quarterly* 193 (March 2008): 154–156.

Johnson, Ian. *The Souls of China: The Return of Religion After Mao*. New York: Pantheon, 2017.

Smithies, James. *The Digital Humanities and the Digital Modern*. London: Palgrave, 2017.

Spiro, Lisa. "'This Is Why We Fight:' Defining the Values of the Digital Humanities." In *Debates in the Digital Humanities*, edited by Matthew Gold, 16–35. University of Minnesota Press, 2012.

Yang, Fenggang. *Atlas of Religion in China: Social and Geographical Contexts*. Leiden: Brill, 2018.

Yang, Fenggang. "The Failure of the Campaign to Demolish Church Crosses in Zhejiang Province, 2013–2016: A Temporal and Spatial Analysis," *Review of Religion and Chinese Society*, 5, no. 1 (2018): 5–25.

Louis Kaplan and Melissa Shiff
From Ararat to Kimberley: Activating Imaginary Jewish Homelands with Augmented and Virtual Reality

Figure 1: Melissa Shiff and Imaginary Jewish Homelands, *Mapping Ararat*, 2012. Augmented reality still image. Ararat Synagogue on the 18th Green of the River Oaks Golf Course, Grand Island, New York

Introductory Overview

Whenever and wherever a virulent and viral form of anti-Semitism has occupied the public discourse in modern history, political efforts and ambitious plans have arisen to locate a haven and a homeland for such persecuted Jews somewhere on the globe. While the story of Theodore Herzl and the political success of Zionism is the most well-known narrative and well-rehearsed script, there were numerous other largely forgotten plans that did not seek a return of the Jews to their ancient abode in Israel in order to posit a Jewish home or refuge. It is from these effaced paths that our historical research and speculative creations begin as we activate imaginary Jewish homelands.

This essay reviews the development, technical specifications, and results (both realized and in process) of two research-creation projects in the digital humanities (DH) that combine new media art practice with academic research[1] and that incorporate the emergent technologies of augmented reality and virtual reality to stage counterfactual histories. It does this by exploring two plans for Jewish homelands that arose in response to specific moments in the history of religious persecution of Jews in modern times. In the first project *Mapping Ararat* (2011–2014), we created an augmented reality (AR) walking tour that takes visitors on an incredible journey that images and imagines what would have happened if Major Mordecai Noah had succeeded in his 1825 plan to transform Grand Island, New York into a "refuge for the Jews." In our current phase entitled *The Imaginary Homelands of I.N. Steinberg* (2015–2021), we have shifted to virtual reality as the preferred technological medium in order to explore Steinberg's valiant attempts to find a refuge for persecuted Jews fleeing Nazi Germany in such far-flung places as Kimberley, Western Australia, Port Davey Tasmania, and the Saramacca district of Suriname against the backdrop of World War II and the Holocaust. We focus here on the so-called Kimberley scheme and how this charismatic Jewish leader almost succeeded in securing a homeland for Jewish refugees in this remote region. Tapping into Steinberg's vast archive located at the YIVO Institute for Jewish Research in New York[2], we are using these source materials to texture the 3-D architectural models of a virtual world constructed in the Unity gaming engine that one can navigate using various platforms.

In contrast to the other essays featured in this collection, it is important to stress at the outset that *Mapping Ararat* and *Imaginary Jewish Homelands* are as much new media arts projects as historical research endeavors. This is underscored by the fact that both were awarded funding in the hybrid research-creation category by the Canadian Social Sciences and Humanities Research Council (SSHRC). The role of the imagination and fantasy is crucial to these projects as they consider counterfactual scenarios ("what if's") in modern Jewish history. Another way of expressing our different focus and emphasis would be to say that instead of using the tools afforded by DH to archive various religious

---

[1] In this regard, our collaborative digital art and humanities project is akin to the work of other research-led practices in the visual arts such as those by Johanna Drucker (at the Graduate School of Education and Information Studies at UCLA) and Victoria Szabo (at the WiredLab! for Digital Art History & Visual Culture at Duke University). The augmented realty artist John Craig Freeman (Professor of New Media, Emerson College) was a collaborator on *Mapping Ararat* while Ultan Byrne (Lecturer in Architecture, University of Toronto) has been involved with both phases of *Imaginary Jewish Homelands*.
[2] The Papers of Isaac Nachman Steinberg (RG 366) located at the YIVO Institute for Jewish Research at the Center for Jewish History in New York consists of twenty-six linear feet of boxes and the Australian component is comprised of hundreds of folders.

sites, we have built sites that were once proposed but that were never realized in actuality in order to stage alternative histories. The creation of an augmented reality walking tour using the software application Layar and the virtual reality tour based on the Unity gaming engine provide us with the new media tools necessary to construct contemporary digital objects at the crossroads of artmaking and humanistic inquiry.

The last decade has witnessed the rapid rise in museum-sponsored and independent artist projects that utilize augmented reality as a means to foster digital cultural heritage. This objective is certainly at the core of the two Jewish cultural memory projects that are discussed in this article. In "Apprehending the Past: Augmented Reality, Archives, and Cultural Memory," Victoria Szabo offers a valuable review of this "confluence of digital technologies with cultural heritage initiatives" (2018, 372) and she assesses how and why mobile augmented reality applications have become the preferred technological means for such site-specific projects in light of their ability to provide "digitally delivered overlays or supplements to existing material environments" (373). These overlays can unlock a repressed historical past in a familiar landscape in a work such as Heidi Rae Cooley and Duncan Buell's AR walking tour *Ghosts from the Horseshoe* (2012). This project is designed to expose the buried and repressed history of slave labor that went into the making of the University of South Carolina campus in Columbia and to promote a new habit of thinking. Cooley and Buell once referred to the *Mapping Ararat* project as "perhaps the closest analog to *Ghosts*" but she notes their major difference as follows: "Mapping Ararat thus imagines what might have been: Ghosts presents 'what was' but has been kept hidden" (2014, 208). However, there is an even more direct comparator in a self-professed DH and cultural heritage project that uses augmented reality (and the Layar app) to activate counterfactual histories in exactly the same way as *Mapping Ararat*.[3] This speculative application is to be found in the cutting-edge work of the Alaskan-based new media artist Nathan Shafer and his *Dirigibles of Denali* (2015-present). Shafer uses augmented reality to reimagine and reanimate the fantastic architectural plans for domed cities in and around Anchorage dating from the 1960's and 1970's that were approved but that were never realized. The result

---

[3] Liron Efrat frames the similarity between these two projects in the section on "Counterfactual Histories" of her dissertation (in progress) on augmented reality art. Efrat writes, "In both projects, AR is used to digitally execute an already approved plan, which eventually did not materialize in physical space." See "Welcome to The Thriving Field of the Real: The Aesthetics of Convergence and REALization in Augmented Reality," Graduate Department of Art History, University of Toronto, Chapter 4.2.1.3.

is the three site-specific and geo-located visions of Seward's Success, Denali City, and Arctic Town.[4]

Mapping Ararat
Site Specificity Goes Digital

In writing about *Mapping Ararat: An Imaginary Jewish Homelands Project* and its significance in the context of DH, Christopher Cantwell pointed out its capacity "not only to recreate lost religious landscapes, but also to imagine religious landscapes that never were."[5] The implementation of the emergent digital technology of augmented reality enables such a speculative enterprise. *Mapping Ararat* constructs an augmented reality walking tour that embeds 3D computer graphics modeled in the Maya and Rhino software programs into the everyday landscape at the very sites where Mordecai Noah plotted and projected his Jewish homeland on the banks of the Niagara River outside of Buffalo, New York. With smartphones in hand, visitors are able to divine, locate, and navigate architectural landmarks (e.g. synagogue) and potent symbols (e.g. flag) that are built to scale. These so-called assets are viewed on the screen of a mobile phone or a tablet device using the publicly available Layar application that relies on the use of geo-locational technology (GPS) to enable a site-specific mapping of Ararat with exact cartographic coordinates. These assets are not in the physical landscape; instead they are housed on a server and inserted into the landscape, so that this fictive Jewish homeland unfolds onscreen at these very sites. In this way, augmented reality moves site-specific installation art into the digital realm providing access to an imaginary dreamscape (as seen through the iPad or the iPhone) where counterfactual visions are superimposed over reality. In posing this parallel universe and mixed reality, Ararat's electronic monuments conjure

---

4 Shafer has gone even further to enlist science fiction writers to compose stories "imagining a world were these cities had actually been built, and an alternate history of Alaska reflecting that reality." See his on-line description "Dirigibles of Denali: The Unbuilt Domed Cities of Alaska Reimagined with Augmented Reality" (http://www.nshafer.com/dirigiblesofdenali/) and his collaborative on-line essay with Patrick Lichty (2016) on the links of this futuristic project to science fiction. "Looking at the *Dirigibles* project as a sort of alternate cultural data visualization of the near future might be an apt metaphor for the speculative fiction that the project is creating."
5 Christopher D. Cantwell, "Mapping Ararat: A New (To Me) Digital Project in the Study of Religion," *Religion in American History* (May 08, 2015). http://usreligion.blogspot.ca/2015/05/mapping-ararat-new-to-me-digital.html.

the Jewish phantoms that are still haunting the contemporary landscape of Grand Island, New York.

Our team takes the view that augmented reality is a fantastic and phantasmatic medium—one that opens up alternatives through which we encounter the ghosts and specters of things that might have been or that still might be yet to come. Here, the mobile camera phone functions not as a transparent window on the world or as a magic mirror reflection but rather as a spectral refraction that points to paths that were not taken but that haunt contemporary reality. The site-specific nature of the tour also resembles a treasure hunt as participants receive a map that marks the places where they must search for and find the augments. Each thumbnail-sized icon has been created from a birds-eye view render of the 3D Maya model. (Fig. 2) The augmented reality walking tour consists of twenty-four visual attractions in total. Each point of interest has an accompanying audio file that provides the virtual tourist with historical information and flights of fancy about the particular sites mixing fact and fiction in a multi-layered soundscape that also serves a pedagogical function.

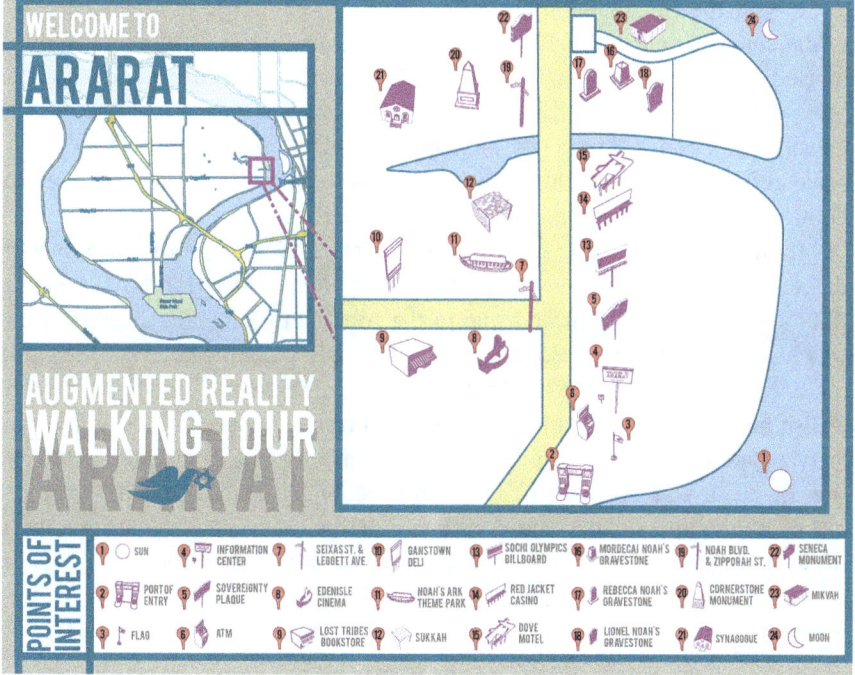

Figure 2: Melissa Shiff and Imaginary Jewish Homelands, *Mapping Ararat*, 2012. Adobe Illustrator graphic. *Mapping Ararat* augmented reality walking tour map

Historical Roots

Who was this dreamer and schemer with a plan to settle the Jews on the border between the United States and Canada? To review the historical record, Major Mordecai Manuel Noah (1785–1851) was the most prominent American Jew of his era. This larger than life personality corresponded with four presidents and once held a diplomatic post as the Consul to the Kingdom of Tunis in North Africa during the administration of James Madison. Based in New York City for most of his career, he was a leading Tammany Hall politician, a successful playwright, a major newspaper publisher, and a visionary who was the first person to propose the creation of an autonomous Jewish homeland in modern times. He called it Ararat and he conceived it as "a City of Refuge for the Jews."[6] In choosing this particular name, the Major revealed his Biblical Noah complex. According to the Bible, Mount Ararat was the resting place for Noah's Ark after the flood. Mordecai Noah thought that his new homeland could serve the same function for Jews around the world faced with Anti-Semitism and religious persecution. Thus, it is not surprising to learn that Noah's initial Ararat proposal coincided with the so-called Hep-Hep Riots against German Jews that took place in the summer and fall of 1819. Furthermore, Secretary of State James Monroe himself recalled Noah from Tunis in 1815 when the diplomat was revealed to be Jewish, claiming that this would "form an obstacle to the exercise of [his] Consular function."[7] This act of religious discrimination constituting both a personal and professional slight had an impact on Noah and led him to seek Jewish political and cultural autonomy via the Ararat scheme.

Grounded in a weighty piece of actual history, *Mapping Ararat* begins with the three hundred pounds cornerstone that Noah ordered from Cleveland, Ohio and that played a pivotal role in the Ararat Proclamation Ceremony that was held at St. Paul's Episcopal Church in Buffalo on September 15th 1825. It is the only relic remaining from Mordecai Noah's ambitious endeavor to create a Jewish homeland on Grand Island and it is currently housed at the Buffalo History Museum where our team made a pilgrimage at the beginning of our research. But it was always our intention to mobilize this cornerstone taking it out of the confines of the museum and situating it as part of the interactive AR walking tour. During our archival research, we also discovered an illustration

---

[6] This phrasing is found on the Ararat cornerstone that was revealed at the Proclamation ceremony that ironically took place at the St. Paul's Episcopal Church in Buffalo on September 15, 1825. We discuss the significance of this relic further below.

[7] Dated 15 April 1815, Monroe's dispatch is cited in J. Robert Moskin, *American Statecraft: The Story of the U.S. Foreign Service* (New York: Macmillan, 2013), 100–101.

published in a book dated from 1841 that depicts the brick and wooden obelisk constructed to house the cornerstone as a mid-nineteenth century tourist attraction.[8] This drawing was then rendered using the proprietary 3D modeling program Maya. While it is generally known for its 3-D computer animation capabilities, we are using it solely as a modeling application. After that, it was inserted into the landscape through a series of steps using the AR browser and software application Layar. Finally, with a smart phone in hand, a visitor sees the monument on Grand Island in its augmented reality form. In this way, *Mapping Ararat* restores the cornerstone and reanimates it for a touristic use in the digital era. Given that the Layar application allows for the taking of screenshots by the visitors in their encounters with the augments, this provides further opportunities for interactivity and the digital documentation of imagined spaces. In this particular screenshot, the photographer asks the tourist to pose in a way so that it seems "as if" he is reading the inscription on the Ararat cornerstone monument when he is actually looking at nothing at all. The result is a virtual photograph that serves as a visual enactment of what history looks like when it has been written in an alternative universe. (Fig. 3)

---

[8] This text reads: "The monument erected by Major Noah is now standing. It is about 14 feet in height. The lower part is built of brick—the upper or pyramidal portion is of wood, and the whole painted white." In John W. Barber and Henry Howe. *Historical Collections of the State of New York* (New York: S. Tuttle, 1841), 154.

Figure 3: Melissa Shiff and Imaginary Jewish Homelands, *Mapping Ararat*, 2012. Augmented reality screenshot. Tourist (Professor Adam Rovner) "reading" Ararat Cornerstone Monument

In developing the augmented reality walking tour, we were very fortunate to find another key historical artifact during our archival research. This one underscores the importance of the cartographic aspect of *Mapping Ararat* as is the case with many DH projects. It is an old map of Grand Island that was published in David H. Burr's *Atlas of the State of New York* in 1829.[9] Burr's *Atlas* provides colored plates for each of the fifty-six counties in New York and the map of Erie County (No. 50) includes Grand Island. In charting the island, Burr's *Atlas* lists Arrarat—spelled with 3 r's—as an actual geographical location. This map and its placement of Ararat on the northeast side of the Island was what enabled our team to root our imaginary Jewish homeland and augmented reality walking tour in a specific physical site. In other words, we are utilizing this area on the map as the location for our AR walking tour. This area is now known as White-

---

[9] David H. Burr, *An Atlas of the State of New York Containing a Map of the State and of the Several Counties* (New York: David H. Burr, 1829).

haven and it is the current home of the Radisson Hotel Niagara Falls-Grand Island.

Jewish Paths Not Taken: Synagogue and Gravestones

Starting from these real historical artifacts, our project moves into more speculative realms to pose the counterfactual question—What if? How might the course of Jewish and American history turned out differently if Ararat had been able to survive and to thrive? *Mapping Ararat* exposes viewers to the contingencies of history by plotting a counterfactual history that plays out this "what if" scenario on the Ararat path not taken. In illuminating this alternative trajectory of modern Jewish history, we are recalling that history is a construct of competing political desires and wills which could have turned out quite differently and that it is written from the perspective of that which prevails. This premise was at the basis of the exhibition: *Where to?* at the Israeli Museum of Digital Art in Holon, Israel in the spring of 2012 that featured *Mapping Ararat* prominently. The exhibition allowed artists and historians to tap into modern Jewish cultural memory and Israeli state archives in order to imagine possible roads not taken for the establishment of a Jewish homeland in light of the contested circumstances of the present.[10] As a new media art practice, augmented reality and its overlaying of the imaginary onto real space offers the technological means to give a vision and a voice to a failed plan such as Noah's Ararat. In this way, *Mapping Ararat* and its use of augmented reality can serve as a model for other digital art and humanities projects that seek to imagine (and experience) what might have been.

Let us now focus on a few of the more successful electronic monuments that we have conjured on the AR walking tour in order to illustrate further the rele-

---

10 This group exhibition featured Michael Blum, Ariella Azoulay, and Yael Bartana among others. The curatorial statement reads: "Through the exhibited works and the historical materials gathered for the exhibition, we suggest reintroducing these forgotten currents and ideas to the public discourse, bringing the 'losers' of history to the center of the stage, and once again presenting the question of Jewish existence as a current problem that remains unsolved." In a political landscape full of anxieties about the sustainability of Zionism based in the holy land or critical of its abuses of power in relation to the Palestinian population, it is easy to see why the *Mapping Ararat* project would resonate in Israel among post-Zionists and others seeking political and aesthetic alternatives. Thus, the curators selected the Ararat virtual synagogue as the on-line banner for the entire exhibition. See Udi Edelman, Eyal Danon, and Ran Kasmy-Ilan, *Where to?*, The Israel Center for Digital Art (2012) at their website https://www.digitalartlab.org.il/.

vance of our project to DH in context of the study of religion. Like the other augments on the tour, this three-dimensional synagogue is built to scale so that one can navigate around it or even go "inside" this particular model when viewing it on one's electronic device. It is important to point out here that *Mapping Ararat* is quite different from most academic uses of augmented reality that deal with the reenactment of events or the reconstruction of historic buildings that have a basis in things that actually existed. The work of DH historian John Bonnett (2003) and the 3D virtual buildings project in Ottawa that taught students how to generate models of historic settlements through the use of 3D modelling software comes to mind as an early example of this trend in the Canadian context. Other more recent examples of historical modeling and reconstruction create AR tourist apps based on archaeological evidence. The many examples constructed by the 3D computer graphics company MOPTIL (Mobile Optical Illusions) serve as a case in point. MOPTIL's apps have been specifically designed for Greek tourism at such world heritage sites as the Acropolis in Athens or the Knossos Palace in Crete.[11] In contrast, *Mapping Ararat* occupies a more hypothetical space given that it speculates and extrapolates from an actual proposal that never came to fruition. The construction of the Ararat synagogue augment offers a good case study of this mode of extrapolation in that it is based on actual architectural designs in upstate New York as well as synagogue designs in New York City from the same foundational period during the first half of the nineteenth century. Jumping to the present, one sees how the synagogue and the imaginary contours of Ararat contest the contemporary use of this particular site, for the Ararat synagogue has been sited at the edge of the eighteenth green of the River Oaks Golf Course in Grand Island. This means that worshippers have to watch out for flying golf balls if they want to go "inside" the virtual structure. This real-life hazard provides an excellent example of the surreal juxtapositions that can ensue when creating tourist attractions on this augmented reality walking tour. In this context, it should be mentioned that we have conducted successful on-site tours with different classes from the University at Buffalo. These guided experiences show the viability of AR walking tours as a pedagogical device that plays to the mobile technological habits of a generation of digital natives who have been raised on adventurous software applications and video games.

Another fascinating juxtaposition with religious overtones involves the construction of the gravestone augments to remember Mordecai Noah and his family

---

11 The company began its operations in 2014 and the complete list of the seven archaeological sites for which they have produced AR tours can be found at the MOPTIL website http://moptil.com/sites/.

members. These have been inserted into the Whitehaven cemetery on Grand Island. This is a decidedly Christian cemetery at the epicenter of where Ararat would have been according to Burr's *Atlas*. With this transplantation, we have repatriated Noah's gravesite relocating it from the Shearith Israel Cemetery in New York to his imagined Jewish homeland of Ararat. In doing this, we also have converted an actual monument into an electronic monument (to recall the digital cultural theorist Gregory Ulmer's formulation[12]). In terms of its construction, the Noah augment copies the 1875 drawing of his actual tombstone made by the artist A.H. Nieto and thus it parallels the fabrication of the cornerstone monument.[13]

In addition to the founder's gravestone, there are two others of this type on the AR walking tour. These are the gravestone augments constructed for Noah's wife Rebecca Jackson and for his youngest son Lionel. The tombstone for Lionel Noah poses questions related to religious faith and belonging and carry us further along the path of Ararat's counterfactual history. These alternative possibilities are raised in the audio track on the walking tour as the narrative accompanying Lionel Noah's gravestone moves between fact and fantasy. The text alludes to the genealogical fact that Lionel named his son Lionel Jr. in an act of Christian assimilation directly opposed to the Jewish practice of naming one's children only after one's deceased ancestors. Coincidentally, it turns out that Lionel Jr. repeated the same gesture in the next generation by naming his son Lionel Jr. too. In divining the Jewish ghosts of Grand Island, the placement of this tombstone in Ararat imagines an alternative history where Jewish naming practices would have foreclosed this possibility and would have led to a very different religious outcome and genealogical record.

The "homecoming" visit of one group of Lionel Noah's descendants to Grand Island in 2014 captures the poignancy of our project. In this field trip to "Ararat", Mordecai Noah's great-great-great-great grandchildren (both of whom are Christian) pose in front of their ancestors' virtual graves. (Fig. 4) In this speculative manner, Noah's actual descendants occupy the space of contested memories and imagine an alternative history for themselves and for this place. Such an image raises the counterfactual question of "what if?" directly and it allows us to peer into the contingencies of history. One senses the affective power and potential cognitive dissonance involved in asking Noah's descendants to embark on the Ararat augmented reality walking tour. Their uncanny presence on

---

[12] See Gregory Ulmer, *Electronic Monuments*. (Minneapolis: University of Minnesota Press, 2005).
[13] This illustration is found in the exhibition catalog edited by Abraham Karp, *Mordecai Manuel Noah: The First American Jew* (New York: Yeshiva University Museum, 1987), 72.

Grand Island raises the Jewish ancestral ghosts that might have been and that continue to haunt this place in the subjunctive mood.

Figure 4: Melissa Shiff and Imaginary Jewish Homelands, *Mapping Ararat*, 2012. Augmented reality screenshot. Visit to Rebecca Noah's gravestone augment by Mordecai Noah's great-great-great-great grandchildren

Virtual Kimberley

Setting the Scene

In order to reimagine and recreate our next Jewish homeland, we recently traveled to a remote and scarcely populated region of Australia to recover a forgotten chapter in modern Jewish history that began when I. N. Steinberg landed at the port of Perth in May 1939. This indefatigable Jewish political leader arrived in Australia seeking a refuge for the persecuted Jews of Europe just four months before the outbreak of World War II and only three years before the implementation of the Final Solution of the Jewish question better known as the Holocaust. He would spend the next four years petitioning the Australian people across the entire continent to save European Jews from the Nazi terror. Since the inception of this phase of our project in spring 2015, we have focused on the Kimberley scheme that was Steinberg's most sustained effort in Australia. Following in his footsteps on a recent field trip to the region, we were based in the town of Kununurra and in the vicinity of Lake Argyle (neither of which existed in his time). These places could only come into being as the result of the Ord River Diversion Dam—an irrigation and hydro-power scheme instituted some decades after the prescient Steinberg already envisioned it in 1939. This breathtaking landscape is a constructed one given that Lake Argyle is one of the largest man-made bodies of water in the world. The success of this fertile plain causes us to think that Steinberg's vision of a Jewish settlement in the Kimberley was not so far-fetched. While Steinberg managed to get the provincial government of Western Australia to agree to his plan in principle rather quickly, his efforts to convince the federal government were put on hold when World War II broke out in the Pacific region. There was also opposition from nativists and anti-Semites who wanted to keep Australia's borders closed to Jewish refugees at the time.

As his political organization's name indicates, Steinberg's ambition was to find "free land" anywhere on the globe. The Freelanders were not interested in following the Zionist path of restoring the Jewish homeland in the contested lands of the Middle East (Israel/Palestine). Instead, their political movement was known as territorialism and its advocates sought to carve out a niche for a Jewish settlement in a place where "the land should be uninhabited or sparsely populated so as to avoid competition with the native population."[14] In contrast to the Zionists, Steinberg did not seek a separate nation-state for the Jews, but rather a semi-autonomous colony where Jews would further their vibrant Yiddish culture while becoming Australian citizens. Shortly after his arrival, Steinberg headed to the Kimberley to meet with the Australian pastoralist and rancher, Michael P. Durack. Heavily in debt on account of a decade of poor beef-cattle prices in the 1930's, Durack was looking to get rid of some of his vast lands and to make a deal with Steinberg for this territory covering an area of some seven million acres that extended across Western Australia and into the Northern Territory and that was equal to the size of Belgium.

Steinberg also was aware of the aboriginal questions raised by his settlement plan and the need for there to be peaceful co-existence in the region. However, we want to recall that *Imaginary Jewish Homelands* draws from an historical archive dated from 1939 to 1943 and therefore decades before the recognition of aboriginal land claims in Australia. While there is an absence of an aboriginal perspective or voices in the I.N. Steinberg archive, he was aware of the need for any potential Jewish settlement to acknowledge and respect aboriginal rights. As he insisted in one of the many memoranda on his proposal, "The rights of the aboriginals in the areas selected for Jewish settlement will be strictly upheld and every facility given them to raise their standard of living in accordance with the

---

[14] "Memorandum prepared by the Freeland League for Jewish Territorial Colonization for the Consideration of the Delegates at the International Refugees Conference at Evian" (1938), 3. Papers of Isaac Nachman Steinberg (RG 366), YIVO Institute for Jewish History, Center for Jewish History, New York.

policy of the country."¹⁵ Nevertheless, Steinberg's plan remained an abstraction and he never got to the actual negotiation stage with the aboriginal people of Miriwoong Country whose lands were colonized by the British Empire and that were later settled by the Durack family. We also want to caution against the imposition of a contemporary post-colonialist perspective regarding these materials and the automatic assumption that the aboriginals of that day would have viewed the Jews as colonizers rather than as refugees with a just cause to settle in the Kimberley. This takes us back to a well-documented heroic historical incident when William Cooper who was the long-time leader of the Koori Australian Aborigines' League protested the "cruel persecution of Jews in Germany" by the Nazis after he learned about Kristallnacht (the night of broken glass) that took place on 9–10 November 1938.¹⁶ About a month later, Cooper led a delegation walking ten kilometers to the German Embassy in Melbourne to submit the petition but he was refused admittance by the Counsel-General. This famous case study serves as an emblem of inter-racial dialogue and co-existence in its recognition of the shared suffering of Jews *and* aboriginals as victims of racial discrimination at the time. Finally, it should be noted that the absence of any aboriginal structures in Virtual Kimberley was a conscious decision that we made because our project seeks to imagine a *Jewish* refuge and refugee camp, but this in no way imagines or means that the aboriginal people would be absent from this massive landscape in the Kimberley. Indeed, we felt that any gesture of "speaking for the other" in this way could be deemed as a type of colonizing gesture.

From Historical Archive to Virtual World

We are developing a Virtual Reality (VR) world using the Unity gaming engine in order to envision what might have been. Virtual Kimberley constitutes an interactive environment that consists of ninety-three structures ranging from refugee tents to larger institutional buildings. The buildings are modelled in the architec-

---

15 "Memorandum of Freeland League on the conditions which a large area of land in Australia should be granted for Jewish Colonization" (1940), 5. Papers of Isaac Nachman Steinberg (RG 366), YIVO Institute for Jewish History, Center for Jewish History, New York.
16 Ron Jontof-Hutter, "Kristallnacht and the Righteous Australian Aboriginal William Cooper," *J-Wire* (9 November 2018): http://www.jwire.com.au/kristallnacht-and-the-righteous-australian-aboriginal-william-cooper/. We also want to thank Jayne Josem, the Museum Director of the Jewish Holocaust Centre in Melbourne, for her valuable insights on William Cooper's valiant protest against the Nazi regime.

tural software program Rhino and textured in 3D Studio Max and Photoshop before they are imported into Unity. VR serves as the perfect medium for this project because it constitutes an imaginary dreamscape that allows for the creation of Steinberg's unrealized visions for a Jewish homeland in the Kimberley region. But instead of texturing the digital models in our virtual world with images of bricks and mortar, we have photographed thousands of documents from the vast Steinberg collection as the source material to overlay the polygon meshes that comprise the buildings of this Imaginary Jewish Homeland. Converting texts into textures, the figure invoked here is one of paper houses for a failed Jewish Utopia evoking the fragility and flimsiness of a project "on paper" that never made it past this initial idealistic stage. In this way, the project moves from the historical archive to the virtual realm allowing the public to see and read the archive on the VR tour and thereby providing interactive access to this imaginary world. In this context, it should be mentioned that this project envisions a number of user experiences utilizing different formats all within the Unity gaming platform. These include a large format projection or a video wall or a head mounted display for a fully immersive experience using the HTC Vive, a virtual reality headset. Through the interactive use of the head mounted display, the user will encounter each building in full scale so that the texts on the exterior and interior walls will be massive and imposing. We are also planning a limited and more directed user-experience on our website, which we plan to launch soon (www.imaginaryjewishhomelands.com).

There are three components to the virtual settlement in terms of the types of architectural structures. The first component is comprised of refugee tents as well as small houses that are fragile in nature. They are textured with newspaper clippings taken from actual articles published worldwide during the years when the Kimberley scheme was being considered. There is a conscious choice here to texture these models with a flimsy material punctured with gaps and holes. The tan color and the weathered condition of the newspapers (that are unfortunately decomposing in the archive) mimics the look of an actual refugee tent. Their fragility is also used to connote the marginal status of the refugee whose dwellings are makeshift and temporary. The second component is comprised of about a dozen institutional buildings as we imagine the types of structures that such a community would have built. These include the I.N. Steinberg House, Barn, Newspaper Factory, Post Office, School, among others. These structures have been selected based on the types and classification of the documents and images that were found in the archive. As one example, let us turn to the building with the word "Memorandum" on its entrance. (Fig. 5) The Freeland League Memoranda building in Virtual Kimberley is textured with the most important memoranda and other official documents from Steinberg's time in Australia in pursuit of the Kimberley plan including memos sent to the Prime Ministers of the day (Robert Menzies and John Curtin). Finally, the third component of the virtual settlement is comprised of what looks to be a small subdivision of residential houses. They are more formed than those in the refugee colony with four walls supporting these structures but every so often one notices that there are gaps and unfinished aspects here as well. This grouping imagines that the refugees would have moved up economically over time as they "settled into" the new land and became Australian citizens (as Steinberg envisioned). To focus on just one of the many examples, the subdivision includes this orange house with English and Yiddish texts announcing a meeting of the Freeland League in New York on the subject "The People of Australia is Ready to Grant a Haven to the Jewish People" featuring guest speakers Sir Norman Angell (who was the winner of the 1933 Nobel Peace Prize) and the American philosopher Sidney Hook.

Figure 5: Melissa Shiff and Imaginary Jewish Homelands, *Virtual Kimberley*, 2017. Unity screengrab with metadata. Freeland League Memoranda Building

The choices made with respect to the architectural styles used for most of the larger buildings are derived from the period of the international style and from functionalist architecture. We looked for buildings in this style to base our designs and these were often modified to meet our needs. For example, the town hall is based on the white functionalist building in Prague called the Manes Gallery that opened in 1930. Meanwhile, the Archive building is based on the Villa Roemer in Hamburg from 1928 designed by Karl Schneider. We are also speculating that Jewish architects who migrated from Germany and Eastern Europe to Tel Aviv (such as Erich Mendelsohn and Zeev Rechter) and who built the so-called White City of Bauhaus-inspired buildings there in the 1930's and 1940's would have emigrated to the Kimberley instead. Here we are imagining a counter-factual history—what if these architects had the chance to design Steinberg's Jewish Utopia? The Melech Ravitch House (Fig. 6) that honors the Jewish poet and adventurer who was the first to consider Australia as a possible place of Jewish refuge in 1933 is based on an unrealized Rechter building that has been modified. This structure houses the Kodak snapshots that Ravitch took on his epic journey as well as their captions and a few of his other writings on the subject.

Figure 6: Melissa Shiff and Imaginary Jewish Homelands, *Virtual Kimberley*, 2017. 3-D Studio Max render. Melech Ravitch House

The project offers an additional dynamic feature that brings it into the world of information studies and library science. If you want to learn more about any document that has been used to texture a model, then you can mouse over it to call up its metadata. We refer to this indexical practice as spatial archiving. The practice of spatial archiving as a mode of information visualization links the historical documents that have been used to texture each model with the data about the document in our archive. Each entry contains such fields as type of document, author, date, source, and comments. In this way, we are deploying a three-dimensional and interactive means in order to display and reveal an archive that consists of two-dimensional documents and artifacts. There are 1271 entries in the current index of the project ranging from state memoranda to personal letters, from newspaper clippings to children's drawings, from manuscripts to published pamphlets and essays, from period maps to the documentary photographs that Steinberg took on his initial expedition to the Durack lands in 1939. There are representative texts from a number of the languages that the polyglot Steinberg knew including English, German, and Yiddish. Every single document used to texture the models has a corresponding Hex color that is attached to it that allows for its retrieval via coding. While sometimes there are hundreds of documents used to texture one model (e. g., The Post Office), the smaller models such as the refugee tents rely on only one source document at times.

In terms of the user experience, one enters the virtual world from a birds-eye view where one is able to see all ninety-three models laid out in a grid-like pattern. The user then can click on any building and fly over to it for closer inspection and exploration. There are three other modes that are possible in the user experience—a ground view, an orbit view, and an inside view. In the orbit view, the buildings are isolated on a white background and one can zoom in and out using a track pad or mouse. This is also the level of experience where one can retrieve the metadata and engage in spatial archiving. Finally, the larger buildings are textured on the inside with different documents that are laid out in montage fashion through a series of rooms. By clicking on the space bar, one can proceed from room to room as well as to pivot so that one can look around inside these rooms (through the use of the arrow keys). The ability to traverse the virtual world relies on the writing of code in the C# computer programming language and this enables the user to control the models and to take them through their scripted actions and animations.

Transforming the Archive into a Diagram

In theoretical terms, it is also possible to think about this project and its traversal of a virtual world in terms of the shift from the "archive" to the "diagram" as theorized by the French philosopher Gilles Deleuze and more recently by the American film and literary scholar, Tom Conley. This is not surprising to learn when one considers the fact that Deleuze has been framed as the philosopher of the virtual. The key passage here is from Deleuze's book on *Foucault* (1986) when he makes the following distinction. "The *diagram* is no longer an auditory or visual archive but a map, a cartography that is coextensive with the whole social field."[17] While the historical archive is fixed and static in the realm of being, the diagram constitutes a mobile space of becoming. As Conley puts it in a recent essay on "Deleuze and the Baroque Diagram", "The archive deals with what was in the realm of the what is, while the diagram begins from what is to project what will be."[18] Moreover, the diagram is always in the realm of the cartographic—a mode between writing and drawing that seeks to be mapped. Thus, "the diagram

---

[17] Gilles Deleuze, "A New Cartographer," in *Foucault,* trans. Sean Hand (Minneapolis, MN: University of Minnesota Press, 1988), 34.
[18] Tom Conley, "The Strategist and the Stratigrapher," in David N. Rodowick, ed., *Afterimages of Gilles Deleuze's Film Philosophy* (Minneapolis, MN: University of Minnesota Press, 2010), 196.

is a map of possibility, of a *devenir* exceeding an 'archive or historical repository of forms."[19]

Our virtual world functions in a similar manner as it transforms the archive into the diagram and as it partakes of a desire to be mapped. Turning texts into textures, it animates and mobilizes the historical repository of forms that constitute the thousands of static documents in the Steinberg Archive at the YIVO Institute for Jewish Studies. As the user-participant traverses this virtual world and encounters its multiple pathways and its lines of flight, the archive is turned into a map of possibilities and trajectories. Furthermore, we have placed our virtual world on the orange part of the map ("Pastoral Map of North Australia and Kimberley") that carves out the borders of Durack's lands adding another cartographic dimension to the project. (Fig. 7) On the topographical level, this also has created a ground of rich earth tones that complement the fiery rock formations found in the Kimberley region. As Conley writes, "Mixing writing and drawing, the diagram is an intermediate shape between an object conceived and an object realized."[20] This is another way to situate virtual Kimberley as diagrammatic—mixing the voluminous writings in the Steinberg archive with the drawings comprised of computer graphics and 3-D architecture models that the user sets into motion by traversing the space and by mapping it. The result is this imaginary Jewish homeland—an intermediate shape between an object conceived (i.e., Steinberg's Kimberley plan) and an object realized using the Unity gaming engine (virtual Kimberley).

---

19 Tom Conley, "Deleuze and the Baroque Diagram," in Angelika Zirker, Matthias Bauer, Olga Fischer, and Christina Ljungberg, eds., *Dimensions of Iconicity* (Amsterdam: John Benjamins Publishing Company, 2017), 153.
20 *Ibid.*

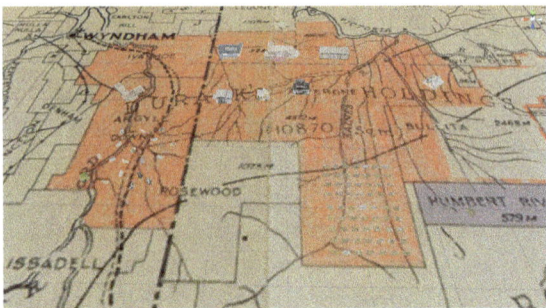

Figure 7: Melissa Shiff and Imaginary Jewish Homelands, *Virtual Kimberley*, 2017. Unity screengrab. Overview of *Virtual Kimberley* structures placed on the orange part of "Pastoral Map of North Australia and Kimberley" carving out borders of M.P. Durack's lands

Post-script

In his posthumously published draft entitled "The Actual and the Virtual," Deleuze states, "Every actual [object] surrounds itself with a cloud of virtuals."[21] If the State of Israel constitutes the actualized event of Jewish national sovereignty, then *Mapping Ararat* (with its implementation of augmented reality) and *Virtual Kimberley* (with its implementation of virtual reality) provide us with a cloud of potentialities (and of unknowing) from which we can only speculate what history might have become. (Fig. 8)

---

21 Gilles DeLeuze and Claire Parnet, "The Actual and the Virtual," in *Dialogues II*, trans. Eliot Ross Albert (New York: Columbia University Press, 2002), 148. If every document housed in the Steinberg archive is to be viewed as an actual object, then Imaginary Jewish Homelands mobilizes their virtual potential whether in the Unity gaming engine, in the cloud, or on the web.

Figure 8: Melissa Shiff and Imaginary Jewish Homelands, *Virtual Kimberley*, 2017. 3-D Studio Max photomontage. Jewish refugee house in the Lake Argyle region of the Kimberley

Selected References

Barber, John W., and Henry Howe. *Historical Collections of the State of New York*. New York: S. Tuttle, 1841.

Bonnett, John. "Following in Rabelais' Footsteps: Immersive History and the 3D Virtual Building Project," *History and Computing* 13, no.2, 2003: 107–150.

Burr, David H. *An Atlas of the State of New York Containing a Map of the State and of the Several Counties*. New York: David H. Burr, 1829.

Cantwell, Christopher D. "Mapping Ararat: A New (To Me) Digital Project in the Study of Religion," *Religion in American History Blog* (May 8, 2015). Webpage: http://usreligion.blogspot.ca/2015/05/mapping-ararat-new-to-me-digital.html.

Cooley, Heidi Rae and Duncan A. Buell. "Ghosts of the Horseshoe, a Mobile Application: Fostering a New Habit of Thinking about the History of University of South Carolina's Historic Horseshoe" in Samantha K. Hastings, Annual Review of Cultural Heritage Informatics (Lanham: Rowman and Littlefield, 2014), 193–212.

Deleuze, Gilles, *Foucault*. trans. Sean Hand. Minneapolis, MN: University of Minnesota Press, 1988.

DeLeuze, Gilles, and Claire Parnet, *Dialogues II*. trans. Eliot Ross Albert. New York: Columbia University Press, 2002.

Karp, Abraham. *Mordecai Manuel Noah: The First American Jew*. New York: Yeshiva University Museum, 1987.

Lichty, Patrick and Nathan Shafer. "AR, Alaska and Augmenting the Circumpolar," in Kevin Hamilton, ed., CAA Conference Edition 2016, Washington, DC, *NMC/Media-N: Journal of the New Media Caucus.* Website: http://median.newmediacaucus.org/caa-conference-edition-2016-washington-dc/ar-alaska.

MOPTIL (Mobile Optical Illusions) Website: http://moptil.com/.

Moskin, J. Robert. *American Statecraft: The Story of the U.S. Foreign Service.* New York: Macmillan, 2013.

Rodowick, David N., ed. *Afterimages of Gilles Deleuze's Film Philosophy.* Minneapolis, MN: University of Minnesota Press, 2010.

Shafer, Nathan. "Dirigibles of Denali: The Unbuilt Domed Cities of Alaska Reimagined with Augmented Reality." Website: http://www.nshafer.com/dirigiblesofdenali/.

Shiff, Melissa, Louis Kaplan, and John Craig Freeman. *Mapping Ararat: An Imaginary Jewish Homelands Project.* Website: http://www.mappingararat.com.

Szabo, Victoria. "Apprehending the Past: Augmented Reality, Archives, and Cultural Memory," in Jentery Sayers, ed., *The Routledge Companion to Media Studies and Digital Humanities* (New York: Routledge, 2018), 372–383.

Ulmer, Gregory. *Electronic Monuments.* Minneapolis: University of Minnesota Press, 2005.

Zirker, Angelika, Matthias Bauer, Olga Fischer, and Christina Ljungberg, eds. *Dimensions of Iconicity.* Amsterdam: John Benjamins Publishing Company, 2017.

Caleb Elfenbein, Farah Bakaari, Julia Schafer
Mapping Anti-Muslim Hostility and its Effects

Introduction: Confessions of a Neophyte

In September of 2015, I (Caleb) was teaching my introduction to Islam course when the story of Ahmed Mohamed became national news. Mohamed, then 14, was arrested for allegedly bringing a hoax bomb to school. It turns out that it was a digital clock that he had built for a school assignment. Nonetheless, Mohamed was led out of school in handcuffs, a hauntingly terrified look on his face. I remember teaching that morning. These are the conditions in which we are studying Islam in the contemporary United States, I explained as I projected an image of officers leading him out of the principal's office in handcuffs. Suspicion of Muslims is sufficient to lead to the arrest of a 14-year-old student fulfilling an assignment. I did not know it at the time, but in retrospect, this was the moment that what became Mapping Islamophobia began to percolate in my mind.

Just two months after Mohamed's arrest, in the context of an increasingly incendiary, anti-Muslim political environment, coordinated November attacks in Paris appear to have unleashed a remarkable wave of anti-Muslim hostility across the United States. While obviously troubling, this in itself was not unprecedented. The attacks of September 11 led to a spectacular rise in anti-Muslim activity, including hate crimes and government sweeps and detentions. Yet political discourses at the time were very different than those circulating in 2015. While government surveillance programs continued well beyond the months immediately following September 11, rates of hate crimes and open harassment of Muslims abated considerably in the space of a few months (even if remaining higher than pre-September 11 levels).

In contrast, in 2015, openly anti-Muslim political discourse appears to have created conditions in which public anti-Muslim hostility thrived, resulting in significant increases in rates of hate crimes and other anti-Muslim activity. It is in the context of these conditions that I happened to attend a workshop in postcolonial digital humanities (DH), which, quite unexpectedly, transformed the way that I think about research and the role that scholars play in public life.

I must admit that prior to this workshop I did not have a particularly good sense of the remarkable diversity of work that falls within the category of DH. In fact, I was attending the workshop largely because I was the incoming director of Grinnell College's Center for the Humanities and I thought that I ought to develop a better understanding of this dynamic, growing field of work. It just so happens that Roopika Rasim, who led the workshop, is a leading voice in postcolonial DH. This subfield of DH emphasizes the social justice possibilities of wedding technology and humanistic inquiry. Among the project examples Dr. Rasim presented was Mapping Police Violence.[1] I am not exaggerating when I say that encountering this website changed the trajectory of my career.

The rise of Black Lives Matter beginning in 2014 and increasing media coverage had made me more aware of police violence against African Americans, but seeing the website, developed and maintained by a research collective that includes Samuel Sinyangwe, Deray McKesson, and Brittany Packnett, gave me an entirely different understanding of this civil rights travesty. The website's landing page presents an interactive animated map with a pin for every death at the hands of law enforcement. A flash highlights individual data points as the map moves through time. Seeing these flashes accumulate over time had an effect on me. It communicated the horror of police violence in a way that news stories alone could not. I have since come to appreciate my response to seeing this website as developing affective understanding—an emotional response to a phenomenon that exists in fundamental relation to our intellectual understanding of that phenomenon.

Presenting data in a way that elicits an emotional response is one way that Mapping Police Violence is humanistic in nature. We cannot reduce this kind of engagement with data as mere "tugging at the heart strings," which suggests a kind of duplicitous tactic. As Sara Ahmed argues in *The Politics of Human Emotion*, emotion is a central component of human sociality, a crucial aspect of how we experience and engage the world around us. Deliberately working at an emotional register draws attention to the limits of rational understanding in capturing the human experience. In the case of Mapping Police Violence, seeing the data points flash across the screen gave me an opportunity to appreciate the cumulative weight of incident after incident of police violence against African Americans in a way that statistics never had.

A second humanistic element of Mapping Police Violence relates to data accessibility. Each data point on the map is clickable, opening a pop-up that contains information about the person behind the pin. These stories underscore that the weight of police violence against African Americans, which the animated map communicates so effectively, accumulates in and through individual lives.

---

[1] *Mapping Police Violence*, https://mappingpoliceviolence.org/, inspired *Mapping Islamophobia* both in its goals and modes of data visualization.

For those who are not directly affected by police violence against African Americans, these two humanistic elements of Mapping Police Violence have the potential to create empathy of a kind that other methods of presenting and engaging data might not.

Mapping Police Violence had a profound effect on me. In addition to helping me understand police violence against African Americans in an entirely new way, I knew right away that the site could serve as a model for a similar project on Islamophobia and anti-Muslim hostility. The resulting project, Mapping Islamophobia, has been the most rewarding work I have done as a scholar.

Mapping Islamophobia draws together a variety of my interests and goals as a scholar. Since I began graduate school, I have tried to imagine how I could produce scholarship meant for a broad audience. As a professor, I see the immense pedagogical value of mapping and data visualization in helping students begin to engage information in creative and accessible ways. As a scholar of Islamic studies, I have long felt a responsibility to Muslim communities. I am well aware that my career is made possible by the lives of those for whom Islam has been and is important in some way. As a scholar of religion, I am fascinated by the conditions in which societies engage in debates and negotiations about the values and goals animating community life, or what I have come to call the conditions of public life.

Of course, I am able to see how the project draws these things together only after having worked on it for almost two years. When I began Mapping Islamophobia, all I had was an idea for a map and a sense that this could be a good way to communicate to a broader public audience. I had no idea how to create a map like the one I imagined. The workshop motivated me to approach Grinnell's curricular technology specialist at the time, Mike Conner, explaining what I hoped I could do. He introduced me to Carto, a very user-friendly web-based mapping platform.[2] He showed me how to create a dataset that would provide Carto with the information it needed to do what I had in mind and helped me begin to think about basic design questions. It was so much more straightforward than I had imagined. This was an empowering discovery. From there I began to think about how to collect data on anti-Muslim hostility. I worked with a research librarian, Dr. Phil Jones, to identify the ideal methods of harvesting data from media sources and experimented with different search strings until I began getting good results. After a brief initial period collecting data, I turned

---

[2] Over time, I have also come to recognize Carto's limitations. Relative to other platforms, it has limited functionality and options for different kinds of data animation and interactivity. Tableau, for example, offers more options for data visualization of data interactivity, but requires much more training to use and maintain.

to data coding. What patterns was I seeing? How could I translate those patterns into categories? I had also begun collaborating with student researchers, bringing my first student aboard within a few months of getting the project off the ground.

By this time, the process of data collection and entry was in place, making it possible to move beyond a few initial entries in creating the first dataset on anti-Muslim hostility. Although we have refined things over time, the basic process has remained more or less the same. Each row of data begins with a news article, typically gathered through news alerts, from which we derive information about the location of an incident, the people involved, the nature of the incident, and relevant details about what happened. We then use Google maps to get geolocation information (down to a building or intersection whenever possible). At times, we have to search for and consult sources beyond our initial article, but it is surprising how much you can pull from one article. Once we have completed the entry we file the article so that we have a record of all sources. Periodically, we upload an updated dataset to Carto, which in turn updates the maps on Mapping Islamophobia.

It is a pretty straightforward though time consuming process—each entry takes about fifteen to twenty minutes, sometimes more when it is necessary to draw on additional resources to identify a precise date or location. With well over 3,000 rows of data as of this writing, you can see why it is necessary to have many people working on the project at once. Even with multiple people, it can be difficult to keep up with current developments while also collecting data on past years.

There is a pattern in this brief account of how I began Mapping Islamophobia: This project has been a collaboration from the start. DH projects require bringing together people with different skillsets to solve a problem. With Mapping Islamophobia, the problem was how to represent and communicate in accessible fashion a social justice problem I saw unfolding before my eyes. I was able to go live with the project within the space of a few months in large part because of the group of people who helped me think through how to solve this problem.

From its very first map (Figure 1), which uses Carto's animation function to show incidences of anti-Muslim hostility as they unfolded over time, the project has grown to include over a dozen maps drawing on two distinct datasets. Most of these maps are interactive. Users can click on each point on the map to learn more about the stories behind the data.

Mapping Islamophobia has continued to be a fundamentally collaborative project beyond the creation of this first map. The addition of a second dataset, which tracks American Muslim participation in public life through various

Figure 1: The project's first map, which uses animation to show changes in anti-Muslim hostility over time.

kinds of community outreach and political activity, emerged out of a question that arose in conversation with the very first student collaborator, Chloe Briney. Early on in the initial stages of data collection, she asked what we should do with the "good stories," examples of cooperation between Muslim and non-Muslim communities in the United States. Having started the project to highlight the place of anti-Muslim hostility in the conditions of public life for American Muslims, I did not have a good answer to her excellent question. It took months of thinking—and further questions about where the "good stories" fit into my work from audience members in early public presentation of the project's findings—to formulate an answer.

These "good stories" now sit beside incidences of anti-Muslim hostility to create a picture of the complexity of public life for American Muslims. Visual representations on Mapping Islamophobia present this picture in an accessible fashion. Users can see trends in anti-Muslim hostility as well as in American Muslim outreach efforts. Having these maps side by side on the site provides an opportunity to see that we cannot understand the heartwarming stories of cooperation and dialogue without reference to the place of anti-Muslim hostility in the conditions of public life for American Muslims.

Mapping Islamophobia is, at base, an educational resource. The site provides users with an opportunity to explore anti-Muslim hostility and American Muslim participation in public life, presenting information in a way that allows them to draw their own conclusions. This is asking a lot of users and may in some ways limit our audience. Nonetheless, we hope this decision leads to quality of engagement with the data for those who are willing to put in the time to exploring Mapping Islamophobia, including teachers and students in secondary school and higher education.

As the project has continued to unfold and grow over time, student collaborators have taken on an increasingly significant role. My co-authors, Julia Schafer and Farah Omer, have each taken primary responsibility for one of the two datasets, allowing me to take a broader view of the project and to develop related scholarly work that analyzes the significant data we have collected. Along the way, they have been essential conversation partners, exploring the implications of decisions we make about how to code data, what data to include and why, and the findings about the conditions of public life for American Muslims that present themselves in the course of our work. In addition to opening space for their immense intellectual contributions to the project, thinking of student contributors as collaborators ensures that Mapping Islamophobia meets or exceeds best practices in DH work.

Article 1 of the Student Collaborators' Bill of Rights, created by staff and affiliates of the UCLA Center for Digital Humanities, states, "As a general principle, a student must be paid for his or her time if he or she is not empowered to make critical decisions about the intellectual design of a project or a portion of a project (and credited accordingly). Students should not perform mechanical labor, such as data-entry or scanning, without pay."[3] I have been fortunate enough to have access to resources to pay all of my student contributors an hourly wage, using a combination of my institutional research funds and resources available through Grinnell's recent Mellon Foundation DH grant. This is especially important for those who are somewhat less involved than Julia and Farah in big-picture decision-making conversations. Still, whatever the nature of their contributions, being able to compensate all student collaborators has helped make Mapping Islamophobia an educational *and* professional experience for them.

In what follows, Julia and Farah will present their own experiences with Mapping Islamophobia, offering important insight into student collaborator experience in the development of a DH project. They present their contributions

---

3 UCLA, "HumTech," http://cdh.ucla.edu/news/a-student-collaborators-bill-of-rights/.

and experiences in the framework of their respective datasets. In so doing, they will provide additional detail about the project and its ongoing work.

Reflections of a Student Collaborator: Julia Schafer

Our first dataset records instances of anti-Muslim hostility across the United States. Anti-Muslim hostility takes a multitude of forms in public life, from anti-Islam graffiti on mosques, anti-Muslim protests, and public campaigns targeting Muslims to local, state, and national political discourse and state-level anti-shariʻa legislation. A critical assumption of our project is that anti-Muslim hostility creates conditions that are harmful to free and voluntary participation in public life. We use a series of interactive maps to visualize the ways in which different expressions of public hate manifest, both geographically and temporally, in the lives of Muslim American communities and across the United States. I have been intimately involved in creating this and managing this dataset. In the process, I have come to understand the value and role of DH in creating scholarship for the public good, some key facets of what goes into creating work meant for a public audience, and the terrain of collaboration in DH work.

As Caleb noted, our goal is to provide data in a way that empowers viewers to develop their own understanding of anti-Muslim hostility in the United States. Toward this end we have carefully curated information about the nature of the events that anti-Muslim hostility inspires, made this data available to users in an accessible way, and made a deliberate decision to refrain from providing in-depth analysis of our data.

A core value of Mapping Islamophobia is public accessibility. The desire to make data more legible to broader publics has presented us as scholars with an opportunity to consider what motivates publicly-facing work. For example, as a student contributor, this was the first time I thought about why and how it might be useful to purposefully omit explicit analysis in scholarly work. We certainly have our own analysis that reflects a deep engagement with the thousands of rows of data we have collected, but we have decided to create a resource that shows data in as accessible a form as possible rather than telling people what the data means. In this sense, Mapping Islamophobia is deeply informed by pedagogical considerations. We want to make sure that students, and others, have the chance to work with the data and come to their own conclusions. These decisions about legibility and access were central to our mission of creating a DH project that contributes to public education for social good.

To facilitate users' direct engagement with data, we created a careful system of collecting, organizing, and coding the information we include in our datasets. Each point of data is recorded and curated to provide the location, date, and nature of an event driven by hostility. Our categories were developed through a series of collaborative discussions over time as our team began collecting and parsing data. The fact of deciding how to code an event is its own analysis, requiring in-depth discussions to make sense of the stories we have found and continue to find through our research process. As contributors we have each been able to contribute to the key questions of what data to include, why we should include it, and what implications its inclusion and presentation will have for how users understand the conditions of public life for American Muslims.

Our map presenting anti-Muslim hostility data by the gender of those affected is illustrative of the implicit analysis in much of our decision-making.[4] When we created this map, which we display below, we noticed that the green dots representing incidents affecting Muslim women were obscured by the other data points depicting incidents that affected Muslim men or Muslims more generally. We felt that this reproduced what we had learned anecdotally—that Muslim women, especially Muslim women wearing head covering of some kind, seem to be less likely to report everyday experiences of anti-Muslim hostility. We wanted our map to counter this trend, and pulling that layer of data forward on the map is one subtle way of doing so. Although we do not talk about that decision on the website, it is an example of how choices we make behind the scenes are in themselves analytical work. Having said that, however, we have made every effort to be as transparent as possible in all that we do.

In fact, one of the first decisions made about the project was to be completely transparent in how we collect and record data. Our entire data set is published online and available to the public. We have also provided direct access to every source used to compile the data in the form of an attached URL. This serves two purposes. First, a key value of publicly facing work is transparency. If we are asking our users to engage our data, we need to be clear about how we have put together the information before them. Second, given some public skepticism around the extent of anti-Muslim hostility in the United States we wanted our data to be as close to beyond question as possible.

Another way we prioritize transparency is through careful recordkeeping. We have documented the major—and a good number of seemingly minor—decisions

---

[4] There is certainly a good argument to be made for more explicit analysis, especially when creating a resource of possible value to policymakers. *TellMAMA: Measuring Anti-Muslim Attacks*, https://tellmamauk.org/, based in the United Kingdom, teams with a range of partners to produce reports drawing on and illuminating its data.

Mapping Anti-Muslim Hostility and its Effects — 257

we have made in the process of developing this project. If anyone has questions about decisions we have made or specific elements of the project, we can answer them in fairly detailed fashion. This has influenced me to be very intentional as I work on the dataset. As I make decisions and keep records of my search process, I am constantly aware of the fact that my actions are public. I have come to see transparency as one element of creating a project with deep integrity. Further elements of creating a project with integrity is care in the data collection and coding process.

To ensure the integrity of the data behind our project, we use data sourced from media outlets with clear editorial oversight.[5] Additionally, when creating entries, we privilege articles that emerge out of reports to civil rights organizations and/or law enforcement agencies. Our team uses online databases to access archives of news media sources that report on instances anti-Muslim hostility. We chose to restrict our data to these types of sources because they provide the most transparency and authority in reporting on cases of hostility. The project is ongoing, and through individual work on the data set, we both maintain up-to-date reporting on the current year and work backwards through time. The work of the project is distributed by calendar year of analysis across multiple research team members, but decisions regarding visualization, coding, and language are made as a team.

Coding the data with consistency is another important element of maintaining the integrity of the project. Compiling and presenting data is essential to the interactive nature of the site, but ensuring that the data is well- and consistently-organized requires a challenging analysis and decision-making process for the

---

[5] We recognize that opting to include only those incidents that receive media attention from outlets with clear editorial processes and oversight limits our dataset. We feel strongly, however, that building on the careful editorial work of local, regional, and national "print" news outlets ensures that we can be very confident in all of our data. In an era of claims around "fake news" and "hate crimes hoaxes," we want our data to be as immune from criticism as possible. We also want to make sure that people using our site can access all of our sources themselves, a key element of transparency. Other projects have made less restrictive choices about data collection. For example, a project out of the New America foundation that came online in 2018, *Anti-Muslim Activities in the United States*, https://www.newamerica.org/in-depth/anti-muslim-activity/, includes data from what their site describes as a wide range of publicly available sources, including news reports and legislation. *TellMama: Measuring Anti-Muslim Attacks*, https://tellmamauk.org/, uses crowd-sourcing and testimonial methods for gathering data. Pro-Publica's *Documenting Hate*, https://projects.propublica.org/graphics/hatecrimes, which seeks to build a general dataset of hateful activity across the United States, uses an approach that combines these methods, including data from publicly available sources as well as a crowdsourcing component, such as reports by those targeted, witnesses, and others with knowledge of incidents.

researcher. The majority of our collaborative group meetings are dedicated to discussion about questions of coding events and the use of language in our project. In this process, both students and faculty have worked closely together to make substantial decisions regarding the (implicit) analysis entailed in coding data and writing descriptions. These discussions are an ongoing feature of the project as new and challenging questions and problems arise from the research process.

In order to fully represent the many expressions of anti-Muslim hostility, we decided to code each event as either a crime against a person, a crime against a place, a bias-related incident, legislation, public speech by a political figure, and public campaigns.[6] We also record the gender of those affected by an incident, the URL or bibliographic record of the media news source, the exact location (using latitude and longitude), and a brief, carefully crafted description of the event. All of this information makes its way directly into how we present our data, from the range of maps on anti-Muslim hostility on our website to the pop-up windows that provide information on the stories behind the data. When developing the descriptions that appear in these windows, we pay careful attention to writing formally and objectively, specifically refraining from any editorial comment. Our aim is to report the data clearly and factually and to promote the legibility and clarity of our project.

Beyond presenting the data in accessible fashion, there are a number of steps we take to support legibility through quality control, or maintaining uniform standards in coding and writing. As an editor of the anti-Muslim hostility dataset, I have worked to maintain an unbiased and formal writing style as a means of ensuring continuity as other student contributors rotate on and off the project. Different writing and working styles are a reality of collaboration, making the role of editor absolutely essential in a project such as ours. We

---

[6] Coding data is a central feature of DH work. It forms the basis for data organization and presentation, providing categories to make it possible for users to navigate the information you're providing. The Southern Poverty Law Center's *Hate Map*, https://www.splcenter.org/hate-map, provides a nice model of how data coding provides a way to filter information. They have organized things according to the ideology of the organizations they are tracking and offer an easily navigable menu for exploring the data accordingly. We were able to establish our six main coding categories fairly early on in the project, in part because patterns in the data quickly began to appear. We also use a second level of coding, which we call "Event (short name)." This coding is slightly more descriptive of individual incidents. Over time, the terms in this second level of data coding began to proliferate, creating inconsistency across the dataset, leading us to consolidate them while maintaining sufficient specificity to serve the intended descriptive purpose. We use drop down menus in our Excel spreadsheets to maintain data entry uniformity in many categories. This is especially useful when many different people serve as contributors.

Figure 2: Users can select one year or multiple years of data.

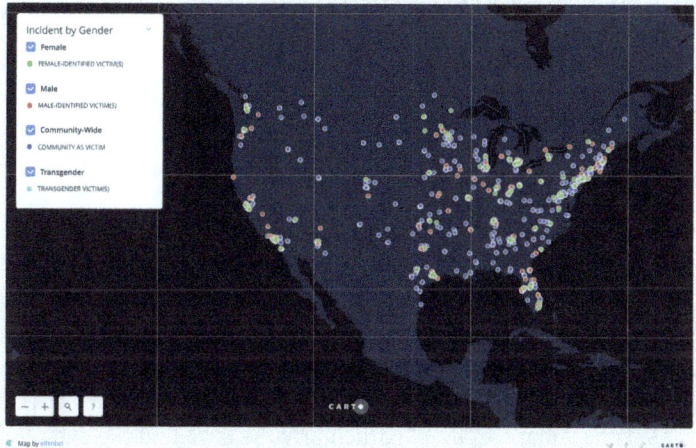

Figure 3: Users can sort the data by the gender of those affected. Many incidents, including vandalism, political discourse, and legislation, affect the entire community and so we code it accordingly.

want our users to engage data as seamlessly as possible. Having as close to a uniform voice as is feasible is an important way to work toward this goal.

Figure 4: Users can sort data by type of activity. The layers correspond to the codes we assign each data point. In this map, we display an example of a pop-up window that provides access to stories behind data points, an essential part of the project.

Reflections of a Student Collaborator: Farah Omer

Our second dataset documents outreach efforts that Muslim Americans do to humanize themselves in the face of anti-Muslim hostility. These humanizing efforts take a multitude of forms, such as mosque open houses, interfaith dialogue, and "Meet a Muslim" and "Ask a Muslim Anything" events. Islamic organizations, local mosques, families, and college students organize them. This dataset serves a critical role in the Mapping Islamophobia project because it illustrates effects of public hate on Muslim American lives, particularly the form and the extent of their participation in public life. I (Farah) have been primarily focused on creating and managing this dataset. In the process, I have come to understand that deciding on the parameters of our datasets, coding the data that drives Mapping Islamophobia, and making decisions about data visualization are themselves forms of analysis, showing that DH work, even when it does not include much explicit analysis, as with our site, is a scholarly endeavor. Moreover, as with Julia and Caleb, I have come to see the unique possibilities that the field of DH provides scholars who want to create publicly facing scholarship that addresses pressing social justice issues.

Prior to working on this project, I did not fully understand the collective burden on Muslim Americans to constantly distance themselves from extremist ideologies and terrorist attacks. Of course, we have all become well acquainted with the script that follows each terrorist attack. Muslim religious leaders, private citizens, and Muslim advocacy organizations alike publicly denounce the attacks, calling them un-Islamic. Conservative media outlets, politicians, and anti-Muslim activists typically ignore these statements, publicly demanding that Muslims prove that Islam does not condone such violence and extremist ideologies. Collecting data on the many ways that Muslims are reaching out to their non-Muslim neighbors across the country has shown us how hollow these calls really are, and our Mapping Islamophobia data visualizations present this evidence in an extremely accessible and compelling way.

Studying the outreach efforts of American Muslim communities in the context of anti-Muslim hostility provides incredible insight into the effects of public hate on the ability of American Muslims to partake in a core value of democratic life in the United States: the ability to freely and voluntarily participate in public life and discourse. Although, as Julia mentioned, we largely refrain from explicit analysis on our website, the user can observe these relationships in the maps. For instance, data visualization allows us to see the correlation between spikes in anti-Muslim hostility and increases in the humanizing outreach efforts of American Muslim communities. Our site explores such correlation and connection in a variety of ways, which Figures 5 and 6 demonstrate.

Figure 5: Mapping Islamophobia includes an animated map showing American Muslim public engagement and humanizing outreach as they unfold over time. We juxtapose this map with its twin map (Figure 1) presenting manifestations of anti-Muslim hostility over time.

Figure 6: Users can engage data about anti-Muslim hostility and American Muslim humanizing efforts on the same map. Mapping Islamophobia includes a series of maps presenting data from both datasets by year. As with other maps on the site, users can access information about the stories behind the data.

Mapping Islamophobia aims to shed light on the incredible pressure on Muslim Americans to exonerate themselves from public suspicion and uses digital tools to make this pressure legible and accessible to a wider audience. Analyzing the data shows that there is a definite correlation between anti-Muslim hostility and Muslim public outreach and engagement, in terms of both when we see ebbs and flows in outreach work and what people doing the work have to say about the impetus behind it.

By drawing attention to the relationship between anti-Muslim activity and Muslim public outreach and engagement, we attempt to bridge the divide between scholarship and activism, creating a space for the two to inform and enrich each other. After all, our data consist of many American Muslims who, through community engagement, want to actively change the perception of Muslims in their own communities. We are indebted to them because their hard work makes our scholarship possible. In return, we hope that some of these same activists can use our interactive, user-friendly maps to inform their work by engaging their audiences in visualizations of the rise of anti-Muslim hostility and efforts of Muslim Americans to counter this phenomenon.

In addition to our scholarly pursuits, we hope that our publicly engaged humanistic work will become a tool for activists who are fighting for this vulnerable population. We are keenly aware of the complexity and the burden of the communities that make our scholarship possible. The technological tools of mapping and data visualization help us make a small contribution to their efforts.

I have found that creating the parameters of the dataset documenting humanizing efforts of American Muslims has been one of the hardest and most rewarding aspects of creating these tools. Caleb, Julia, and I have spent a tremendous amount of time discussing what and who to include and as well as the ethical and analytical implications of the collaborative decisions we make. Do we include professional speakers who advocate for interfaith dialogue? Or is their status as professional advocates at odds with the project's goal of showcasing the burden of everyday Muslim Americans? If we opt to not include such actors, are we erasing important work, skewing the reality of American Muslim outreach efforts? Do we include individuals who engage in outreach efforts not to humanize Muslims, but simply to practice the tenants of their faith that encourage charity and generosity? These are some of the questions we have continued to wrestle with in our work. These decisions are in themselves acts of analysis and, as such, are a foundational element of our scholarly work.

Decisions about the parameters of this dataset—about who and what to include—are directly tied to questions about coding the data, which is itself also a deeply analytical process. For instance, in the humanizing public life dataset we have a category for political activity. This covers electoral work, such as Muslim Americans running for office, organizing voter-registration campaigns, and sponsoring fundraisers for political candidates. When considering adding this category, Caleb and I had a long discussion about the potential benefits and consequences of including Muslim American candidates in our dataset and what we hoped the category would convey. We agreed that including such candidacies is important for two central reasons.

First, we wanted to include American Muslim candidates for public office because running for office is such a core element of participatory democracy. After all, the dataset is about American Muslim participation in public life. Second, including Muslim candidates in the dataset enables us to show how their rates of candidacy have changed over time, plunging after 2001, remaining quite low for over fifteen years, and entering a potentially new phase of increased rates of candidacy in response to the incredible growth of public anti-Muslim hostility since 2015.

Yet we are also aware of the added burden of being a Muslim candidate for office amidst this apparent increase. We found that Muslim candidates often struggle to be seen as equal citizens who are running for office simply because they want to improve the lives of their potential constituents. Instead, coverage tends to emphasize their Muslim identity and they are seemingly required to prove their allegiance to "American values." Therefore, after many discussions, we agreed to include them in the dataset only when their religion is a part of their media coverage. We have found that at present we have not had to exclude more than a few candidates as a result of this decision, certainly not enough to skew results. This is an important finding in our project. Running for office is itself an instance of humanizing outreach for American Muslims. Thus, although we code such candidacies as "political activity," the stories behind the data show that the nature of American Muslim candidacies is shaped by the conditions of public life, much like the broader story that Mapping Islamophobia tells. Public anti-Muslim hostility affects what voters expect to hear from Muslim candidates, raising important questions about their capacity to freely choose how to present themselves and discuss their platforms.

The data we have collected for this dataset shows quite clearly that humanizing outreach efforts far outnumber instances of political activity. While the humanizing nature of American Muslim candidacies for public office illustrates the often blurry boundaries of the codes that datasets require, coding them as "political activity" helps us show how anti-Muslim hate affects the nature of American Muslim participation in public life. Data visualization illustrates the relative frequency of humanizing outreach and political activity with a force that simply presenting percentages would not. Hopefully this helps generate affective understanding for the viewer. Highlighting political activity as a separate category, represented by a distinct color in our maps, makes this possible. The coding decisions we make do not occur in a vacuum. They are tied to what we are trying to communicate to users and, importantly, *how* we hope to communicate it.

Figure 7: Users can select layers of data to explore the different kinds of humanizing work in which American Muslims have engaged over time.

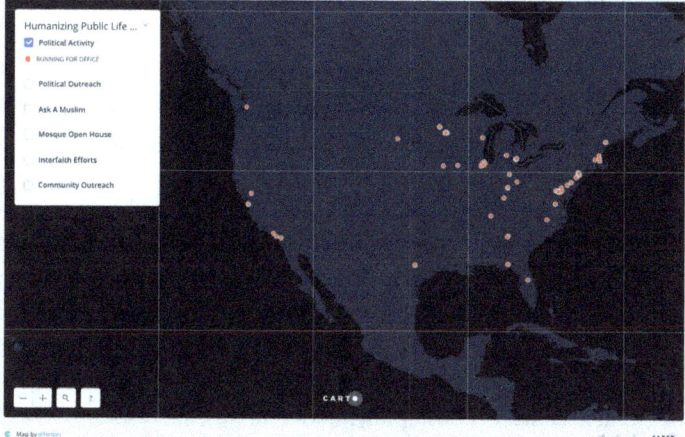

Figure 8: Sortable data allows users to see the incredible discrepancy between the scope of humanizing outreach (evident in Figure 7) and the numbers of American Muslims running for political office over time.

Reflections Together

As you can see, each of us brings different skills, experience, and sensibilities to this project. Our collaboration has made this project much richer than it would be otherwise. We have all learned a lot about DH as the project has unfolded. This process has also occasioned important conversations about what it means to do public scholarship. The project's emphasis on a pressing civil rights issue is a significant element of how we understand the purpose of public scholarship. But equally important is our recognition of our debt to the community we are seeking to serve, our attempt to be as transparent as possible in all that we do, and our decision to create an educational resource. These are some of the core values of DH as we have come to understand them. It has been a wonderful process to discover these values in and through creating a DH project, moving from neophyte status to advocates of the value of this kind of work as a mode of accessible publicly engaged scholarship.

We have already seen one example of how our publicly engaged DH work can reach audiences that more traditional forms of scholarship might not. Jason Harshman, a professor at the University of Iowa College of Education, has integrated Mapping Islamophobia into his social studies teacher-training courses and professional development workshops, and Alisa Meggitt, a teacher at North Central Junior High in Iowa City, Iowa, draws on the project to open conversations about global citizenship with her students. In both cases, the content *and* the form/medium of Mapping Islamophobia have been important to the way these educators have drawn on the project in their teaching.

Throughout this essay, we have emphasized that Mapping Islamophobia presents information in a way that encourages users to explore the data and generate their own insights and conclusions. Clearly, from deciding on dataset parameters to coding and making decisions about data visualization, we are making a range of analytical decisions behind the scenes. It is our hope that these decisions leave sufficient room for users of Mapping Islamophobia to do their own work. At the same time, we are also very aware and self-conscious of the fact that we are telling a story, a story that we think is a pressing social justice concern. This is a difficult balance. We acknowledge this balance by including a small measure of explicit analysis accompanying one map. Users would have to spend a fair amount of time exploring the Mapping Islamophobia website before discovering this particular map, "Hate's Effects: An Analysis of Collected Data." Even still, we preface the map with a prompt to the user, asking them to consider the data for a few moments before moving on to our brief analysis.

In Retrospect: Reflections of a Neophyte Project Director

Nearly four years of working on Mapping Islamophobia has provided plenty of opportunity to consider what I might have done differently. There are so many decisions that go into getting a project off the ground. Very few of these decisions are set in stone, yet changes do become more complicated and time consuming over time. I have considered on a number of occasions, for example, whether the site's data management and visualization platform, Carto, is the best fit for the project over the long term. I have also reconsidered some of the categories I have put in place to code and organize data.

Reflecting on the big picture, however, these are not the things that rise to the top of the list of things I would do differently, particularly if the question is what, on the basis of my own experience, I would advise others to keep in mind as they consider starting a DH project. What rises to the top of the list are considerations about project sustainability. A senior colleague warned me as I began working on Mapping Islamophobia that the web is (among many other things) a graveyard of once-promising DH projects. The best way to avoid this fate is to consider sustainability from the first moments of your project, which in retrospect I wish I had done more intentionally. Thinking about sus-

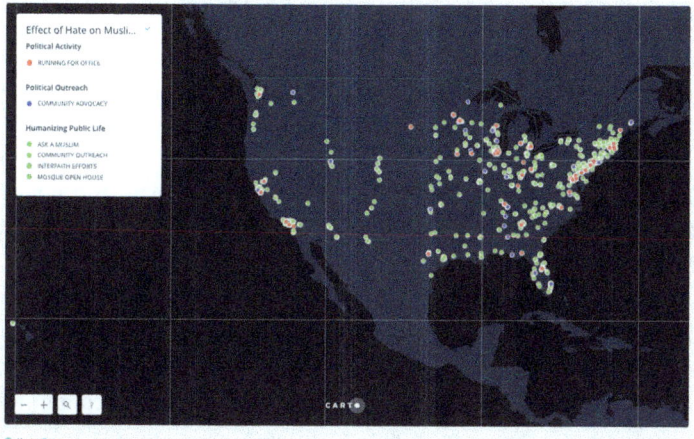

This map gathers "humanizing" work into one broad category (represented on the map by green dots)—activities that seek to or have the effect of humanizing American Muslims for the broader American public—and compares it with political activity and outreach undertaken by American Muslims (represented on the map by red and purple dots). Please spend a few moments considering what the data visualization suggests about the relative frequency of these kinds of activities, and then scroll below the map to see our analysis.

The green dots on the map above represent a variety of activities, a good number of which have the express purpose of humanizing American Muslims for the broader American public. This includes open mosque events, "ask a Muslim" events, interfaith initiatives, and public presentations about Islam. Sometimes community outreach efforts also consist of activities that American Muslims do simply because they are striving to be good people—distributing water in Flint, Michigan, opening mosques to shelter people after natural disasters, operating soup kitchens—but by virtue of today's climate these efforts also signify something more than that. We have lumped all of this work into one category on this map so that we can illustrate the disparity between the amount of humanizing work and political activity in which American Muslims engage. We do not think this disparity is a coincidence.

We think it is an effect of anti-Muslim hostility in public life. American Muslims are spending an incredible amount of time and energy doing humanizing community outreach, and we cannot help but wonder if this makes it less possible to engage in public life in different ways, like running for office and participating in policy debates around our most pressing domestic and international issues. A good number of the red dots on the map, which represent American Muslims running for office, reflects the efforts of individuals running for office multiple times, making the disparity between green (humanizing work) and red/purple (political activity and outreach) even more notable. The number of American Muslims running for political office is on the rise, though it is still well below the numbers seen in the years leading up to 2001.

This is of course just one possible effect of anti-Muslim hostility in public life. The fear that results from the threat of routine harm is clear enough in first-person accounts of American Muslim life in the contemporary United States. Yet in important ways the disparity we see in the data we present in this map likely grows out of such fear, motivating people to pursue certain kinds of public activities over others. In a society that prides itself on voluntary and free participation in public life, this strikes us a tragic feature of contemporary American life.

Figure 9: We present what we think are our most significant findings in a map that is nestled in the site rather than highlighted as a key element.

tainability includes being clear-eyed about how the project relates to your overall research agenda, developing a good sense of what kinds of support and collaborations you'll need (and have access to), and being open to a mode of doing research that departs from the "solo researcher" model that remains the norm in many fields.

If you are considering a DH project, one of the first steps to take is to check to see what technical support your institution can provide over the long term. As colleges and universities encourage more and more graduate students and faculty to pursue work in DH, it is essential to ask what kind of support your project will receive once you have moved beyond the initial, and often very exciting, start-up moments. The more successful your institution is at encouraging DH, the more demands will be made on your staff collaborators as they support an ever-expanding roster of projects. This means that technical changes, like moving data management and visualization platforms, may feel like insurmountable challenges—very few of the technical or content-related decisions are set in stone, but as the project moves along, you will likely need technical expertise and other kinds of support to make them.

Ongoing project support is particularly important if your initiative, like Mapping Islamophobia, is an open-ended endeavor.[7] Mapping Islamophobia does not have an obvious end point. When I started the project, I did not really understand what it meant to embark on an open-ended DH project, especially one whose goal was, at least in part, to serve as a resource for activists and policymakers. I started it because it felt important to do so. However, I did not truly stop to think where this project fit into my long-term research horizon. About two years into the project, a senior colleague advised me that in order for Mapping Islamophobia to "count" as fully as possible in salary and promotion reviews, I would need to publish about the project and its data in peer-reviewed settings. This seemed reasonable enough at the time. Now that I have committed significant time to doing so, however, I have real questions about the project's sustainability.

My undergraduate student collaborators have been a welcome and essential element of Mapping Islamophobia. Still, the time they have to contribute has its limits, not least of which is how quickly cohorts move through their undergraduate careers. When you consider the time it takes to train students, even with significant support from staff collaborators in DH, undergraduate collaborators may cycle through the project more quickly than you can replenish the pipeline of those ready to contribute. Project sustainability requires that you consider such factors. As I focused more time on publishing around Mapping Islamophobia, I began to depend more on student collaborators to collect and enter data and for project management. Managing and editing this work still takes time, and so balancing the role of project director with other parts of your research agenda—not to mention teaching and service—can be quite challenging. Having student collaborators, and being a good mentor to them, requires a significant investment in time.

Moreover, sustaining an ongoing project according to best (ethical) practices requires that you compensate your student collaborators. Just as you need to in-

---

[7] Not all, or even most, DH projects are ongoing in this way. A wonderful example of a discrete DH project is Kayla Wheeler's "Mapping Malcolm's Boston: Exploring the City that Made Malcolm X," *Mizan Project* 30 Nov. 2017, http://www.mizanproject.org/mapping-malcolms-boston/. Aspects of the project may continue to develop over time. Nevertheless, because it draws on what is ultimately a limited dataset project sustainability looks different than for projects with ever-growing datasets.

quire into what kind of ongoing technical support you will have, long-term sustainability requires that you be clear about what kind of financial support your institution can provide for student compensation that probably exceeds what would normally be available for research assistants. If your institution does not or cannot provide this kind of support, identifying external sources of funding becomes all the more important. Of course, this takes significant time as well.

I have enjoyed, and am enjoying, developing and maintaining Mapping Islamophobia. As I reflect on the past years, I wish that I had taken more time to talk to people already doing DH work to ask for their advice. Gathering information more intentionally from experienced DH practitioners would have helped me consider big questions about where the project fits into my research agenda—and professional goals more generally—and how to sustain the project over time even when I turned my attention to other things, as inevitably happens in our professional lives. Digital humanities projects, like many other kinds of research projects, will change and grow, and having a clear sense of your goals, project scope, and long-term sustainability, even around things as basic as site maintenance, if your project has a more delimited temporal and/or substantive scope than Mapping Islamophobia, will help make the experience all the more fulfilling.

Our description of Mapping Islamophobia shows that the project unfolded through an organic process. We learned as we went. It is one of the things that has made working on Mapping Islamophobia such a joy. I have learned more than I could have imagined about a huge range of things—data collection, data coding, data visualization, not to mention all that working on the project has helped me see about anti-Muslim activity in the United States and the tremendous work that Muslims communities have undertaken to provide a gracious model of public life. Much of this learning has resulted from the collaborative nature of DH work. This has helped me see that while project sustainability does depend on clarity regarding your own goals and horizons, it also depends on embracing a way of learning and researching fundamentally rooted in community.

It is not surprising that some of the projects that have most influenced the way I think about DH are sustained by collaboratives. Mapping Police Violence, which I discuss earlier in the chapter, has a multi-person planning committee in place to run the project. Another amazing example of social justice-oriented DH, the Anti-Eviction Mapping Project, also depends on an extensive leadership and contributor collective. Had I studied other projects more intentionally I may very well have created a very different model for sustaining Mapping Islamophobia. While very few of us will embark on projects as ambitious as these amazing examples, their structure helps me reflect on my own experience and brings something into clear relief. Passion will help get your project off the ground. Your collaborative relationships will be what truly makes them possible in the fullest sense, hopefully in a way that will sustain them over the long term.

Selected References

"A Student Collaborators Bill of Rights," UCLA HumTech (8 June 2015): https://humtech.ucla.edu/news/a-student-collaborators-bill-of-rights/.

"Anti-Muslim Activities in the United States." New America Foundation. https://www.newamerica.org/in-depth/anti-muslim-activity/.

"Documenting Hate." ProPublica. https://projects.propublica.org/graphics/hatecrimes.

Sinyangwe, Samuel and DeRay McKesson. *Mapping Police Violence*. https://mappingpoliceviolence.org/.

Southern Poverty Law Center. "Hate Map." https://www.splcenter.org/hate-map.

"TellMAMA." Faith Matters. https://tellmamauk.org/.

Wheeler, Kayla. "Mapping Malcom's Boston: Exploring the City that Made Malcom X," *Mizan Project* (30 November 2017): http://www.mizanproject.org/mapping-malcolms-boston/.

Part IV: **Issues**

Christopher R. Cotter and David G. Robertson
Critique and Community: Podcasting Religious Studies

Introduction

The *Religious Studies Project* (RSP) began in May 2011 when the authors of this chapter met in the Students' Association bar at the University of Edinburgh and decided to record a few audio interviews with scholars passing through the local RS seminar series. Formally launching in January 2012, it has become a truly international collaborative[1] enterprise, and is currently sponsored by the *British Association for the Study of Religions* (BASR), the *North American Association for the Study of Religion* (NAASR), the *Australian Association for the Study of Religion* (AASR), and the *International Association for the History of Religions* (IAHR). In September 2017, The Religious Studies Project Association—the non-profit organisation behind the scenes—gained charitable status as a 'Scottish Charitable Incorporated Organisation' (SCIO). Our newly minted constitution[2] outlines the purposes of the organisation as follows:

1. To disseminate contemporary issues in the academic study of religion/s ('Religious Studies') to a wide audience, and provide a resource for students engaged in such study, their teachers, and interested members of the public.
2. To provide engaging, concise, reliable and accessible points of entry to the most important concepts, traditions, scholars and methodologies in the contemporary study of religion/s, without pushing a confessional or apologetic agenda.
3. To pursue these aims principally through the maintenance of the *Religious Studies Project* website,[3] and supporting an associated editorial team in

---

[1] Note the 'collaborative'. We humbly and gratefully acknowledge the work of editors past and present: Katie Aston, Ella Bock, Sammy Bishop, Helen Bradstock, Sidney Castillo, Thomas J. Coleman III, Louise Connelly, Breann Fallon, Daniel Favand, Cole Gleason, Hanna Lehtinen, Martin Lepage, Knut Melvær, Kyle Messick, Raymond Radford, Venetia Robertson, Jane Skjoldli, Per Smith, Jonathan Tuckett, and Kevin Whitesides. We also acknowledge all of those who have contributed as interviewers, interviewees and respondents, unfortunately too numerous to list here, but a full list can be found at http://www.religiousstudiesproject.com/persons.
[2] Religious Studies Project, "Constitution," https://www.religiousstudiesproject.com/contributors/.
[3] Religious Studies Project, http://www.religiousstudiesproject.com/.

the production, dissemination and archiving of regular audio podcasts and written features, in the maintenance of a social media presence and email list, and in editorial duties for the associated journal *Implicit Religion*.[4]

To this end, we have produced over 250 podcasts, roughly 30 minutes each, with leading scholars on cutting-edge theoretical, methodological, and empirical issues in Religious Studies, in combination with regular response essays, which reflect on, expand upon, or critique our podcast output. For our purposes, a podcast is understood as 'audio [or, increasingly often, video] content available on the Internet that can be automatically delivered to your computer or MP3 player.'[5]

All our new podcasts now come complete with a written transcription, some with video as well as audio, and all are now also released through our *YouTube* channel,[6] as well as through *iTunes* and other podcast feeds. The website also features a weekly digest of opportunities (jobs, funding, calls for papers, etc), roundtable discussions, book reviews and other occasional publications, and provides a forum for discussion which is augmented by our lively social media presence. By March 2018, listeners had downloaded our podcasts over 490,000 times, with new podcasts averaging over 1,000 downloads in their first week, and our first ever podcast—with James Cox on "The Phenomenology of Religion"—now boasts over 7,700 downloads. The website receives over 150,000 hits per year, and we are currently followed by over 4,800 accounts on *Facebook*, and 4,300 on *Twitter*. In 2016 the first 'RSP Book'[7] was published—*After World Religions*—which expanded on a podcast of the same name, itself a response to a second interview with James Cox, on the World Religions paradigm.[8] We also began collaborating with Equinox to produce and transform the journal *Implicit Religion* following the death of founding editor Ed-

---

4 *Implicit Religion*, https://journals.equinoxpub.com/index.php/IR. Produced in collaboration with the RSP since 2016.
5 Michael W. Geoghegan and Dan Klass, *Podcast Solutions: The Complete Guide to Podcasting* (Berkeley, CA: Friends of ED, 2005), 5.
6 Religious Studies Project's YouTube Channel, https://www.youtube.com/channel/UCypfitkcldmX1CpAqCp7PKw.
7 Christopher R. Cotter and David G. Robertson, eds., *After World Religions: Reconstructing Religious Studies* (London: Routledge, 2016).
8 This book was the culmination of a *Facebook* interaction with Russell McCutcheon. At the time of writing, the authors have never met Russell in the 'meat world', but have participated in a number of collaborations with him and Alabama colleagues which would not have been possible without the kinds of online interaction he describes in his chapter in the present volume.

ward Bailey. Following initial sponsorship from the BASR, we now receive additional funding from the NAASR and IAHR, as well as maintaining relationships with the *Australian Association for the Study of Religions*, the *European Association for the Study of Religions*, and other organisations. In addition to these sponsorships and some revenue from advertising, we have recently started a *Patreon* campaign[9] and added a donations button to our homepage, in the hope that the RSP can become part of the solution to the exploitation of labour (particularly the labour of early career academics) so prevalent in the contemporary, neoliberal academic marketplace.

Our focus in this chapter shall be upon the podcasts themselves. First, we address the question *"why produce podcasts at all"?* Second, we discuss the practicalities of podcast production—from technical aspects of recording, editing and disseminating, to more structural issues of managing data and organising an international team of volunteers. Finally, we turn to some of the challenges and criticisms we have faced along the way, and reflect upon where we can go from here ... beyond interview-style podcasts, and beyond Religious Studies.

Why Podcasts?

This is not the place to go into the history and technical detail of podcasting in general.[10] For the uninitiated, however, a podcast is a form of episodic audio content published on the Internet and designed to be listened to using an *iPod* or other personal audio device. Podcasts emerged during the 2000s, but have reached maturity with the explosion of smartphone ownership since 2010, and at time of writing, 24% of Americans access podcasts at least once a month, rising to 31% among 25–54 year olds, with use continuing to grow year-on-year.[11]

Back in 2012, we could already see a number of distinct advantages to the format, particularly when we thought about our own consumption of the medi-

---

[9] Religious Studies Project's Patreon page, https://www.patreon.com/projectrs.
[10] But, see: Peter Ractham and Xuesong Zhang, "Podcasting in Academia: A New Knowledge Management Paradigm within Academic Settings," *Proceedings of the 2006 ACM SIGMIS CPR Conference on Computer Personnel Research 2006*, April 13–15 2006, Claremont, California, USA; Raymond Mugwanya, Gary Marsden and Richard Boateng, "A Preliminary Study of Podcasting in Developing Higher Education Institutions: A South African Case," *Journal of Systems and Information Technology* 13:3 (2011): 268–285.
[11] Edison Research, "The Podcast Consumer 2017," *Edison Research* (Apr. 2017). http://www.edisonresearch.com/wp-content/uploads/2017/04/Podcast-Consumer-2017.pdf.

um. Podcasts had provided us with: company when engaged in lonely, solitary tasks; a feeling of community; a personally-curated, 24/7 radio station on topics of interest; and an accessible point of entry into a variety of topics from film criticism and politics, to comic books, true crime, and classical music. Given that we both spent a significant amount of time engaging with podcasts, where was the podcast for our chosen discipline, the academic study of religion? Sure, we were aware of a couple that were out there—and more have since come onto the scene[12]—but we felt, at the time, that the existing output was poorly promoted, with recordings that were too long and abstruse and often with a thinly-veiled agenda. Nothing seemed to fit what we were looking for in an RS podcast. So we started recording the podcasts we wanted to hear—interviews with top scholars drawing on their most interesting research to converse about the fundamental methodological and theoretical issues in the field in a relaxed, concise and perhaps a little irreverent manner.[13]

The podcast format has a number of distinct advantages—for members of the public, students of all levels,[14] and more established academics. It democratizes knowledge and humanizes its production by giving listeners the chance to hear academics talking naturally, and offering an introduction to topics somewhere between a *Wikipedia* entry and a full-length book. A lot of material can be covered in half an hour, yet this can be digested at the listener's own pace, in their own time and space, again and again if necessary. Although podcasts lack the discursive element of traditional lectures, they can be used in a 'flipped classroom' model, and students find them particularly useful for revision.[15] Furthermore, regardless of our position in the field, we all have to focus our reading, and a podcast can help fill in some of the inevitable blanks, and facilitate listeners' keeping on top of the latest research as well as current perspectives on older scholars and themes.

In an era of departmental streamlining and closure, and with increasing isolation and stress brought on by the marketization of education and by limited budgets for conference participation, regularly listening to a podcast can provide a vital connection to the world outside the confines of one's institution that can be academically stimulating and provide a sense of community and common

---

12 See Michael J. Altman, "Podcasting Religious Studies," *Religion* 45:4 (2015): 573–584.
13 See Christopher R. Cotter and David G. Robertson, "Unlocking the Ivory Tower: The Religious Studies Project One Year On," *BASR Bulletin* 121 (Nov. 2012): 11–12.
14 See Ian Blair, "P.E.A.R.S. The Religious Studies Project and Undergraduates," *BASR Bulletin* 126 (May 2015): 14.
15 McGarr, O. "A Review of Podcasting in Higher Education: Its Influence on the Traditional Lecture," *Australasian Journal of Educational Technology* 25:3 (2009): 309–321.

purpose that might be lacking in one's immediate environment. And, similarly, given the increasing pressure for academics to relate their research to public interest, and to make sure their research is accessible for said public and has 'impact', recording a podcast is a simple and efficient way to disseminate research freely and accessibly to thousands of interested listeners in perpetuity. Much as Plate argues about *Massive Open Online Courses* (MOOCs) in this volume, podcasts 'can offer a critical intervention into the understanding of religion', helping us to 'push beyond academic insider-speak' and shape broader societal discourses.

All of this is not to say that simply recording a single podcast will have impact. A synergy of fortuitous factors worked in our favour, including timing, relatively light workloads, an unsaturated market, the support of the BASR and some senior colleagues,[16] and the availability and enthusiasm of a large number of interviewees, interviewers, authors and more. We also had a plan: there is no point launching enthusiastically into recording a couple of podcasts with departmental colleagues, only to have the venture fizzle out in a few weeks. Get some recordings in the bag. Build up a following. Take a break if you need to. But—importantly—make sure that when you establish a regular pattern of output that you stick with it. There is something particularly galling about that one webpage that was set up back in 2014 as the public face of Research Project X, but which was never updated and which served as nothing more than a half-hearted nod to the impact agenda before a retreat back behind the protective walls of the ivory tower. We selected a weekly schedule, with a long break during the summer, though that is not the only model. Our colleagues at the University of Alabama have successfully established an alternative pattern—releasing seven podcasts between March 2017 and March 2018—that is 'occasional', aiming to 'reflect the goings-on in the department and the larger field.'[17]

We quickly adopted an attitude of 'don't wait to be given permission', and this attitude has pervaded RSP output to this day. The point was not to merely replicate existing academic structures and outputs, but to complement, challenge and expand upon them. Indeed, it is unclear whether we would have been able to build anything like the resource we did had we been bound by a department or institution, because of the issue of justifying the cost in staff

---

[16] In particular George Chryssides, Dominic Corrywright, James Cox, Carole Cusack, Graham Harvey, Hannah Holtschneider, Bettina Schmidt and Stephen Sutcliffe were supportive in the very early days, and Russell McCutcheon later became instrumental in the institutionalisation and diversification of the project.
[17] Michael Altman, "We Have a Podcast," *Culture on the Edge*, 2 Mar. 2017, https://religion.ua.edu/blog/2017/03/02/we-have-a-podcast/.

time and resources for each episode, slow-moving checks and balances, and the inbuilt conservatism of institutional structures. Having built up a reputation, however, it is encouraging to see these existing academic structures engaging with RSP outputs in the form of citations and entries on course syllabi at universities from Alabama to Chester, Turku to Sydney, and, indeed, at Edinburgh,[18] as well as some creative and innovative engagements, such as that explored by Michel Desjardins at Wilfrid Laurier University in Canada, who built RSP podcasts

> into his (required) MA Method and Theory course. In groups, students selected 'the podcasts that attracted them the most' and then, along with other activities and short papers, facilitated class discussion on them. The result? A 'method and theory course that reflected student interest and instructor knowledge'. In this particular instance, the students seemed to enjoy 'having choice, they found it easy to listen to the podcasts, and it was impossible for people not to have opinions about what they'd heard.'[19]

These and other engagements with RSP podcasts reflect the logics of the contemporary Higher Education environment in which access 'to knowledge and ownership of knowledge [...] is no longer a marker of privilege and academic status,' where it 'is beholden on teachers to show and share how they are co-users of open access information'.[20] We passionately believe that the careful use of podcasts helps in no small way to mitigate the pitfalls of the 'transfer theory' of teaching[21] and develop a media- and resource-rich environment suited to the vagaries of the twenty-first century. But if that addresses the 'why' question, how might one even begin?

How Do We Podcast?

The project launched in late 2011 with £200: £100 from each of us. £100 went to setting up a *WordPress* website to host the podcast, and the remainder was spent on a digital recording device. We decided on a *Zoom H2*, a self-contained battery-operated unit with an excellent microphone. This would remain our primary de-

---

[18] See David G. Robertson, "Teaching Matters: Using Podcasts in Tutorials," *BASR Bulletin* 129 (Nov. 2016): 17–18.
[19] Jack Tsonis, Christopher R. Cotter and David G. Robertson, "The Religious Studies Project... In Teaching & Learning," *BASR Bulletin* 124 (May 2014): 11–12.
[20] Dominic Corrywright, "Landscape of Learning and Teaching in Religion and Theology: Perspectives and Mechanisms for Complex Learning, Programme Health and Pedagogical Wellbeing," *DISKUS: The Journal of the British Association for the Study of Religions* 14 (2013): 1–20.
[21] Dennis Fox, "Personal Theories of Teaching," *Studies in Higher Education* 8:2 (1983): 151–163.

vice for the first three years of the project, and in fact we still use it today in certain circumstances. It can produce great results in a quiet environment, but does pick up a lot of background noise if a quiet area cannot be found.

We recorded the interviews and the episode introductions separately, as we still do today. Initially, when we were doing all the interviews ourselves, we would simply alternate so the person not presenting the interview recorded the introduction. Once other interviewers joined the team in 2013, we began presenting the introductions together, taking on the role of hosts. Separate introductions are needed because interviews do not necessarily go out in the order in which they are recorded, in part to ensure variety of interviewer, location, and subject matter, but also because some interviews find a respondent more quickly than others. From the very start, we introduced a regular written response expanding, reflecting upon or critiquing the podcast, as an encouragement to others to continue the conversation, to provide an opportunity for up-and-coming scholars, and to establish the idea that there is no single, definitive 'RSP view' on a topic. It also underlined that the podcasts are part of a larger project—the RSP as 'hub', rather than 'show'. As previously mentioned, over time RSP publications have grown to include an edited book, a journal and a number of articles, including the one that you are reading.

To edit the episodes together, we used *Audacity*, an Open Source audio editor which is available on all platforms.[22] While not as fully featured as some editors (called DAWs in the trade, for Digital Audio Workstation), it has excellent noise reduction, is easy to use and extremely reliable. Again, we continue to use this to this day, although now we have a team of audio editors. We have also added a mastering stage through the online service *Auphonic*, which standardizes and optimises the audio, as well as automating some of the publishing process.[23] *WordPress* handles standard audio formats already, so we only had to add a simple (free) audio player—*PowerPress* by *Blubrry*[24]—to embed the podcasts into the website posts. Syndicating them to *iTunes* (historically the home of the *iPod* and therefore of podcasts, and still arguably the major library) requires setting up a specific RSS feed for podcast output, as well as some additional descriptors, keywords and so on. This can be done through *iTunes* itself, but *PowerPress* helpfully takes care of all of this, as well as ensuring that the feed is picked up by other podcast apps, and providing a free basic statistics service, shortcodes, and more.

---

22 *Audacity*, https://www.audacityteam.org/download/.
23 *Auphonic*, https://auphonic.com/.
24 *PowerPress*, https://create.blubrry.com/resources/powerpress/.

As the team of interviewers grew, it became increasingly important to standardise and improve audio quality. The plurality of voices on the RSP is a strength, but it also presents problems, as interviewers are not generally trained in audio production, so simple things which affect recording may not occur to them. Background noise, for example; while the gentle ambience of the city or of a park is not an issue, audio from a cafe may be almost unusable due to the high-pitched noise of cutlery and glassware which cuts across everything else, and must be removed manually—a time-consuming and difficult process which can often mean cutting quality content for audio reasons. Moreover, interviewees are academics, trained as researchers rather than media pundits, and so will turn away from the microphone mid-sentence, shuffle their notes next to the mic, or tap a pen or watch on the table while they speak, any of which potentially ruining the interview. With a weekly turnaround, it has not generally been possible to scrap an interview completely, or re-record it—although we have done this on occasion.

Audio quality is the single most common, and, indeed, highest profile complaint, that the project has received, with Mike Altman noting in journal *Religion* that a 'major flaw in RSP's podcasts is the production quality'.[25] Thus, much money and time has been spent to improve it, although this also presented challenges. In 2015, we had enough sponsorship to be able to purchase some professional-quality equipment, including microphones and a digital interface, but it simply was not viable to provide this for all our interviewers. Nor was this rig enough to record roundtables or panels with multiple speakers. Our solution at present is to purchase special clip-on mics[26] for our interviewers which work with any device, including a smartphone, iPad or laptop—these can give excellent results (although this is just as dependent upon background noise as our *Zoom H2* is), but for a much more affordable price. In addition, we will use our professional rig (now expanded with a mixing desk and additional mics for use in roundtables, etc) whenever possible—for example, if attending a conference, this will be set up and available for interviewers to use.

The expanding team created other systemic challenges. Audio was coming in from multiple sources, in multiple formats, and all large data files. We needed a way to store these files, and to share them. We tried using a private file server

---

[25] Altman, "Podcasting Religious Studies," 578.
[26] MOVO PM20 Dual-Headed Lavalier Condenser Microphones, which can be purchased through Amazon. Two institutional factors drove this decision as much as audio quality—plug-and-play compatibility with recording devices (including smartphones, as Apple have a different connection from other brands), and that it can be delivered worldwide. We do not want to be paying as much to mail as we paid for the actual hardware.

on our website, hidden from the public, but it quickly became unwieldy and glitchy[27] and we switched to using *Google Drive*. We also found that our respondents were having trouble with accessing the sometimes unusual file types and with their large size (many institutional mailboxes won't accept attachments larger than 25mb, for example), so we developed a system which automatically converts submitted audio to a simple .mp3. We have developed other forms of automation too: interviewers submit their interviews through an online form, which informs the editors and contains the information needed to create the text that accompanies the podcast when published, and the *Auphonic* mastering stage uploads the final audio to the Wordpress site and YouTube.

As our editorial team is spread across four continents,[28] another challenge has been creating an organisational system to handle it. Email threads quickly became unmanageable, and it was clear that a lot of work was being duplicated by the two Editors-in-Chief. With our then Managing Editor Daniel Favand and Webmaster Knut Melvær, we developed a cloud-based system using *Trello*, a project management tool, and *Google Docs*, in which each interview comes into the system via the online form, has the appropriate team members assigned to it, and is finally scheduled after finding a respondent and being edited. This has streamlined the production process enormously, which has helped us identify problems, as well as making it easier for new editors to join the team.

These systematic developments build upon the work of all the editors who have contributed over time, and the time saved on day-to-day tasks enables us to spend that time on pursuing the more ambitious aims of the RSP. It also means that, if needed, the RSP could continue without the founding Editors. Moreover, it means that we have established an open source, cloud-based infrastructure for others to use in the future to create similar—or preferably different! —public-facing initiatives. Indeed, we have spoken about this in seminars on the digital humanities (DH) at the University of Chester and the Open University, and intend to make this knowledge publicly available in future.

---

27 Powered by the open source *Pydio* (www.pydio.com). The issues with this platform seemed to be caused by our server, however.

28 Europe, Australasia, North America and South America. We are actively working on Asia and Africa. Antarctica will be a challenge, however.

A Rocky Road

As the preceding discussion has indicated, producing this born-digital, Open Access resource has not been all plain sailing, and we have been on the receiving end of a number of important criticisms over the years.

Going beyond criticisms of our audio quality, it was occasionally pointed out to us that our podcasts might be problematic for people for whom English is not a first language, or indeed for those with hearing impairments. One step that we took to address this was ceasing our established practice of recording our podcast intros and outros in various pubs and bars, and refraining from beginning each episode with niche pop culture references, as we would often do in the early days. Although we do still maintain a level of irreverent humour (as particularly evidenced by our annual 'festive midwinter special'), we decided that a bit more professionalism on our part would reduce the opportunity for things to be 'lost in translation'. We have also begun to transcribe new podcasts (the back catalogue is still a work in progress), which means that they can now be more easily cited and utilized in the classroom, and this also softens some of the barriers surrounding spoken English. These transcriptions are not cheap, however. Although we do receive enough funding from our headline sponsors to cover basic running costs, innovations such as transcriptions, new software and hardware, all cost money-hence the funding drive mentioned in the introduction. As Plate argues in his chapter in this volume, the popular rhetoric that MOOCs (and podcasts) are 'free' ignores the (often highly significant) costs in time and (financial) resources for those who produce them. We want to be part of the solution to the systemic exploitation of the 'free' labour of (junior) academics, and our ambition would be to be able to pay a fair rate for the work carried out by our large team of volunteers 'for the good of the discipline'. The first months of 2018 are a case in point, where a number of technical issues caused by our former hosting provider, and then compatibility issues with our new provider, disrupted our output for a number of weeks, and involved a significant amount of labour to restore functionality. Such events cannot be predicted, but without the 'stability' that comes from being attached to a university or other institution, even small injections of capital make a significant difference. And returning to the language issue, we are certainly willing to entertain the possibility of publishing podcasts in other languages in future, but that will require editorial team members and transcribers who have a sound knowledge of those languages, and might also raise issues of translation for our existing audience. However, we are working closely with the EASR and IAHR to expand the global regions covered in the substance of podcasts, and also to develop a better geographical spread of interviewees, interviewers and editorial team members.

This leads us to another group of important criticisms that we have received over the years surrounding, for example, the spread of topics covered, the prevalence of white males in our output, the featuring of the occasional controversial individual or argument, and the lack of a 'traditional' peer review process. Given our situatedness as two white, relatively privileged, relatively heterosexual, 'British'—or, rather, Scottish and (Northern) Irish[29]—men, who have been closely associated with the Religious Studies system at the University of Edinburgh system for over a decade, and who have very specific research interests, it is somewhat unsurprising that—despite best intentions—RSP output has fallen foul of these critiques. Yet these are critiques that we take seriously, and it is worth explicitly considering them here as they highlight issues that others may face in similar ventures, as well as in the field in general.

A pithy example is our annual 'Christmas special'. In frequently referring to this 'non-denominational festive midwinter special' as our 'Christmas' special, we have certainly adopted the hegemonic discourse of our social and historical context. We don't produce special comedy episodes to celebrate Yom Kippur, Diwali or American Independence Day, but neither is there a tradition of producing such specials in mainstream UK media. One could, of course, challenge such a hegemony, but our annual special is not a matter of religious observance—it is not 'about' Christmas. We have resources to produce one special per year, so it makes sense to publish it during the most widely-observed public holiday in the regions in which the vast majority of our audience live.

Turning to our more 'standard' output, we have on a few occasions taken the contextually 'easy' route and released a podcast which didn't quite live up to our implicit editorial line, and we did once release a podcast with an individual not known to us whose work turned out to be somewhat controversial in another geographical region.[30] Furthermore, our first foray into publishing a more 'traditional' research paper on the website resulted in one reader commenting:

---

29 See Christopher R. Cotter, "You're Greek? Well..., I'm (Northern) Irish, Kind'a ... " in *Fabricating Identities*, Russell T. McCutcheon, ed. (Sheffield: Equinox, 2017): 34–41.
30 See Philip Deslippe, "Stretching Good Faith: A Response to Candy Gunther Brown," *The Religious Studies Project*, 29 Jun 2017, https://religiousstudiesproject.com/2017/06/29/stretching-good-faith-a-response-to-candy-gunther-brown-philip-deslippe/.

> If RSP is going to publish 'research articles' representing the academic study of religions then the mss. should I think go through the normal process of peer review to help authors submit the best possible version before publication.[31]

On the latter point, part of the very purpose of the RSP was to provide an alternative to those institutional processes—such as lengthy periods of peer review—that, though important, slow everything down and stifle debate. At the time, Chris argued that our 'less formal approach allows the authors to take comments on board from members of the academic community and revise the text accordingly [...] to produce the best possible text' yet acknowledged the 'need to balance our preference for informality and accessibility with the standards of the academy.' In essence, if not in name, this was an example of 'post-publication peer-review.'[32] Ultimately, however, the issue was sidestepped when we began to direct outputs based on 'original research' to *Implicit Religion*. On the former point, we have now tightened up our 'vetting' procedure, informally assessing potential contributors based on where they work/study, what/where they have published, who else has worked with them, what conferences have they presented at, and so on. Where questions have been raised, we have asked around trusted colleagues, or have simply tried to raise these questions in the podcast itself (to greater or lesser success). Further, we have clarified our intellectual angle through the development of our constitution (see above), and the reformulated rubric for *Implicit Religion*, which takes a broad scope and showcase

> analyses of material from the mundane to the extraordinary, but always with critical questions in mind such as: why is this data boundary-challenging? what do such marginal cases tell us about boundary management and category formation with respect to religion? and what interests are being served through acts of inclusion and exclusion?[33]

On the 'diversity' front, it is certainly true that we have hastily convened roundtable discussions with arrays of panelists that would be difficult to describe as 'diverse'. The makeup of our very first roundtable discussion on 'The Future of

---

[31] Michael Stausberg and Knut Melvær, "What is the Study of Religion/s? Self-Presentations of the Discipline on University Web Pages," *The Religious Studies Project*, 6 Dec. 2013, http://religiousstudiesproject.com/2013/12/06/what-is-the-study-of-religionsself-presentations-of-the-discipline-on-university-web-pages/.

[32] Jane Hunter, "Post-Publication Peer Review: Opening Up Scientific Conversation." *Front. Comput. Neurosci.* 6 (2012): 63.

[33] *Implicit Religion* index, https://journals.equinoxpub.com/index.php/IR.

Religious Studies' in March 2012[34] provoked one visitor to comment 'Perhaps the photographs are misleading or I've missed something, but why are the discussants all white and male?' Five years later, a similar issue was raised surrounding another roundtable discussion[35] which once again featured an all-white panel. A simple lack of resources is partly to blame (including time and money to fund travel, etc), as is a need for timely and topical content. Faced with a choice between a less-than-ideally representative roundtable or no roundtable at all, we have generally opted for the former. But we are well aware that, although we have definitely been improving in this regard, this excuse cannot be used in perpetuity. Nevertheless, while our output may seem overwhelmingly white to listeners in other geographical contexts, the demographics of the critical study of religion in the UK are, as yet, predominantly white. And while the criticism of gender inequality may certainly be leveled at some of our podcasts, it is certainly not the case for our output overall.

A more cynical response here might be to ask 'who made us the police of Religious Studies?' We've been producing a free resource for over five years in our 'spare time' with very limited resources, so of course there are going to be omissions, of course things will slip through the net, and of course we will (unintentionally) repeat and reinforce some of the inequalities that plague the field (globally and in our UK context). People are always welcome to point out our omissions and to contribute themselves or suggest others we might approach. People are always welcome—indeed encouraged—to carry on the discussion on our website and social media feeds, on the podcast itself and in other fora. We think in particular here of the fantastic and varied discussion triggered by our 2016 interview with Teemu Taira which was picked up by a least five other scholarly blogs, and snowballed into a highly fruitful and, at times, heated dialogue.[36] But whilst there might be some truth in this cynical response, we are keenly aware that we have a great deal of responsibility. We had this responsibility when we started (even though we might not have realised it), but this is par-

---

34 Christopher R. Cotter, Ethan Gjerset Quillen, David G. Robertson, Liam Sutherland, Jonathan Tuckett and Kevin Whitesides, "Roundtable: What is the Future of Religious Studies?" *The Religious Studies Project*, 21 Mar. 2012, http://www.religiousstudiesproject.com/podcast/roundtable-what-is-the-future-of-religious-studies/.
35 Emily Clark, Finbarr Curtis, M. Cooper Harriss, Rachel Lindsey, Craig Martin, Derek Nelson and Brad Stoddard, "Six Scholars Discuss the Dissertation to First Book Process." *The Religious Studies Project,* 20 Mar. 2017, http://www.religiousstudiesproject.com/podcast/six-scholars-discuss-the-dissertation-to-first-book-process/.
36 See here for a list of these: Teemu Taira and Breann Fallon, "Categorising 'Religion': From Case Studies to Methodology," *The Religious Studies Project*, 19 Sep. 2016, http://www.religiousstudiesproject.com/podcast/categorising-religion-from-case-studies-to-methodology/.

ticularly the case now, given our position of authority in the field, our recently acquired charitable status, and the fact that we are sponsored by some of the highest bodies in RS. It's not just our reputation that's on the line anymore.

Although we might be irreverent, we do take things seriously, and we are trying to become more proactive than reactive. We also have a commitment to the principles of academic freedom and our editorial team will not veto content which questions or tests established ideas or received wisdom, develops or advances new ideas, or presents controversial or unpopular points of view. For example, a lively comment thread following one response essay prompted Chris to reply that

> although we generally try and keep a relatively tight reign on the critical and non-confessional (for want of a better term) nature of the content on this site, this is much more the case in terms of the podcasts. We see these responses in particular as sites for listeners— and this includes the respondents—to react to the podcast, and to facilitate exactly the sort of debate that is now occurring in this comments section.[37]

Indeed, our focus is much broader than most academic publications, both geographically and thematically, so the range of opinion on the RSP is always going to be challenging. However, such material must be presented in a collegial and respectful manner, and consistent with the principles of equality and diversity adopted by our editorial team's host academic institutions. Controversies have been few and far between and we like to think that when something has gone awry and problems have been pointed out that we have been gracious, understanding, and attempted to move forward in a manner that will preserve the existing ethos of the RSP whilst incorporating the critique, learning from it, and putting measures in place to ensure things are different in future. And we now have a much larger editorial team and board of trustees to hold us to account. But there will always be more to be done ...

Conclusion

We've come a long way from those days in the pub with our trusty *Zoom H2*, and some might lament this institutionalization and 'routinization of charisma',[38] yet

---

[37] See Race Mochridhe, "Theologies That Cannot Be: A Response to the RSP Interview with Dr. Caroline Blyth," *The Religious Studies Project*, 9 Mar. 2017, http://religiousstudiesproject.com/2017/03/09/theologies-that-cannot-be-a-response-to-the-rsp-interview-with-dr-caroline-blyth/.
[38] Max Weber, *The Theory of Social and Economic Organization* (New York: The Free Press, 1997).

we see it as simply the inevitable, logical, and indeed appropriate process which any well-meaning, public facing DH venture will undergo. At the same time, we are acutely aware that we caught lightning in a bottle with the RSP, so the issue now is to decide what we want to achieve with it.

The name 'The Religious Studies Project' was deliberately chosen to be ambitious. Our field is at a crossroads: departments are being squeezed due to cuts and the neoliberalisation of the academy; the subject is being 'balkanised' into departments made up of multiple area-studies scholars with little interest in cross-cultural comparison or theoretical issues, and—sometimes—an apologetic agenda;[39] religion is a more prominent aspect of public and political discourse than it has been for decades, yet our analysis is not being sought or heard.[40] Our larger Project, then, is to get Religious Studies—social-scientific, non-confessional, critical Religious Studies—the voice it deserves.

The public remains largely ignorant of what Religious Studies *does*; we can help to change that. We believe that these topics are intrinsically interesting, and we know that a person talking naturally about a subject they are passionate about is always engaging. Too few of us know how to go about it, however, as these are not skills we are typically trained in, and moreover the current academic climate rewards us for work aimed only at our peers and all but inaccessible to the public, in journals, conferences and committees (notably, the Research Excellence Framework (REF) in the UK).[41] The RSP has built a platform for scholars to put forward research for free and in a way that anyone can understand—which after all should be a central concern for a public-funded intellectual. Moreover, by focusing on intersections with other subjects in the humanities, we can give those scholars the tools to explore that avenue, and promote the relevance of Religious Studies in the social sciences more broadly.

But is our model the only one? Can podcasts be used in other ways? What might the future of podcasting be in Religious Studies? Can podcasts *be* scholarship? Or are they a medium in which we talk *about* scholarship? Back in 2015, Mike Altman conducted an initial review of the uptake of podcasting within Religious Studies, observing quite rightly that much like 'the classic academic monograph or journal article, [existing RS] podcasts still rely on language and argu-

---

39 Aaron W. Hughes, *Islam and the tyranny of authenticity an inquiry into disciplinary apologetics and self-deception* (Sheffield: Equinox, 2015).
40 Titus Hjelm, "Understanding the New Visibility of Religion," *Journal of Religion in Europe* 7 3:4 (2014): 203–22.
41 See Jonathan Tuckett, "Shall We Play the Game?" *The Religious Studies Project*, 22 Mar, 2018, https://religiousstudiesproject.com/2018/03/22/shall-we-play-the-game/.

mentation, even if they are 'heard' instead of read'.[42] Some of our podcasts, such as our roundtable discussions, conference diaries, book review episodes, live episodes and compilation episodes have begun to push the format beyond the standard 'tell me about your latest book' format, but in each of these areas there is much room for improvement. Apart from recording in front of a live audience, in what other ways could the audience—in the room or digesting the podcast in their own time—be brought into the process? Future endeavours might make much more use of audience participation, in written, audio and video form, through the podcast channel itself, or on associated websites and blogs. For example, portions of this paper were presented as part of the University of Edinburgh Religious Studies research seminar, which was audio recorded, transcribed and released via the RSP in March 2018,[43] meaning that the paper has received multiple peer reviews, from multiple audiences, and been digested in person, via transcript, and via earbuds before making it to the final version you are reading.

There is great potential for podcasts to be rapidly convened to comment on current news and events, both within the field itself, and more broadly relating to 'religion' around the globe—much as we did when the initial results of the 2011 UK census were published.[44] Our experimental 'compilation episode' model—most recently on the sociology of religion beyond the secularization thesis[45]—could be fully embraced, with topics planned well in advance, making use of audio archives and scholarly correspondents worldwide to produce critical primers on key topics and up-to-date, multifaceted commentaries on the most pressing issues of the day. More broadly, scholars might wish to explore the production of academic audiobooks, working across multiple languages, innovative excursions into comedy, drama, music and soundscapes, or the production of more 'documentary-style' episodes, or any combination of the above. Further-

---

42 Altman, "Podcasting Religious Studies," 579.

43 Christopher R. Cotter, Stephen Gregg, Suzanne Owen, David G. Robertson and Steven J. Sutcliffe, "The BASR and the Impact of Religious Studies." *The Religious Studies* Project, 12 Mar, 2018, https://religiousstudiesproject.com/podcast/the-basr-and-the-impact-of-religious-studies/.

44 George Chryssides, Christopher R. Cotter, David G. Robertson, Bettina Schmidt, Beth Singler and Teemu Taira, "Podcast: Religion in the 2011 Census," *The Religious Studies Project*,14 Dec. 2012, http://www.religiousstudiesproject.com/2012/12/14/podcast-religion-in-the-2011-census/.

45 Christopher R. Cotter, Carole Cusack, Grace Davie, Jonathan Jong, Kim Knott, David G. Robertson, Paul-François Tremlett, Joseph Webster, Linda Woodhead, "New Horizons in the Sociology of Religion: Beyond Secularization?" *The Religious Studies Project*, 12 Dec. 2016, https://religiousstudiesproject.com/podcast/new-horizons-in-the-sociology-of-religion-beyond-secularization/.

more, there are a growing number of RS-related podcasts out there, most of which have creative commons licenses,[46] meaning that this content can be utilized to varying degrees by others, allowing collegial commentary, critique and collaboration across podcasts, across the globe. Yet, all of these innovations would require significantly more time and resources than are currently at our disposal. Indeed, as Mike Altman concludes:

> The podcast is service to the field—academic icing on the cake of real research (read books and articles). Producing high quality, well-researched, clearly articulated, and accessible podcasts [...] requires the same amount of energy and resources as quality books, chapters, and articles. Until tenure and promotion committees and academic administrations are willing to recognize a truly academic podcast as scholarly research they will probably never happen.[47]

Epilogue

Thinking beyond podcasting and Religious Studies, what can others take from this chapter? There is an important difference of approach between the RSP and traditional academic platforms. Had we sought perfect audio, an ideal web interface and perfectly diverse participants from day one, the project would probably never have happened, and certainly not keeping to a weekly schedule. Like *Facebook*'s original motto, 'Move fast and break things', we use an iterative model where we try a lot of things, and improve on what is working as we go along. In this way, our publishing model is closer to journalism or software development than traditional academia, but this may be an approach that academia needs to embrace. That one perfect journal article behind a paywall belongs to another age, and arguably serves only publishing houses.

If you want the public to listen, they have to be able to hear you.

Selected References

Altman, Michael J. "Podcasting Religious Studies." *Religion* 45:4 (2015): 573–584.
Blair, Ian. "P.E.A.R.S. The Religious Studies Project and Undergraduates." *BASR Bulletin* 126 (May 2015): 14.
Cotter, Christopher R. "You're Greek? Well..., I'm (Northern) Irish, Kind'a ... " In *Fabricating Identities*, edited by R.T. McCutcheon, 34–41. Sheffield: Equinox, 2017.

---

46 https://creativecommons.org/.
47 Altman, "Podcasting Religious Studies," 582.

Cotter, Christopher R. and David G. Robertson. "Unlocking the Ivory Tower: The Religious Studies Project One Year On." *BASR Bulletin* 121 (November 2012): 11–12.

Cotter, Christopher R. and David G. Robertson, eds. *After World Religions: Reconstructing Religious Studies*. London: Routledge, 2016.

Corrywright, Dominic. "Landscape of Learning and Teaching in Religion and Theology: Perspectives and Mechanisms for Complex Learning, Programme Health and Pedagogical Well-being." *DISKUS: The Journal of the British Association for the Study of Religions* 14 (2013): 1–20.

Edison Research. "The Podcast Consumer 2017." *Edison Research* (April 2017). http://www.edisonresearch.com/wp-content/uploads/2017/04/Podcast-Consumer-2017.pdf.

Fox, Dennis. "Personal Theories of Teaching." *Studies in Higher Education* 8 (2) (1983): 151–163.

Geoghegan, Michael W., and Dan Klass. *Podcast Solutions: The Complete Guide to Podcasting*. Berkeley, CA: Friends of ED, 2005.

Hjelm, Titus. "Understanding the New Visibility of Religion." *Journal of Religion in Europe* 7 (3–4) (2014): 203–22.

Hughes, Aaron W. *Islam and the Tyranny of Authenticity: An Inquiry into Disciplinary Apologetics and Self-Deception*. Sheffield: Equinox, 2015.

Hunter, Jane. "Post-Publication Peer Review: Opening Up Scientific Conversation." *Front. Comput. Neurosci.* 6 (2012): 63. doi: 10.3389/fncom.2012.00063.

Mugwanya, Raymond, Gary Marsden, and Richard Boateng. "A Preliminary Study of Podcasting in Developing Higher Education Institutions: A South African Case." *Journal of Systems and Information Technology* 13 (3) (2011): 268–285.

Ractham, Peter, and Xuesong Zhang. "Podcasting in Academia: A New Knowledge Management Paradigm within Academic Settings." *Proceedings of the 2006 ACM SIGMIS CPR Conference on Computer Personnel Research 2006*, April 13–15 2006, Claremont, California, USA. https://www.researchgate.net/publication/221644207_Podcasting_in_academia_a_new_knowledge_management_paradigm_within_academic_settings.

Robertson, David G. "Teaching Matters: Using Podcasts in Tutorials." *BASR Bulletin* 129 (November 2016): 17–18.

Tsonis, Jack, Christopher R. Cotter, and David G. Robertson. "The Religious Studies Project… In Teaching & Learning." *BASR Bulletin* 124 (May 2014): 11–12.

S. Brent Plate
Public Pedagogy: MOOCs and Their Revolutionary Discontents

My main interest in digital humanities (DH) has to do with the ways they open up scholarly research to a broader public in new ways. That's not their only use, but that's the form of the digital that has most interested me. I appreciate the ways museums are creating innovative audio guides to their exhibitions, the ways historians are working with GIS and web-based work to chart a past that people can see and interact with (see Quintman and Schaeffer, this volume), the ways online publications and podcasts (see Cotter and Robertson, this volume) are allowing research to be more open access, and the ways open webinars allow interested parties to sign up, listen, and comment.

At the same time, I value *experience* in the undergraduate classroom, the face-to-face encounters, field trips, and collaborative projects that come from brick-and-mortar pedagogy. In my small liberal arts college, I have been experimenting with teaching about the fullness of "sensual religion." I can pull off some pedagogical gymnastics because my students and I can jump in a van and head down the road to a Vietnamese Buddhist Temple to smell incense, a Bosnian mosque to share an iftar meal during Ramadan, and a Church of God in Christ to hear the sonic percussions of gospel music. I became convinced that there was no other way to teach about the spiritual senses than some variation of a brick-and-mortar classroom, supplemented by field trips.

So, when the dean of my college contacted me and asked me to contribute a MOOC (Massive Open Online Course) to the Harvard-MIT non-profit venture called edX, I began with a dose of skepticism. Teaching a "regular" class about the senses was tricky enough. Doing it online seemed not only counter intuitive but perhaps even foolhardy. Reservations aside, I took on the challenge, seeing it as a novel way to contribute to the public understanding of religion, and do that through the material and sensual avenues that I've researched. The result was a course entitled "Spirituality and Sensuality" that launched on edX in Spring of 2015.[1]

My brief account here gives an overview of MOOCs in general, paying attention to the promises and perils of this form of public education, and then recounts my experiences teaching in this environment, before ending with a peti-

---

[1] Even the title of the course was different from anything I'd offer in a credit-bearing college course. The title was for publicity, for attracting a larger swath of people.

tion for scholars of religion to find ways to be involved in these productions of higher education that are open to the general public.

History and Promises of the MOOC

MOOCs evolved from correspondence courses and distance education. The difference is that MOOCs take advantage of online digital platforms to provide text, video, and audio, and that they are, as the title implies, "open" to people not formally pursuing degrees. Many of them are free for anyone, anywhere. Fuller histories of the education system are readily available online, so I will just give a quick set up here.[2]

The major push in the new generation of MOOCs began in 2011 when Stanford University put three computer science courses online. Three of today's biggest providers, edX, Coursera, and Udacity, all started up in 2012. By 2013 there was already overhyped criticism that the moment for MOOCs was over, but enrollment numbers tell a different story. In 2018, over 101 million "students" (I want to use that term loosely) around the world signed up for at least one course, 20 million of them for the first time ever. In 2014, there had only been 17 million students registered worldwide. That's an increased usage of nearly 600% over those four years. In 2018 there were over 900 universities offering a collective total of 11,400 courses.[3] And Udacity announced their revenue more than doubled from 2016 to 2017, earning $70-million in 2017, and making a billion-dollar company.[4] Within this massive growth, credential-sequence courses are a growing area, meaning that the courses actually count toward a certification, and sometimes degree, of one type or other. The world's largest provider, Coursera, now offers an online MBA program and charges $22,000 for it. In 2019, *Forbes* reported that Coursera is now worth over $1 billion, particularly after signing a deal with the Abu Dhabi School of Government to train up to

---

2 See Paul Stacey, "Pedagogy of MOOCs," *The International Journal for Innovation and Quality in Learning*, 3 (2014): 111–115. The Wikipedia entry for "Massive open online course" is also useful for an overview and links.
3 Statistics about MOOCs in this paragraph come from Dhawal Shah, "Year of MOOC-Based Degrees: A Review of MOOC Stats and Trends in 2018," *Class Central*, 6 Jan. 2019, https://www.classcentral.com/report/moocs-stats-and-trends-2018/.
4 Heather Somerville, "Udacity, with Eye to Eventual IPO, Says Revenue More than Doubled in 2017," *Reuters*, 27 Feb. 2018, https://www.reuters.com/article/us-udacity-revenue/udacity-with-eye-to-eventual-ipo-says-revenue-more-than-doubled-in-2017-idUSKCN1GB2E2.

60,000 employees in computer skills. The greatest growth in participants is coming from India, China, Mexico, and Brazil.[5]

But many MOOCs are free of charge and open to anyone, anywhere. Some providers are non-profit (e. g., edX, Khan Academy) and offer free materials, while others are for-profit (e. g., Coursera, Udacity) with varying pay structures. Many of the free courses have an optional "certificate" available, meaning that if you do passing work you receive an actual certificate, but no college credits. For a field like Religious Studies, such a certificate is fairly meaningless, and yet 180 people paid $60 to get a certificate for passing my course.

Many of the promises of MOOCs are couched in socio-economic terms: that there are hundreds of thousands of people on waitlists for community colleges in California, that MOOCs break down the elitism of higher education systems, and that hundreds of millions of people in the global south can have access to quality learning, all repeating the "anyone, anywhere" mantra. There is a lot to this—the fact that Stanford, MIT, and Harvard are working together to offer free educational content to the world is significant in and of itself—though some of the rhetoric smacks of hyperbole. It's difficult to see how the model can be sustainable without great amounts of private and public funding. Most of the courses remain cheap or free for the students enrolled, but they come with a large bill for those who create and maintain them, as I'll outline below. The largest contributors to funding the courses have been groups such as the Bill and Melinda Gates Foundation, MacArthur Foundation, National Science Foundation, as well as top-tier universities such as Stanford, Harvard, MIT, Caltech, University of Pennsylvania, and the University of Texas at Austin. But there's no telling what happens if the fickle nature of philanthropy sees another shiny object to run after.[6]

Yet it's also hyperbole to criticize MOOCs by saying they can't shape the education of a global public in positive ways. While the largest MOOC providers have been US-based, in the past few years FutureLearn in the UK (run through The Open University) and XuetangX (founded by Tsignhua University in China) have emerged in the top five service providers, further indicating the

---

5 Discussing such issues in the slow moving field of academic publishing means all these numbers will be changing rapidly. For now, see Susan Adams, "Online Education Provider is Now Worth More Than $1 billion," *Forbes*, 25 Apr. 2019, https://www.forbes.com/sites/susanadams/2019/04/25/online-education-provider-coursera-is-now-worth-more-than-1-billion/#469a618030e1.

6 See the important critique, "The Creeping Capitalist Takeover of Higher Education," by Kevin Carey, vice president for education policy and knowledge management of the centrist think tank New America. *Huffpost*, 1 Apr. 2019, https://www.huffpost.com/highline/article/capitalist-takeover-college/.

global reach and interest in MOOCs. The MOOC enrollments alone should make us pause and challenge many of us to get involved. Seen alongside the massive interest in TED talks,[7] informational podcasts,[8] DIY YouTube videos,[9] The Teaching Company's Great Courses DVDs,[10] and even the tremendous number of annual museum visits,[11] there is clearly an ongoing public interest in continuing education outside traditional higher education. Among the subjects in which great swaths of the public have an interest is religious studies.

Teaching Religion through a MOOC

For most MOOCs, on most platforms, content is king. Courses are set up to teach specific skills, or gain a knowledge *about* something, mostly about technology or business. The most massive of the MOOCs teach code, business tactics, or "how to" speak better/think better/write better. In 2016, around 20% of all offerings were in the social sciences or humanities (about equal numbers in each) which, considering the larger state of higher education, isn't actually that low. Then again, these courses are not as enrolled as ones that teach code, foreign languages, or technology.[12]

According to one key clearinghouse of MOOC offerings, Class Central (www.class-central.com), there are 69 courses in "Religion" available at the time of writing the final draft of this chapter in late 2019. There were only 34 courses in religion when I first started writing this in late 2017. Seven of those are run by edX through Harvard's Religious Literacy Project, which has a

---

[7] In 2015, TED talks were being watched 1 billion times a year.
[8] Podcast listening has continued to rise, and estimates suggest that 21% of the U.S. population listened to a podcast in 2016. Andrew Meola, "Podcasts Are Becoming More Popular Among Listeners and Advertisers," *Business Insider*, 6 June 2016, http://www.businessinsider.com/podcasts-are-becoming-more-popular-among-listeners-and-advertisers-2016-6.
[9] The top DIY YouTube channels each have millions of followers and hundreds of millions of collective views.
[10] The Teaching Company has produced over 500 courses since it began in 1990, and reports total sales in twenty-five years of 14 million copies, claiming $100 million per year. See Nevin Martell, "Before YouTube and online classes, there were the Great Courses," *Washington Post*, 3 Sept 2015, http://wapo.st/1KXa4hs?tid=ss_mail&utm_term=.62f3e7fd2f3d.
[11] The American Alliance of Museums (AAM) reports that there are over 850 million museum visits per year in the United States. See (AAM), "Museum Facts," http://www.aam-us.org/about-museums/museum-facts.
[12] Statistics about MOOCs in this section come from Dhawal Shah, "By the Numbers: MOOCS in 2016," *Class Central*, 25 Dec. 2016, https://www.class-central.com/report/mooc-stats-2016/.

"World Religions Through Their Scriptures" program—Diane Moore has done much of the work there. A number of religion courses are now available in Arabic, Russian, French, Italian, and Hindi.

Nonetheless, several courses offered through Udemy, for instance, retain some very problematic approaches (a course on "Wicca and Witchcraft for Beginners" offers the chance to "cast spells" and "summon elementals, dragons, ancestors and deities"). There are a number of courses available with "religion" as a theme, though these are mainly from philosophical or psychological perspectives, with religion relegated to an intellectualist set of doctrines. Indeed, a vast majority of all MOOCs under a broad umbrella of "religious studies" approach religion through textual or philosophical methods. (Not listed in Class Central's list is Anthony Pinn's excellent "Religion and Hip Hop Culture" which was run in 2015, and is still available at edX.) In short, at present there are not a lot of MOOC offerings in religious studies, though that is changing, and FutureLearn has recently made available a number of courses in conjunction with European universities of Nottingham, Groningen, and Trinity College Dublin.

With many MOOCs, there is often little interest in independent critical thinking, or connective cognitive operations that require reflection and seeing from new perspectives, or, and this is my interest, creating embodied pedagogical practices. Many of the available courses, even in humanities and social sciences, are still dominated by the "sage on the stage" model: just set up a video camera and let the professor do her/his lecture, complete with chalkboard scribblings. This works if content is the thing being disseminated. The differences here are often noted through the nomenclature "xMOOC" and "cMOOC" with x indicating the more traditional pedagogy centered around a professor, test-taking, and individual learning, while c is for connectivity and aims toward community learning and networking.

With the new wave of MOOCs there has been a noticeable uptick in the cMOOC variety, and indeed all the religion courses I've visited (all created in the past 5–6 years) have worked on models that emphasize shorter videos (typically ranging from 2–15 minutes) sandwiched with readings and activities. In these newer models, professors of the courses are pictured in their offices, or around a table with students, or interviewing other "experts" on topics, but they also walk through museums or stand in front of a green screen as charts, images, and information are generated behind them. At other times, all that is heard is a voiceover while images of religious practices and spaces are shown on screen, or animated scenes play out visual representations of the things being described. The audio and visual components have been highly edited and the resulting videos look more like a documentary film than a class lecture.

Videos are then layered along with readings, links to online charts, histories, and stories, and usually some type of quiz/evaluation based on multiple choice. Finally, and this is a key part that I'll discuss further, there is a strong participation component.

My MOOC: From Professor to Director

In 2014, Hamilton College received a grant from the Andrew Mellon foundation in the amount of $250,000 to establish a set of MOOCs.[13] Much of the original grant was established to encourage liberal arts colleges to be part of online education movements, and to chart some possible new directions for MOOCs. In 2015, Hamilton and Colgate joined Davidson College and Wellesley College to form a consortium of liberal arts colleges teaching MOOCs, and share resources.[14] There was an interest on the part of edX, which had mainly been run through R1 institutions, along with a handful of top-tier liberal arts colleges, to think through what a liberal arts approach to education might offer the format of MOOCs. The dean of Hamilton at the time, Patrick Reynolds, stated the goals of Hamilton's offerings to be "public scholarship, educational outreach, and better understanding of the applications of online educational technology."[15]

When I finally agreed to take the time to create and run a MOOC, I wanted to be sure we were doing something different than the sage on the stage, and it took some research to find a good model. I finally came across Harvard's "Tangible Things" course (through edX),[16] and began to see ways to create interactive proj-

---

[13] The grant was instigated through the work of Hamilton's then-President, Joan Stewart, and implemented chiefly by David Smallen, Hamilton's Vice President for Libraries and Information Technology, with a lot of assistance from Lisa Forrest, Director of Research and Instructional Design in the library. I should note that my ability to create the course was largely made possible because I teach at a well-endowed college, and even with this, the course was only producible with the Mellon grant and the impetus of the college's interest in combining technology and pedagogy.
[14] See Jeffrey Young, "New Consortium's Mission: Improve Liberal-Arts Teaching Online," in *Chronicle of Higher Education*, 12 May 2015, http://www.chronicle.com/blogs/wiredcampus/new-consortiums-mission-improve-liberal-arts-teaching-online/56621?cid=at&utm_source=at&utm_medium=en.
[15] Vige Barrie, "Hamilton Joins Consortium Focused on Online Pedagogy," *Hamilton College*, 12 May 2015, https://www.hamilton.edu/news/story/hamilton-joins-consortium-focused-on-online-pedagogy.
[16] "Tangible Things: Discovering History Through Artworks, Artifacts, Scientific Specimens, and the Stuff Around You," https://www.edx.org/course/tangible-things-discovering-history-harvardx-usw30x-0.

ects, incorporate text and image onto webpages in ways that didn't have to look like a textbook, embed links to related materials, and finally do some video work that was engaging, showed movement, and not simply a recorded lecture.

While the structure and syllabus of a traditional college class is more or less up to the individual instructors themselves—as long as they fit departmental curricular needs, etc.—a MOOC can only be created through the effort of a large group of people, most of whom are not in any way "experts" on the content. I worked closely with 3–4 people in audio-visual and web design production, and another half dozen who helped with copyright, liaised with edX, and offered pedagogical insights. Further, six students who had taken previous college classes with me were asked to serve as "teaching assistants," and they monitored activity during the course, responding to the discussion prompts, and thus giving a strong sense of interactivity to the course in spite of hundreds of people commenting. Adding to the list of those involved was the dean and other administrators who ran the grant and reported back to Mellon.

As the work went on, I began to feel more like the writer and director (and occasional "star") of a film, and less like a professor. I thought up most of the content, but when it got time to create the format to deliver the content we had to think collectively, and I had to work *with* a crew of people to provide material. Once I acknowledged that I was more of a director who needed to get along with a staff of others, and not the solo sage on a stage, things got a lot easier and allowed me to think in new ways about the content and format.

With the help of an amazing team of ITS staff, we set out to construct one of Hamilton College's first MOOCs, "Spirituality and Sensuality: Sacred Objects in Religious Life."[17] Much of it was based on related topics I researched and wrote about in my book *A History of Religion in 5 ½ Objects*,[18] as it attempted to work through a material culture approach to religion. Working with a key videographer, we set up cameras and filmed in bakeries, cemeteries, churches, museums, forest trails, synagogues, labs, and even a classroom or two. From there we interviewed neuroscientists and pastors, bagel bakers and rabbis, American Indian cultural center directors and archeologists, and created artsy poetry readings. And then we borrowed from the treasure trove that is the internet: more interviews, musical performances, artworks, scientists on TED, cooks on NPR's Science Friday, and maps to help keep our geo-bearings. With help from ITS staff

---

[17] Online at "Spirituality and Sensuality: Sacred Objects in Religious Life," https://www.edx.org/course/spirituality-sensuality-sacred-objects-hamiltonx-relst005-5x.
[18] S. Brent Plate, *A History of Religion in 5 ½ Objects: Bringing the Spiritual to its Senses* (Boston, MA: Beacon Press, 2014).

members who ran the HTML in the background and tweaked and tweaked some more, we got up and running.[19]

By the beginning of the seven-week course in Spring of 2015, around 5,000 people had registered, hailing from over 130 countries. This included over 400 people from India, over 100 from both Brazil and Mexico, nearly 100 from Turkey, and many more. Half came from the United States. About half were between 26–40. A little over half reported they were female, while 1/3 reported male, a small segment said "other," and the rest did not report anything. Almost 75% had a college degree or higher.[20]

As we began the course the registered students introduced themselves in discussion sections: hundreds of people from all over the world uploaded photos, told stories, and offered their takes on ways objects have sensually triggered experiences and memories in their own lives. As we got going into the assignments and responses, there was some disagreement, and differences in language and tradition, but overall there was a strong sense of working together as a group, as reported by a good number of participants. I know not everyone who registered felt such a collective effervescence, but it was striking to me how many connections were made in spite of my initial skepticism of working online.[21]

Similar points are echoed by Harvard's Jennifer Aileen Quigley and Laura Salah Nasrallah in response to a MOOC they offered in 2014 on "The Letters of Paul" through edX. Through their class, they found a large majority of the students reported a strong connection to other students. In a reflection essay on teaching a MOOC, Quigley and Nasrallah conclude:

> The MOOC in particular and online education in general need not be primarily a concession to the expense and challenge of bringing students into one place. It can be a way to accom-

---

[19] The "team" included the hard work of a number of people, and I want to particularly mention the ITS staff of Ted Fondak, Forrest Warner, Bret Olsen, and Lisa Forrest, and my teaching assistants Molly April, Jasmin Thomas, Carrie Cabush, Sophia Henriquez, Molly Root, and Danielle Rodrigues.

[20] One element of online courses is that there are endless amounts of data available about students and their use of the course materials. I can note that there have been 674,596 "clicks" in course content, that the mean time students spent in the course was 1.1 hours, and that out of the 7,000 people who eventually registered over the last two and a half years a little over 900 have engaged at least half the class. That final number may be surprising, but engagement rates across MOOCs are very low in relation to the number of people registered, usually 10% or less.

[21] In spite of many metrics, questionnaires, and other responses, the analysis is difficult to assess. One could say that having only a 10% participation rate (see previous note) makes this a failure, but when we are talking about hundreds of people we might, in the "glass half full" mode, think it was quite successful.

modate a hunger for knowledge, the longing for continuing education *and broad communication and community.*²²

A brick-and-mortar course may or may not create connection, and neither will a MOOC necessarily. But the challenge remains, and there are ways to create different kinds of connections and communities.

The pedagogical challenge of connectivity is furthered in the challenge to make courses embodied. For all the promises of the interactivity of New Media, digital media remains sensually impoverished, relegated to the audio and visual registers. Responding to this, for our course we created discussion and writing prompts that aimed to get people up off their chairs and into their kitchens or out on the sidewalks to re-experience the world around them in fresh forms, and to reflect on how these basic experiences related to religious life and tradition. Some did, some didn't. Enough did. And as we worked toward the halfway mark, I began to sense the shifts toward an embodied pedagogical practice, allowing students to see and feel how "religion" is not all about thoughts in the head or texts to be read.

So, for instance, the third week's topic was "smell and incense." I gave a seven-minute introductory video to the week by making comments about an ancient Aztec censer that is housed in Hamilton College's Wellin Museum. There were readings from Exodus (God's recipe for incense in the temple), sections of Susan Harvey's *Scenting Salvation* and Annick Le Guérer's book *Scent*, as well as a segment from my book. Then we turned to interview a local Presbyterian pastor who uses incense in a moving ritual for remembering the dead (admittedly, not a very "Presbyterian" thing to do), followed by a poem evocative of scent, and then a short audio clip by a neuropsychologist on the olfactory system. Once the science, history, poetry, and devotional dimensions were worked through, we made a discussion prompt that aimed to stimulate students' own experiences with scent. They were asked to respond to the following:

> Go to your kitchen and find three types of spices. If you had to describe the smells to someone with anosmia ("smell blindness"), how would you describe them? What emotions come to the fore as you smell these spices? Are they difficult to describe with words?

---

22 Jennifer Aileen Quigley and Laura Salah Nasrallah, "HarvardX's Early Christianity: The Letters of Paul: A retrospective on online teaching and learning," available at: https://www.academia.edu/34517001/HarvardXs_Early_Christianity_The_Letters_of_Paul_A_retrospective_on_online_teaching_and_learning.

Responses included creative lines like "Cinnamon is like anticipation" or "Vanilla—balloons, party hats, laughter, hugs." A student in Buffalo lived by a General Mills factory and the air that evening smelled like Cheerios. Another talked about her field research in Ethiopia and wrote about her observations of a coffee ceremony there.

Then we worked toward a final assignment for that section which had them conducting,

> a brief interview with another person to discuss scent. Perhaps a local rabbi, imam, priest, or pastor, asking how they use smells in religious services. Or, interview a chef, a baker, a perfume store worker, or an incense shop. Or, talk with a grandparent about their cooking recipes and the spices and smells used. Provide a 2–3 paragraph report on the use of scents in important environments near you.

Again, the responses were broad ranging and creative, from talk with family members—which led to the place of smell and memories—to bakers—which led to comments about "happiness"—to clergy—which led to comment on the conditioned bodily responses by practitioners. Through both of these assignments, students had to reflect back on the medium of written language as a conveyor of information, even when it is notoriously difficult to describe smell in words.

Note that neither of these prompts, nor any of the others in the course, intended to get students to regurgitate readings, memorize a list of scents and olfactory structures, or to offer up some doctrinal comment on scent in religious traditions. There was little "content" that students could walk away with, which may have driven some students away from the course, as some likely had hoped for a list of "what religions believe" type of approach. Instead we aimed to nudge students to rethink the world around them, to realize how the senses operate in basic ways in our everyday life, and then over the course of the MOOC, conceptualize how religious traditions operate through the senses, above, beyond, and before there is any recourse to beliefs, texts, or doctrines. In this way, I saw the course as a critical intervention into public understandings of religion, to push beyond the Protestant-centrism of what religion is.

Once my course was finished, it was "archived," meaning that all the content continues to be available, for free, to anyone who wants to use it. (Available at: https://www.edx.org/course/spirituality-sensuality-sacred-objects-hamiltonx-relst005-5x.) One only needs to register with edX. Three and a half years after my course ran, there are over 8000 people enrolled, and there are still about 20–30 people per week watching videos or engaging readings to this day. All of the discussion boards have been turned off, and so the collective engagement is missing, but the videos and readings can be accessed. One of my hopes was that these modules could be available for others teaching traditional classes, to have a range of resources to use.

MOOCs and the Public Understanding of Religion

As a link between scholarship and the general public, MOOCs are best seen as situated along a line of educational enterprises in the United States, from lyceums to Chautauquas to Mortimer Adler's Great Books. In a 2013 article in the *Chronicle of Higher Education*, Jonathan Freedman saw MOOCs for their public value, while realizing that while the knowledge imparted is wholly "middlebrow," they have a deep value in a history of public education:

> MOOCs are just the latest incarnation of bringing watered-down versions of culture, knowledge, and learning to a mass audience. What we see as the courses' flaws may well be their strengths, and they have the potential to carry those strengths to a broader audience than ever before. Problems arise only when we think of MOOCs as university courses rather than as learning for the masses.[23]

Freedman goes on to suggest that MOOCs "can be thought of as part of the university's contribution to society as a whole, serving as a form of adult education."

I think this is a useful, pragmatic way to imagine such MOOCs: as a middlebrow "contribution to society" through continuing education. Indeed, most critiques of non-credit bearing MOOCs are that they aren't as "rigorous" as real college classes. The obvious response is simply to say: they aren't college courses. As noted, there is an uptick in credit-bearing courses for MOOCs in general, but we need to be able to distinguish the credit-bearing from the non-credit bearing and find the impact of each. There are MOOCs that are credit-bearing and cost-evoking, and these comprise one important manifestation of the digital medium. And there are other MOOCs, like mine, that are free and open to the public and should be compared not to a college class, but to Chautauquas and TED talks, and perhaps even museum visits.

There is, for better and worse, a push across institutions of higher education to offer more online courses, for credit, for paying students. These need a different structure and set of standards than the kind of public course I taught. But I also think any online course, for credit or not, should use the technologies at

---

[23] Jonathan Freedman, "MOOCs: Usefully Middlebrow," *Chronicle of Higher Education*, 25 Nov. 2013, http://www.chronicle.com/article/MOOCs-Are-Usefully-Middlebrow/143183/.

hand to create something beyond the mere transmission of information. The task before us is to utilize digital technologies for a much more robust and thoughtful understanding of contemporary and historical religious life as embodied, as intersecting with other components of human life, and lived out through mundane existence.

The non-credit MOOC, as I see it, can offer a critical intervention into the public understanding of religion. These courses offer an opportunity for some of us to push beyond academic insider-speak and create spaces for comprehension of this odd, contested realm we call religion. Within that, I have continued to see part of my own research and outreach evolve as I try to get beyond the beliefs-doctrines-texts approach that dominates much of the public understanding of religion, and toward body-based practices. Sensual engagement, I have continued to argue, is necessary to begin to understand the comings and goings of religion. MOOCs might yet rise to the challenge of getting us out of our heads and "bring us to our senses."

Selected References

Carey, Kevin. "The Creeping Capitalist Takeover of Higher Education." *Huffington Post.* 1 April 2019: https://www.huffpost.com/highline/article/capitalist-takeover-college/.

Freedman, Jonathan. "MOOCs: Usefully Middlebrow," *Chronicle of Higher Education*, 25 November 2013, http://www.chronicle.com/article/MOOCs-Are-Usefully-Middlebrow/143183/.

Martell, Nevin. "Before YouTube and Online Classes, There Were the Great Courses," *Washington Post*, 3 Sept 2015, http://wapo.st/1KXa4hs?tid=ss_mail&utm_term=.62f3e7fd2f3d.

Plate, S. Brent. *A History of Religion in 5 ½ Objects: Bringing the Spiritual to its Senses.* Boston, MA: Beacon Press, 2014.

Quigley, Jennifer Aileen and Laura Salah Nasrallah. "HarvardX's Early Christianity: The Letters of Paul: A Retrospective on Online Teaching and Learning." Posted at https://www.academia.edu/34517001/HarvardXs_Early_Christianity_The_Letters_of_Paul_A_retrospective_on_online_teaching_and_learning.

Stacey, Paul. "Pedagogy of MOOCs," *The International Journal for Innovation and Quality in Learning*, 3 (2014): 111–115.

Young, Jeffrey. "New Consortium's Mission: Improve Liberal-Arts Teaching Online," *Chronicle of Higher Education*, 12 May 2015,
http://www.chronicle.com/blogs/wiredcampus/new-consortiums-mission-improve-liberal-arts-teaching-online/56621?cid=at&utm_source=at&utm_medium=en.

Wendi Bellar and Heidi A. Campbell
Building Social Sites of Collaborative Research: A Case Study of the Network for New Media, Religion, and Digital Culture Studies

When we talk about digital humanities (DH), the conversation usually aligns itself around a certain set of principles—transparency, interdisciplinarity, collaborative connectivity—rather than around a specific set of theories or methodologies.[1] It was in this spirit of these DH values that the Network for New Media, Religion, and Digital Culture Studies (i.e. the Network) was conceived, constructed, and continues to operate. Indeed, the study of religion and new media, now more formally described as Digital Religion studies, is inherently interdisciplinary, engaging scholars from various backgrounds such as Architecture, Human Computer Interaction, Communication, Religious Studies, and Political Science.[2] A proliferation of scholarly work on Digital Religion across disciplines provided the impetus to create a network that would connect diverse, international scholars, and provide the necessary tools and resources to answer questions related to the ways in which religion is practiced, and evolves, in online and offline spaces that are simultaneously connected.[3] The construction and growth of the Network played a vital role in the development of the growing subfield of Digital Religion, by providing a space to present research, interact with diverse scholars, and create new understandings. As much as Digital Religion is a study of the third space, the place in between and betwixt the online/offline dichotomy where new religious meanings and practices emerge, the Network seeks to serve as a

---

[1] Matthew Kirschenbaum, "What is Digital Humanities and What's it Doing in English Departments?," in *Debates in the Digital Humanities*, Matthew K. Gold, ed. (Minneapolis: University of Minnesota, 2012), 3–11; Lisa Spiro, "This is Why We Fight: Defining the Values of the Digital Humanities," in *Debates in the Digital Humanities*, Matthew K. Gold, ed. (Minneapolis: University of Minnesota, 2012), 16–35.

[2] H. A. Campbell, ed. *Digital Religion: Understanding Religious Practices in New Media Worlds* (London: Routledge, 2013).

[3] H. A. Campbell, "Surveying Theoretical Approaches Within Digital Religion Studies," *New Media & Society* 19, no. 1 (2017): 15–24.

https://doi.org/10.1515/9783110573022-016

third space for which our understandings of Digital Religion deepen and are refined by the sharpening of interdisciplinary collaboration.[4]

The Network serves the goals of interdisciplinary collaboration and transparency in various ways, including an interactive bibliography, online scholar's index, news feed, blog page, links to publications and press about research associated with the site, and an online toolbox with various resources for research. This chapter provides a case study of the construction of a collaborative research site through an examination of the mission of the Network, its tools and resources, and two research studies that were collaboratively conceived and executed through the site. The first section explores the conception of the Network, provides a walkthrough of the site as a whole with descriptions of the various tools and resources available, and provides discussion about how the various resources support an interdisciplinary and collaborative environment. The second section introduces two research studies that explored religious mobile applications and religious memes, and discusses the development and use of digital databases as a method for collaborative research. Finally, the chapter ends with a discussion on how the Network, and other sites like it, can continue to advocate and facilitate more interdisciplinary and collaborative Digital Religion research.

Just as interdisciplinarity, collaboration, and creativity are at the core of DH, The Network mirrors these qualities in the context of Digital Religion by creating a repository of knowledge that crosses international boundaries and disciplinary fields, connecting diverse scholars in ways that promote knowledge collaboration, and documents, and makes visible, knowledge through the use of databases that provide new and experienced scholars alike an entry point into the study of new phenomenon related to Digital Religion. The goal of this chapter is to explain how the Network achieves these goals of collaborative research sites, while also looking to the future for new and better ways to study and understand how religion is understood, practiced, and reshaped in digital spaces.

---

4 S. Hoover and N. Echchaibi. "The 'Third Spaces' of Digital Religion," paper presented at the Center for Media, Religion, and Culture, University of Colorado Boulder, 2012. Retrieved from http://cmrc.colorado.edu/wp-content/uploads/2012/03/Third-Spaces-Essay-Draft-Final.pdf.

Background on the Network for New Media & Digital Culture Studies

> When I graduated with my PhD, I owned the whole literature. There were a couple of books and about three binders full of journal articles. And now, 15 years later, I can't even keep up because people are interested in how religion and new media are interacting in area studies, in political science, in psychology ...

The quote above is from a video on the Network homepage features Heidi Campbell, a researcher in the field of Digital Religion and the founder of the Network. Her statement speaks to one of the needs that led to the creation of the site—the proliferation of scholars interested in studying of religion and new media and the increase in studies being published about the phenomenon in a variety of journals. Campbell initially created the Network on a Wiki platform in 2009 as a way to collect and share resources with other scholars interested in the studying religion and the internet, before moving it to a website. Until the creation of the online site in 2010, there were many other sites that focused on a particular department or scholar's work in the field, but there were no sites with the intent of showcasing leading research in the field as a whole, connecting scholars across international and disciplinary boundaries, and providing an online forum to facilitate collaboration. The Network, which was funded by a grant from the Evans/Glasscock Digital Humanities Project at Texas A&M University, is open to any students, scholars, or independent researchers who are interested in, and doing work at, the intersection of new media and religion.

Members of the Network are given the ability to add to the interactive bibliography, access the scholar's index, and post news about events, publications or other interests related to the Network. They can also request the opportunity to submit a guest blog. There is a vetting process; those interested in becoming a member must submit a request to create a new account and provide a biographical statement including the reason for wanting to join the Network. Network facilitators independently vet each member. In this sense, the Network maintains its source credibility, which is an important characteristic of successful online communities.[5] The vetting ensures that members are connecting with real students and scholars who have a vested interest in providing credible and reliable information.

---

5 M. Ma and R Agarwal, "Through a Glass Darkly?: Information Technology, Identity Verification, and Knowledge Contribution in Online Communities," *Information Systems Research* 18, no. 1 (2007): 42–67.

On the homepage, digitalreligion.tamu.edu, a list of tabs is displayed including: Home, About, Bibliography, Scholars Index, News, Blog, Publications and Press, Researcher's Toolbox, and Contact Us. The Home and About page orient members to the overall goals and mission of the site, which is to "offer an interactive space for researchers and others wishing to learn" more about the growing field. The other tabs provide interactive platforms on which to explore the leading research in the field as well as to connect with scholars who are studying similar themes in Digital Religion, such as identity, authority, community, and authenticity.[6] In addition to the main tabs on the homepage, there are also buttons for Facebook, Twitter, and an RSS feed where members and those interested in the Network can interact through social networking sites. The next section delves into these tabs, which provide key resources and features that lend themselves to a collaborative environment.

Knowledge Collaboration through Key Resources

There are four resources on the site that lend themselves to knowledge collaboration: the interactive bibliography, the scholar's index, the news feed, and the blog. Knowledge collaboration refers specifically to "offering, adding to, recreating, recombining, modifying and integrating knowledge" and is a key feature of successful online communities.[7] Each of the resources listed above specifically add to the collaborative knowledge of the group through various tools such as searching, tagging, messaging, and exporting.

The interactive bibliography (see Fig. 1) fits the definition of knowledge collaboration best because it is co-created by facilitators and members of the Network. Members can add their own or others' publications to the bibliography. Currently, the bibliography boasts more than 550 entries that have been categorized and are searchable across authors, titles, type of publication, year, and keywords. The search function can be used by news scholars to find key articles on topics as varied as gender, 3D environments, and Zen, or by more established scholars to keep up with current research in key areas such as authority, identity, and community. In addition to being searchable from a variety of variables, the entire bibliography, or results from a specific search, or even just a single entry, can be opened in Google Scholar or exported to BibTex, RFT, End Note, XML, or

---

[6] H. A. Campbell, ed. *Digital Religion: Understanding Religious Practices in New Media Worlds* (London: Routledge, 2013).
[7] S. Faraj, S. L. Jarvenpaa, and A. Majchrzak, "Knowledge Collaboration in Online Communities," *Organization Science* 22, no. 5 (2011): 1224.

RIS file format for easy formatting of works cited pages. Because digital scholarship is inherently interdisciplinary, creating repositories of scholarship that cross disciplinary fields and conversations is essential. The interactivity of this particular repository provides a unique collaboration among dispersed members with access to, and expertise in, a diverse set of literatures.

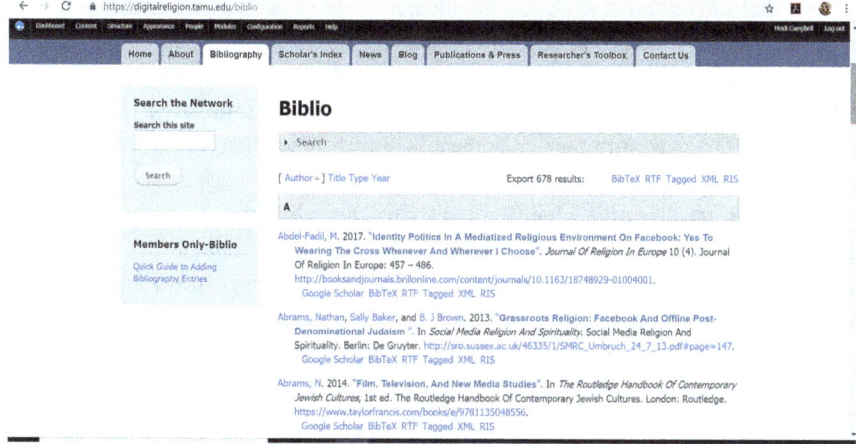

Figure 1: Screenshot of the interactive bibliography on the Network web site.

The scholar's index also instigates connections among scholars by creating a centralized location of information on the Who's Who of the Digital Religion world. Currently, there are more than 250 scholars with active profiles on the Network. Members have access to view each scholar's entry in alphabetical order, and can also edit their own profile to update information as necessary. Scholars are listed by name, affiliation, and keywords on the main index page. When a specific scholar's name is selected, a new page opens up with more information including links to the scholar's web presence, a biographical statement, and the length of membership in the network. Members also have access to a link that allows them to send a private message to the scholar. In this way, the Network provides the tools necessary to start conversations, make connections, and hopefully develop research collaboration among scholars. Although collaboration is a primary objective of digital scholarship, finding partners that are both skilled in DH work and in their respective fields can be a challenge. The Network's scholar's index takes some of the guesswork out of identifying key scholars in the field, and out of figuring out which email address might actually reach the scholar. When scholars on the Network receive a private message, they are notified through a Network-generated email at their listed address, which is kept private.

Members also have the ability to create and share new knowledge through the news and blog pages. Every page of the site has an "Add Content" section at the bottom. All members have the ability to post a news item by clicking the link at the bottom of any page. On the news page specifically there is a guide on how to add news items to the page, which walks users through the technicalities of adding, saving and editing information. The Network specifically solicits announcements of events, calls for papers or panels, news releases of new research publications, or popular press articles about the intersection of religion and new media. In this way, members are able to inform others, and keep abreast of, interesting happenings within the field. Additionally, if there is a new issue that warrants more discussion than a news items, members can request to be a guest-blogger and will be given special access to post a blog entry. This is another area where scholars can take advantage of the collaborative environment to read about and discuss important topics related to Digital Religion. Part of a collaborative research site is sharing information and being in conversation with other scholars; the news and blog pages provide scholars a chance to create, share, and discuss important research findings and contributions to the field.

The Network uses four main resources to build a collaborative environment in which to find, share, inform and discuss important topics among interdisciplinary scholars in the field of Digital Religion. The interactive bibliography and scholar's index provide the foundation by allowing members to co-create an extensive bibliography across academic disciplines. If a student or a new scholar is interested in learning more about the field, they can find established members with whom to form relationships. At the very least, the site provides a central hub where scholars interested in the intersection of religion, new media, and digital culture can find relevant publications, connect with key scholars, and learn and share important news and discussions with each other. However, the collaboration does not end with knowledge; the Network also provides an online collaborative research environment through its researcher's toolbox.

Online Collaboration Through the Researcher's Toolbox

As the field of Digital Religion has expanded, many key scholars in the field have called for new theoretical and methodological tools to examine religious understandings and practices in digital contexts.⁸ In the same sense, DH have relatively recently exploded with various computational tools and methods through which to explore and understand culture in various forms. Because the work of DH is to document knowledge in ways that scholars can build upon, the researcher's toolbox seeks to provide scholars not only with information, but also directly to research tools. The toolbox is a repository of links to research resources including training guides, interdisciplinary perspectives on approaches to studying new media and religion, ethical guidelines for doing digital research, and links to outside databases and tools that scholars can use. In addition, the toolbox houses two databases that are geared toward the study of religious mobile applications and religious memes respectively, to which all members will soon have access. The following sections focus on the collaborative creation of the databases and the resulting publications from each project.

Religious Apps Database

In 2012, after the Network website was publically launched, Campbell submitted another grant proposal to the newly formed Institute for Digital Humanities, Media and Culture at Texas A&M University to develop additional resources to help scholars study a relatively new phenomenon in Digital Religion: religious mobile applications (http://digitalreligion.tamu.edu/religious_app). While earlier studies have explored the use of mobile phones in religious life, relatively few articles have broached the topic of religious apps.⁹ This research project

---

8 H. A. Campbell, "Surveying Theoretical Approaches Within Digital Religion Studies," *New Media & Society* 19, no. 1 (2017): 15–24; M. Lövheim and H. A. Campbell, "Considering Critical Methods and Theoretical Lenses in Digital Religion Studies," *New Media & Society* 19, no. 1 (2017): 5–14; S. Hoover and N. Echchaibi. "The 'Third Spaces' of Digital Religion," paper presented at the Center for Media, Religion, and Culture, University of Colorado Boulder, 2012. Retrieved from http://cmrc.colorado.edu/wp-content/uploads/2012/03/Third-Spaces-Essay-Draft-Final.pdf.
9 see B. Barendregt, "Mobile Religiosity in Indonesia: Mobilized Islam, Islamized Mobility and the Potential of Islamic Techno Nationalism," In *Living the Information Society in Asia,* ed. E. Alampay (Singapore: Institute of Southeast Asian Studies, 2009): 73–92; H. A. Campbell, *When Religion Meets New Media* (London: Routledge, 2010); H. Horst and D. Miller, *The Cell Phone: An Anthropology of Communication* (Oxford: Berg, 2006); A. Roman, "Texting God: SMS and Religion in the Philippines," Paper presented at the 5th International Conference on Media, Religion and Culture, Stockholm, July 2006. Retrieved from http://www.freinademetzcenter.org/pdf/Texting%20God%20SMS%20and%20Religion%20in%20the%20Philippines.pdf; L. Togarasei, "Mediating the Gospel: Pentecostal Christianity and Media Technology in Botswana and Zimbabwe," *Journal of Contemporary Religion* 27, no. 2 (2012): 257–274; Relatively few articles had broached the topic of religious apps *see* R. Torma and P. Teusner, "iReligion," *Studies in*

sought to create a unique methodology and research tool that would allow interdisciplinary researchers to study a variety of religious apps. Specifically, the proposal was to build an app database, housed on the Network site, which would provide access to religious apps from a variety of religions that would be categorized into different themes surrounding religious understandings and practices. While this database would obviously speak to specific research questions within the field of Digital Religion, it also has broader relevance to researchers who are researching other types of apps as well by identifying key technological affordances and features that may be found across app categories. In connection to DH, the construction of the database and its resulting publications also provide an interesting case on which others looking to create collaborative sites can build.

The religious app database (see Figure 2) project developed over six different stages. The first of which was to identify and collect information on the various types of religious apps that were available to users. Four researchers, including Heidi Campbell, the primary investigator, and three graduate students from Texas A&M University, Brian Altenhofen, Wendi Bellar and James Cho, were each assigned an iPod or an iPad with which to collect and store religious apps. The search began with common keywords such as 'religion' and 'spirituality' and for apps associated with the five major world religions (i.e. Christianity, Judaism, Islam, Hinduism, Buddhism).[10] While the researchers were collecting apps on their devices, they were also building the app database from the ground up by inputting specific information about the apps. This included information such as the name of the app, the name(s) of the app developer(s), contact information of the app developer(s), cost, version, and the app classification listed on iTunes.

Stage two involved trying to uncover common themes across the religions so that apps within certain categories could be compared and contrasted with one another by researchers. This involved offline meetings where the researchers met and discussed the common themes they were seeing as well as wrestled with the difficulty of placing an app into only one category. Essentially, the team came to

---

World Christianity 17, no. 2 (2011): 137–155; R. Wagner, "You are What you Install: Religious Authenticity and Identity in Mobile Apps," In *Digital Religion: Understanding Religious Practices in New Media Worlds* (London: Routledge, 2013): 199–206; S. P. Wyche, K. E. Caine, B. K. Davison, S. N. Patel, M. Arteaga, and R. E. Grinter, "Sacred Imagery in Techno-Spiritual Design," In *Proceedings of the SIGCHI Conference on Human Factors in Computing Systems* (April, 2009): 55–58.
10 H. A. Campbell, B. Altenhofen, W. Bellar, and K. Cho, "There's a Religious App for That!: A Framework for Studying Religious Mobile Applications," *Mobile Media & Communication*. 2, no. 2 (2014): 154–172.

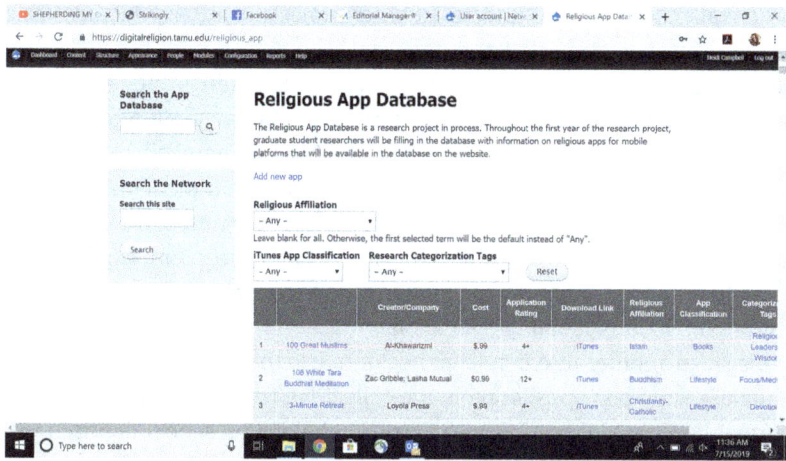

Figure 2: Screenshot of the Religious App Database on the Network web site.

the conclusion that it would be impossible to develop mutually exclusive categories, and found that many apps fit into one or more categories in different ways. However, there were 11 distinct categories, such as prayer, focus/meditation, sacred texts, religious social media, and religious apps for kids, that emerged from analysis of almost 500 religious apps. These were separated into two parent categories: "apps oriented around religious practice" and "apps embedded with religious content."[11] These parent categories help distinguish the focus of the app, that is, whether the app seeks to reproduce a religious practice in the mobile context, or if the app seeks to provide information, resources, and activities that are not necessarily tied to traditional religious practices (e. g. playing a religious game).

Stage three, developing methodology, and stage four, creating content analysis tools, took place concurrently. The researchers went back through the database of religious apps and applied the categories to all 488 apps, while at the same time extrapolating data to be used in the content analysis. This was done by importing the data into spreadsheets and conducting manual counts of categories and app information items. At the same time the research team was developing the methodology and conducting the content analysis, some in-

---

[11] H. A. Campbell, B. Altenhofen, W. Bellar, and K. Cho, "There's a Religious App for That!: A Framework for Studying Religious Mobile Applications," *Mobile Media & Communication*. 2, no. 2 (2014): 104.

dividual members of the team undertook other projects using the database as a starting point for data collection and analysis.

Stage five involved the research team analyzing the findings and formulating category definitions and basic characteristics. These findings are reported in a 2014 *Mobile Media & Communication* article written by Campbell, Altenhofen, Bellar, and Cho. The study's findings showed that relying on the iTunes classifications as a starting point to identifying religious apps was unreliable. Therefore, the new categories were proposed as starting off point for interdisciplinary researchers interested at examining the relationship between mobile media and religion through religious mobile applications.

At this stage another research team began to use the religious app database for a study of sacred text apps found in the collection, with the intent to compare and contrast how each of the five major world religions presented digitized versions of the texts along with specific digital tools with which to study them. For this study, a new coding sheet was created and the information found within the apps themselves was analyzed for specific technological affordances and the ways they represented the religious messages and practices. Currently, this article by Tsuria, Bellar, Cho and Campbell is forthcoming in the *Journal for Contemporary Religion*.

Bellar also utilized the database to identify and analyze Catholic and Islamic prayer apps in her dissertation.[12] The database was instrumental in collecting a representative sample that fit within the prayer category as defined in Campbell et. al's article.[13] In the first phase of Bellar's study, the prayer apps were examined for the ways in which technological and religious affordances combined to create a mobile prayer environment, and how that environment was being explained and marketed to users from the iTunes app descriptions. In the process, more Islamic prayer apps were identified and added to the database so that future scholars will have access to even more apps. Currently, the database houses more than 500 religious apps from various faiths and categories.

The sixth and final stage of the app project and database recently came to a close. The goal was to make the database available to all Network members so that researchers will have access to the religious app data, be able to add to update the database, and also work collaboratively with the data on future research

---

[12] W. Bellar, "iPray: Understanding the Relationship Between Design and Use in Catholic and Islamic Mobile Prayer Applications," Unpublished doctoral dissertation. College Station: Texas A&M University, 2017.

[13] H. A. Campbell, B. Altenhofen, W. Bellar, and K. Cho, "There's a Religious App for That!: A Framework for Studying Religious Mobile Applications," *Mobile Media & Communication*. 2, no. 2 (2014): 154–172.

projects. Currently, the app database is open to members by request. This phase of the Network project speaks to the visibility and transparency values inherent within the DH. Researchers will be able to find and explore specific examples used in the publications, as well as provide critique and inspire conversation surrounding the categories and conclusions from other previous works. As of today, 13 different religions are represented across the more than 500 apps in the database. There is a real need for collaboration to continue this work so that new insights and areas of inquiry are gained to create a clearer picture of mobile religion.

The religious app database project relied heavily on the value of collaboration during all six stages, from beginning to end. Similarly to the ways in which recent Digital Religion scholars focus on the blending and blurring of the online/offline dichotomy, the database collaboration spanned the boundaries of both online and offline environments. It is important for DH scholars to understand the ways in which acting within these environments enable or constrain collaborative projects. Constant self-reflection, on the part of the researchers as individuals and as a team, is a necessary component to successful collaborations within online communities. The religious meme database project also inhibits these qualities, as well as shows the possibilities of collaboration with undergraduate students.

Religious-Oriented Meme Database

Two years after the religious app database was constructed, a new database focused on religious memes began to take shape. In the fall of 2014, a group of undergraduates in a Religious Communication course at Texas A&M taught by Heidi Campbell began collecting and analyzing religious internet memes. Religious internet memes have been defined as "memes circulated on the internet whose images and texts focus on a variety of religious themes and or religious traditions."[14] Under Campbell's supervision, the undergraduates used this definition to identify specific sub-categories of religious-oriented internet memes. The requirement for including a meme in the sample was that it used specific meme stock characters (e.g. success kid or scumbag Steve) to discuss religion,

---

[14] W. Bellar, H. A. Campbell, K. J. Cho, A. Terry, R. Tsuria, A. Yadlin-Segal, and J. Ziemer, "Reading Religion in Internet Memes," *Journal of Religion, Media and Digital Culture* 2, no. 2 (2013): 3–39.

or how it represented a unique religious meme theme or character created to speak to specific beliefs and/or practices of different religious groups.

After each religious meme case study was selected, descriptions of each meme type and their origins were entered into the collaborative database housed on the Network (see Fig. 3). Elements of each meme that were uploaded to the database were also recorded included "religious affiliation," "forms of meme humor," "meme genre" and "religious meme frame" (Religious-Oriented Meme Database, n.d.). Through this analysis, and the help of information found at knowyourmemes.com, information on more than 150 unique memes about religion and their derivations have been collected on the database. As of this writing, members on the Network can search the database specifically for religious affiliation and type of meme humor. They can also add a new meme or meme derivatives, which are variations on a particular type of meme (http://digitalreligion.tamu.edu/religious-oriented-meme-database).

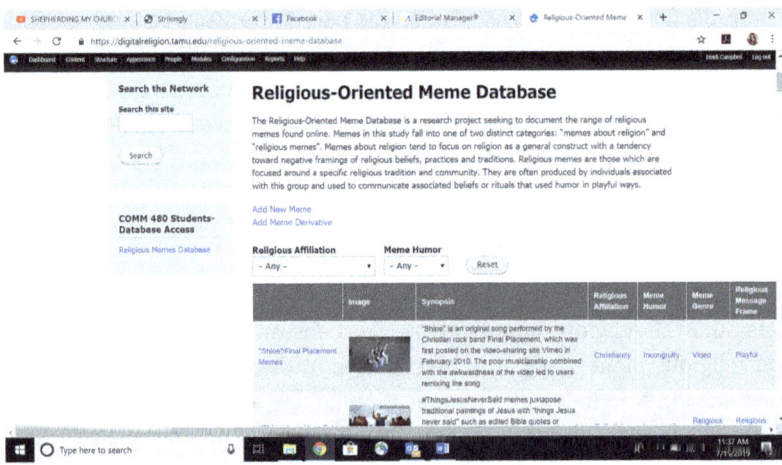

Figure 3: Screenshot of the Religious-Oriented Meme Database on the Network website.

Collaborative work within the religious-oriented meme database has resulted in one publication and another currently under review. Aguilar, Campbell, Stanley, and Taylor, propose that these religious-oriented memes speak not only to conceptualizations of religion within popular culture, but also how these memes express peoples' everyday, shifting, flexible conceptualization of their own religious identities.[15] The research team identified six common genres

---

15 G. K. Aguilar, H. A. Campbell, M. Stanley, and E. Taylor, "Communicating Mixed Messages

and five common religious frames used within religious memes from a sample of 78 original memes that are housed on the Network database. Currently, a new research team is using the database and helping expand contents through a study focused on American-based religious and political memes, and how they present a unique form of American Civil Religion discourse.

Similarly to the religious app database, members of the Network have access to explore the memes currently recorded in the database, add new religious memes to the database, and download information about the memes for use in collaborative research projects. This type of work showcases DH methods (i.e. database design) for exploring a humanities based subject (i.e. Digital Religion). It is also reflective of the value of transparency within DH, in that all of the data collected and analyzed can be viewed and used for future research. Essentially, these databases become a storehouse of knowledge that can be refined, reshaped, and reused in new forms of Digital Religion scholarship.

Building databases in which to store, categories, and analyze digital artifacts is one tool with which DH scholars can collaborate. The religious app database and the religious meme database are two case studies that are designed to foster collaborative research. That collaboration grows stronger when researchers share access to databases so that others can learn from, recreate, and expand our knowledge and understandings of religious information and practices. Innovations in online collaborative tools will help push the field of DH and Digital Religion further by allowing researchers access to specialized knowledge that can be reshape and recreated through democratizing affordances found in digital environments. The next section explores ways in which the Network can continue this mission and impact the field of Digital Religion.

The Future of the Network for New Media, Religion, and Digital Culture Studies

The Network for New Media, Religion, and Digital Culture Studies mirrors the understandings and elements of successful online collaborative sites as it situated within a context of developing new theoretical insights and methodological tools for exploring Digital Religion. The goals of the Network are to create a centralized location of key works in the field, to connect students and scholars across international and disciplinary boundaries, and to develop new collabora-

---

about Religion Through Internet Memes.whoa," *Information, Communication & Society* 20, no. 10 (2017): 1499.

tive tools with which to study digital religious phenomena. It achieves these goals chiefly through an interactive bibliography, an index of scholars in the field, and a toolbox of research resources that are open to Network members.

Limitations do exist in any form of online collaboration. First, the website relies solely on the scholars themselves in order to keep their profiles complete and accurate. Therefore it is necessary to request that scholars update their pages periodically. Whether they update their pages or not is hard to track. Second, the Network must maintain its position as a central hub by attracting new scholars and those outside of academia to take part in the conversation, and by allowing more access to the research databases to outside individuals and groups. Members need to feel agency in not only getting and offering information, but also in recreating, recombining, and integrating that knowledge. One way this can be achieved is through greater access to the data housed on the Network.

Additionally, the Network should take steps to extend transparency and visibility not only to students and scholars in the field, but also to the public at large. The Network has taken steps in partnering with news outlets such as the Religious News Service (RNS), which publishes news stories related to religion and spirituality in an online news format. Through this relationship, the hope is, as the site notes, to "raise public awareness of the unique research being undertaken in new media and religion studies, and translate these important findings to a popular audience." Communicating with the public is, and should continue to be, a value that is cultivated within digital scholarship. Partnering with news services is one such way to ensure that the value is honored.

In DH, and in successful online communities, it is important that a group of people work individually and collectively to achieve similar goals. The Network makes its mission clear and invites scholars at all levels to participate in the community, to start new conversations, and develop new digital tools to push the field forward. It does this in the spirit of transparency, interdisciplinary, and collaborative connectivity —values that are shared within DH and digital research as a whole.

Selected References

Aguilar, G. K., H. A. Campbell, M. Stanley, and E. Taylor. "Communicating Mixed Messages about Religion Through Internet Memes.whoa" *Information, Communication & Society* 20, no. 10 (2017): 1498–1520.

Barendregt, B. "Mobile Religiosity in Indonesia: Mobilized Islam, Islamized Mobility and the Potential of Islamic Techno Nationalism." In *Living the Information Society in Asia*, edited by E. Alampay, 73–92. Singapore: Institute of Southeast Asian Studies, 2009.

Bellar, W., H. A. Campbell, K. J. Cho, A. Terry, R. Tsuria, A. Yadlin-Segal, and J. Ziemer. "Reading Religion in Internet Memes." *Journal of Religion, Media and Digital Culture* 2, no. 2 (2013): 3–39.

Bellar, W. "iPray: Understanding the Relationship Between Design and Use in Catholic and Islamic Mobile Prayer Applications." Unpublished doctoral dissertation. College Station: Texas A&M University, 2017.

Campbell, H. A. *When Religion Meets New Media*. London: Routledge, 2010.

Campbell, H. A., editor. *Digital Religion: Understanding Religious Practices in New Media Worlds*. London: Routledge, 2013.

Campbell, H. A. "Surveying Theoretical Approaches Within Digital Religion Studies." *New Media & Society* 19, no. 1 (2017): 15–24.

Campbell, H. A., B. Altenhofen, W. Bellar, and K. Cho. "There's a Religious App for That!: A Framework for Studying Religious Mobile Applications." *Mobile Media & Communication*. 2, no. 2 (2014): 154–172.

Faraj, S., S. L. Jarvenpaa, and A. Majchrzak. "Knowledge Collaboration in Online Communities." *Organization Science* 22, no. 5 (2011): 1224–1239.

Hoover, S. and N. Echchaibi. "The 'Third Spaces' of Digital Religion." Paper presented at the Center for Media, Religion, and Culture, University of Colorado Boulder, 2012. Retrieved from http://cmrc.colorado.edu/wp-content/uploads/2012/03/Third-Spaces-Essay-Draft-Final.pdf.

Horst, H. and D. Miller. *The Cell Phone: An Anthropology of Communication*. Oxford: Berg, 2006.

Kirschenbaum, M. "What is Digital Humanities and What's it Doing in English Departments?" In *Debates in the digital humanities*, edited by Matthew K. Gold, 3–11. Minneapolis, MN: University of Minnesota, 2012.

Lövheim, M., and H. A. Campbell. "Considering Critical Methods and Theoretical Lenses in Digital Religion Studies." *New Media & Society* 19, no. 1 (2017): 5–14.

Ma, M. and R Agarwal. "Through a Glass Darkly?: Information Technology, Identity Verification, and Knowledge Contribution in Online Communities." *Information Systems Research* 18, no. 1 (2007): 42–67.

Roman, A. "Texting God: SMS and Religion in the Philippines." Paper presented at the 5th International Conference on Media, Religion and Culture, Stockholm, July 2006. Retrieved from http://www.freinademetzcenter.org/pdf/Texting%20God%20SMS%20and%20Religion%20in%20the%20Philippines.pdf.

Spiro, L. "This is Why We Fight: Defining the Values of the Digital Humanities." In *Debates in the Digital Humanities*, edited by Matthew K. Gold, 16–35. Minneapolis, MN: University of Minnesota, 2012.

Togarasei, L. "Mediating the Gospel: Pentecostal Christianity and Media Technology in Botswana and Zimbabwe." *Journal of Contemporary Religion* 27, no. 2 (2012): 257–274.

Torma, R. and P. Teusner. "iReligion." *Studies in World Christianity* 17, no. 2 (2011): 137–155.

Wagner, R. "You are What you Install: Religious Authenticity and Identity in Mobile Apps." In *Digital Religion: Understanding Religious Practices in New Media Worlds*, 199–206. London: Routledge, 2013.

Wyche, S. P., K. E. Caine, B. K. Davison, S. N. Patel, M. Arteaga, and R. E. Grinter. "Sacred Imagery in Techno-Spiritual Design." In *Proceedings of the SIGCHI Conference on Human Factors in Computing Systems* (April 2009): 55–58.

Russell T. McCutcheon
Learning to Code: Digital Pebbles and Institutional Ripples

> Well I started out
> down a dirty road
> Started out
> all alone ...
> "Learning to Fly"
> (Tom Petty and Jeff Lynne [1991])

Knowing Enough to Make You Dangerous

Yes, I know enough about Dreamweaver 3, released in 1999, to make me dangerous; for while I didn't really know how to do all that much with it, even once I attained my own level of mastery (i.e., I just knew what I needed to know), upon first writing this chapter, in the summer of 2017, I saw that version 17 was available—that's right, 17. So I'm a little behind the curve. But given that I was trying to build a Department website while learning it on my own, back in the very early 2000s—consulting "how to" books and scrounging off of the knowledge that a very few friends had already accumulated for themselves, while also doing the various other things that a new Department chair might do in a struggling Department—I'd like to think that I was actually ahead of the curve. For some of us were doing what we now call digital humanities (DH) well before it was considered a distinguishable thing that deserved a name of its own.

But this story is not simply about a Department website, of course; as I learned from the late Jonathan Z. Smith's writings, the things that scholars discuss are best understood as an e.g. of some larger topic, and so focusing too closely on any single example may trick us into thinking that it somehow stands on its own, moves of its own steam, and is therefore inherently interesting or ob-

---

Note: I appreciate that the editors not only invited this contribution but also stood by it when the anonymous outside readers reported that it was not scholarly enough. While analysis and conclusions about the field as a whole are certainly included, the starting point of this chapter includes reflections on a time when, and a situation that, long predates the thing today commonly called "the digital humanities," in hopes of prompting readers to understand the developments undoubtedly chronicled in detail in other chapters as having a beginning and a history, not to mention practical effects that impact the profession and its members' careers as well as the settings in which they carry out their work.

https://doi.org/10.1515/9783110573022-017

viously valuable. So, while none of us should trust our own memory (let alone those of our research subjects), I do recall that all along the new Department website that I was building on evenings and weekends wasn't simply about having a decent site; instead, it was but a node in a concerted effort to revive an entire Department—one that, back in 2000, was on the brink of losing its major and either becoming what we call a service Department or, perhaps, closing entirely.[1] So that investment of time and energy (and, yes, some Department resources as well—but far less than you might think, so long as one has the good will and energy of a committed faculty) into building a content-rich website was just the first of many interconnected digital initiatives tackled over the years by the members of our Department; and although each project was motivated by a specific need, they all eventually pulled the same wagon: reinventing a Humanities Department at a time when they were (and continue to be) under considerable attack.

Of course such a reinvention is never complete; as any good social theorist would likely tell you, groups are always on the brink of demise.[2] But despite being ongoing, even to this day, these interconnected digital projects have been quite successful, inasmuch as they have not only played a prominent role in people across the U.S. and around the world taking an interest in this thing called the Department of Religious Studies at the University of Alabama but, more importantly, those same people trying on for size some of the things that we've successfully been doing in Tuscaloosa—and then having successes of their own.[3]

In fact, that I was invited to contribute to this volume strikes me as persuasive evidence that something that we did in Tuscaloosa worked; for while my

---

[1] This was due to its inadequate number of majors and thus BA graduates—the latter being the coin of our particular corner of the higher ed realm; on our situation, back then, at the University of Alabama, see McCutcheon 2015.

[2] Because their ranks are in a constant state of flux, members are always in need of injecting new energy and new resources into them (e.g., to recruit and initiate new members), making what the late Gary Lease once characterized as a social formation's emergent phase a constant, inasmuch as any group's so-called dominant phase is just the result of a continual state of successful reinvention. See Lease, "The History of 'Religious' Consciousness and the Diffusion of Culture: Strategies for Surviving Dissolution," *Historical Reflections/Reflexions Historique* 20 (1994): 453–79.

[3] While I wouldn't want to over-estimate our effect, a casual look at other Departments' presence online makes evident, at least to me, that people have been paying attention (such as the time the American Academy of Religion's head office contacted us for advice on the video equipment it needed to begin a new initiative of their own). My hope is that this chapter provides further incentive for those who understand their own Department's identity and existence not to be on autopilot—a view held by a surprisingly large number of faculty who seem to take the units in which they do their work for granted.

own research is hardly characterized by the forms that one might reasonably imagine DH might take in our field today, the collective digital work carried out in our Department, by a variety of actors (from faculty and staff to guests, students, grads, and especially our various undergraduate student workers), strikes me as a fitting example of how assorted digital tools affect more than just an individual faculty member's research and teaching.

So the goal of this chapter, much like our reinvention itself, is both modest and ambitious. Initially, it is merely to outline just what we did do over the past twenty years, making evident not only why web-based/digital projects suited what was originally a very small and regionally marginal B.A.-granting unit, but—thinking back to its more ambitious goals—also how they helped us not just to save a Department but helped to position it to triple the size of its tenure-track/tenured faculty and, in the Fall of 2017, launch a new M.A. in which *these very digital tools figure prominently.* So this chapter's aim turns out to be rather more aspiring than simply a list of projects; instead, in the midst of doing that, it also makes an argument for why a faculty member who sees their "serious" research and teaching to lie elsewhere might also learn to code, or at least learn enough about Dreamweaver (let alone Fireworks, Photoshop, Final Cut Pro, some FTP programs, Garage Band, WordPress, Python, etc.) to be just a little dangerous. For although we all earned a Ph.D. by focusing on mastering a narrow specialty, and some of us think that being a professor entails each of us diving even deeper into our own particular rabbit hole, many of us have likely found that our professional success—along with the success of the institutions that employ us and in which we teach our students—also hinges on acquiring some rather unexpected and collaborative skills along the way.

The Digital Pebbles

As already indicated, the first thing on our list was a new website—one of what I'll simply characterize as the various pebbles that, over the past two decades, we've successively tossed in the water in our effort to get this Department back on its feet. Although 2001 was certainly not the earliest years of the web, it was well before Facebook let alone the very term "social media"—in fact, Facebook's precursor, Facemash, wasn't created by then Harvard sophomore Mark Zuckerberg until the Fall of 2003 and it wasn't until 2006 that Facebook opened its membership ranks to almost anyone (instead of just those with a .edu email address). But even though we were still using hardcopy phonebooks and printed course catalogs, the almost limitless footprint of the online world was already pretty apparent.

So we reasoned that if we were trying to recruit students (whether as majors or just to enroll them in our lower-level Core courses—enrollments that so dominate most Religious Studies Department's credit hour production that they can't be ignored), let alone assure the administration (which included a brand new Dean of the College of Arts & Sciences—a college with many Departments far larger and thus more consequential than our own) that it had made a wise decision hiring an outside Chair instead of just closing the Department, reclaiming its space, and distributing its tenure-track faculty and staff to other units, then, in relatively short order, we needed something that someone in another building, either next door or far across campus, could see and which would help to provide them with evidence that their investment was already paying off. (I have in mind, here, other faculty and Department chairs, the Dean, and the Provost/Vice President for Academic Affairs, not to mention our faculty themselves, since finding a faculty member's page is more than likely the first place someone will start when looking for information on them or while trying to contact them.) So while assessing the state of our physical facilities, reviewing faculty productivity reports from the year before, becoming familiar with the budget, learning to work with the staff, deciding what Department procedures needed to be invented, meeting and advising students, etc., it was obvious to this incoming Chair that the website needed attention, for the few pages that it then had relied on so many different styles and fonts and image sizes, not to mention resolution, that it painted an unflattering picture of a unit that was trying to instill a rather different image in people's minds.

So step one was a short term fix: reassigning a faculty member—Kurtis Schaeffer, in fact, who was then in just his second year as an Assistant Professor but who went on to become the longtime Chair of the Department of Religious Studies at the University of Virginia—from one semester's regular courses; he was someone who, despite being a specialist in medieval Tibetan Buddhism, was also willing to learn enough about creating a website to take a stab at adapting a basic university template to our needs and then populating it with consistent and thus professional content. Step two was then commissioning people on campus who did this for a living to come up with a design that might set our Department apart, which included a logo (but what our campus's recently-established branding office now tells us we need to call "a graphic element"). The site they came up with—a stark black and white design that we felt would age well, and which included a stylized version of the late-19th century building in which we were located (an image that intentionally avoided the collection of diverse religious symbols that so many Departments in our field opt for when devising an identifying emblem of some sort)—lasted for many years and was only redesigned in 2017–18; in fact, it was a redesign that prompted us to reconsider the work now done by a website, since our former site, inasmuch as it predated social media, had to serve far more purposes than do sites today. For example, now we just put up a blog post instead of making a new html page filled with photos from the most recent student event or public lecture—but I'm getting ahead of myself.[4]

Although the website was our only online presence for many years, because it predated the arrival of Learning Management Systems (LMS) on campus we were able to gain some of that eventual functionality by having a password protected folder created on the site, thereby allowing us to move reserved readings online as PDFs. (At that time they were retained as hard copies in the main library, which students would photocopy on their own). Actually, we did this several years prior to the rest of the institution catching up. (Moral of the story: there's a nimbleness to the digital world, if you're willing to take advantage of it.) So although our site's "secure" folder (filled with each class's photocopied articles/readings) and our "pdf" folder (filled with public-facing syllabi or flyers) have now been replaced by centralized technologies at the university (aside: what we gain from IT centralization comes at the price of what we inevitably lose from that very centralization), back then they introduced an ease into at least one aspect of our jobs and also into the lives of students. For although few were then using laptops in classes and iPads and smartphones had yet to be invented, and thus no one was relying exclusively on digital readings, the Department was able to take a significant step away from expensive textbooks and their continual (and sometimes suspect) replacement by new editions.[5]

Then, in early 2009, the Department created a page on Facebook; I'd joined back in the summer of 2005, mainly to stay in touch with family back in Canada. In fact, I can still see the August 5, 2005, post from one of our majors at the time:

> You, Dr. M, are the first of my many professors past and present to venture out into the previously uncharted territory known to modern man as thefacebook.com; I salute your bravery, sir.

---

4 See Russell T. McCutcheon, "Why I Blog," in *"Religion" in Theory and in Practice: Demystifying the Field for Burgeoning Academics*, McCutcheon, ed. (Sheffield, UK: Equinox Publishing, 2018) for more background on the choice of a scholar to blog—a venue seen by some as either passé or beneath the dignity of a serious scholar.

5 In fact, at the close of the semester when I first drafted this chapter, as I reviewed student evaluations for all of our classes, comments from an undergraduate seminar relying completely on PDF readings posted online stood out, with a student thanking us for not making them pay for "a $200 textbook" (as the student phrased it).

But social media hadn't taken off all that much until a few years after that and it was only then that we realized that this was also a place for the Department to have a presence, to more effectively communicate with our own students and also with our graduates. (Seeing Facebook as a place to convey an image of the Department to people beyond our campus wasn't something we initially thought much about, to be honest. But now we do, of course.) Not long after that our undergraduate student association established it's own Facebook presence. We didn't start up a Twitter account for the Department (which happened in May 2014) until Michael Altman joined the faculty in 2013–14 and persuaded us that we needed to think of the different audiences that each social media site helped us to reach. (He also organized a contest for students to name my Twitter account, since he also convinced me to join.) What's interesting to note, however, is that the student association had been tweeting, based on our students' own initiative, since July 2011—but we obviously hadn't paid the proper attention to the signals that they had sent us by joining up on their own.[6] So this marked the first time that we started thinking far more intentionally about social media, especially reaching audiences beyond those who already felt some affinity for us (such as current students and our grads).

This had long been apparent at the website, of course, which by then had grown to something like 400 pages (making them became a bit of an after-hours hobby—the sort of work that we never really tend to track *as* work), many of which provided substantive information on the field and which were aimed at people far from Alabama (as well as providing our faculty with plenty of resources, such as searchable online catalogs for the Department's library and fairly large video/DVD collection). But we had not yet really thought through social media as a complement (or replacement?) to the Department's site—a site which, being so large, had by then admittedly become a bit of a task to manage and keep up-to-date. So, we did some thinking about what to name the new Twitter site; we were surprised to learn that @studyreligion (notably in the imperative) had not been taken; it was not only nicely in keeping with a motto we'd come up with over a decade before, Studying Religion *in* Culture,[7] but it also made a strong statement that the Alabama experiment might to have implications in other Departments. After all, ours was hardly the only one in the coun-

---

[6] Students continue to send signals, such as the current generation's migration away from Facebook, which challenges us to continue to be relevant on social media platforms while recognizing that older alums are still on such sites as Facebook.

[7] The italics were meant to convey the distinctive approach we were then hoping to develop at Alabama, in contrast to the, in my view, problematic Tillichian model associated with the more common nomenclature of religion *and* culture.

try in which major/graduate numbers had been declining. But, as we looked around, it seemed we were among a rather small number that were taking an entrepreneurial, and intentionally Department-wide, approach to social media as a way to do something about it. And so @studyreligion was born and in October 2015 it made its first appearance on Instagram as well.

The story so far: a complete website redesign and revision, accomplished initially by faculty as service to the Department, led to enhanced features that impacted our teaching, and which was later complemented by Facebook (2009), Twitter (2014), and Instagram (2015).[8]

At this point we also need to take into account the May 2012 launch of our Department's blog, during Ted Trost's tenure as Department chair (2009–2013); set-up and first managed by our colleague Steven Ramey, it's initial aim was to provide a place for conversation about the theme of that year's lecture series: the relevance of the Humanities.[9] But it soon grew beyond that as more faculty, and then students, grads, and even invited guests, began posting pieces that they'd written especially for this venue, on a wide variety of topics, either briefly summarizing current faculty research or applying class content to understand day-to-day life, politics, and pop culture.[10] This was how WordPress first came into our orbit—a software used in our College's technology office (called etech, and distinguished from the university's central Office for Information Technology); so, with their assistance to customize the theme and provide just a little more functionality, we added an active blog to our social media presence, making even more evident that the days of a website handling all of our online duties had passed; for, as noted above, adding a quick WordPress blog post immediately

---

8 Side note: the University of Alabama's Twitter account is eager for content that promotes the campus, so not long after our Twitter account was launched it became clear that they'd happily retweet posts, especially interesting photos of campus. Given that the Department had long ago gotten into the habit of regularly changing the cover pictures on our Facebook accounts—after all, not just content posts and links communicate activity on a page—it wasn't much effort to start tweeting pictures periodically at the university and appending the #TodayatUA hashtag. The result? While it's unclear exactly what it is worth to a unit, the university and its followers certainly know that this one Department exists.
9 The first post on the site was Stephen Ramey, "What is the Purpose of Education?" *Studying Religion in Culture*, 7 May 2012, https://religion.ua.edu/blog/2012/05/07/what-is-the-purpose-of-education/.
10 The blog (https://religion.ua.edu/blog/) has had well over 150,000 hits since then, with over 1,000 posts on the site, one garnering 2,000 hits in a single day, and another over 3,000 visits in total. We find that it has been an excellent vehicle for continuing to engage students after they've graduated, either by how faculty and current student posts solicit their comments on social media or their willingness to write guests posts.

following an event, containing photos and captions, maybe even an embedded video, and which then automatically hits Facebook and Twitter, was far simpler than html coding in Dreamweaver.

And then, about the same time, three of the faculty members (Steven Ramey, Merinda Simmons, and myself) joined up with colleagues elsewhere (Craig Martin, Monica Miller [who later left the group], Leslie Dorrough Smith, and Vaia Touna) to form a research collaborative that we named Culture on the Edge; with the help of funding from Trost, who was chair at that time, a working meeting at the University of Alabama took place, a common book project was launched, and the group set to work trying to think through the application of Jean-Francois Bayart's work on identity formation (notably his *The Illusion of Cultural Identity* [2005]) to the study of religion. The reasoning was that if, as all of us agreed we ought to, we dropped the pretension that religion is an autonomous and privileged domain and, instead, presumed it to be but one more mundane social site, then what might the work of a scholar of religion look like? A year later, in May 2013, the group launched a blog of its own (hosted by the University of Alabama) and, for the next three years, amassed a pretty intensive series of pithy but intellectually thick posts, written for a wide audience, in which the group's approach was exemplified and explored.[11] Although, strictly speaking, not a Department initiative, Touna eventually joined our faculty (in the Fall of 2015), bringing the UA contingent of the group to four, by which time the move from studying identity as a stable quality to studying (as Bayart phrases it) "acts of identification," had made headway throughout other parts of the Department—both in our teaching and our writing. "Classification is a political act" some of our undergraduate students then started responding, on social media, to news stories they'd find and post online, tagging the Department, influenced by what they were learning in our classes. And while those classes certainly drew on detailed historical or rich ethnographic information, this material was seen not as self-evidently interesting but, instead, used as a pedagogical means to another end, i.e., as noted in the opening of this chapter, as an e.g. of some wider process in which social actors, embedded in contexts not of their making, made moves and engaged in contests. Coming as it did just over a decade after our reinvention began and resulting from initial conversations between Simmons, Ramey, and myself—three colleagues, all at different career stages—concerning things we had in common despite the different areas in

---

11 The blog (https://edge.ua.edu/), which all along had a pedagogical tone (hoping to reach newcomers at whatever level), is still active but, as the group expanded, has been reinvented as a peer review blog working with early career scholars who find the group's approach useful.

which we did our work, Culture on the Edge nicely exemplifies where I'm going in this chapter, for the form our work takes in the online world has had substantive and long-lasting effect on the content of our research and our teaching.[12] And the fact that a variety of thematically-related posts from the group's blog turned into a series of small books, aimed at the classroom (each with an original introduction and substantive afterword by the volume editor), makes the benefits of cross-pollinating from the digital world evident once again.[13]

But again, I'm getting ahead of myself.

Returning to the various digital pebbles that we've tossed out there, there's also Vimeo: in the summer of 2012, to kick off the new semester, we posted our first "welcome back" video to that new account, borrowing the recording equipment from a student media lab on campus. The aim was to surprise returning students, a little, or at least a little more than just putting up different bulletin board content in anticipation of their arrival in August—not realizing that, as with the other digital initiatives, our intended audience would, eventually, reach far beyond our campus. I drove a student worker (Andie Alexander, now working on her own Ph.D. at Emory) around town, with her leaning out the passenger side window, filming local landmarks, and she then learned enough about iMovie to edit it, give it an older sepia look, and pair it with John Sebastian's old theme from "Welcome Back Kotter" (on TV from 1975 to 1979 and John Travolta's big break). Faculty names were listed as if we were co-stars and the Dean's name appeared in the credits as the executive producer. Eventually, the Department purchased its own camera, tripods, and mics (a Zoom mic and a lapel mic), started regularly filming our guests' public lectures, doing some interviews with them, and creating a variety of short series featuring our students and faculty (looking around almost any prof's office there's likely an interesting story associated with each trinket on their book shelves …). Each summer we also make a new welcome back video—usually featuring any new faculty member we've hired. And, as with previous social media ventures, we again learned that the success of any one venture is linked to how a coherent, collective wagon can be pulled by them all, e.g., a new video is embedded in a blog post and it then hits Facebook and Twitter.

---

[12] The Culture on the Edge site, now with 1,000 posts, has had 277,000 hits since going online in May of 2013, with the single most visited post receiving 1,900 hits in one day and the all-time most popular one chalking up a total of 10,500 visits; see the preface to Touna 2018 for more on that site.

[13] Two of the volumes were edited by myself and one by Steven Ramey; the books appear in Vaia Touna's book series, "Working with Culture on the Edge" (published by Equinox of the UK) and are entitled: *Fabricating Origins*, *Fabricating Identity*, and *Fabricating Difference*.

Now, as I briefly noted earlier, Altman has moved us into the latest social media direction with the Study Religion podcast (hosted at SoundCloud and also available for download on iTunes).[14] With the tenth episode now in production (the seventh, of which we were quite proud, focused on a variety of scholars' memories of Jonathan Z. Smith [1938–2017]), we have high hopes for the site inasmuch as it makes a statement about the Department but in the context of the field at large (both in the U.S. and beyond); for although the work of students and faculty are certainly featured, guests to campus appear as well, along with conversations on topics that range widely and which take place far from our home in Manly Hall—again, making evident, we hope, that the approach we've developed here at the University of Alabama pays off when applied elsewhere, inasmuch as we might learn something new about seemingly mundane, culture-wide processes of social formation if we approach our topic in just this or that manner.

There's still many challenges ahead of us, of course, so I don't want to get lost in the list of digital initiatives that I've just put before readers. To name but one such challenge: the faculty involved in these initiatives—more or less involving everyone in the Department, in varying ways, but with Richard Newton (who joined us in August 2018) now taking the lead, especially on producing short, engaging videos for social media that promote our courses and which are aimed at incoming students—have active research and publishing programs of their own, not to mention their teaching and supervision of students. Even recognizing the self-beneficial role played by carrying out Department service one still must take into account that the management of these social media sites takes time and effort (especially throughout the summer when faculty normally are not engaged in performing all that much service). How to ensure that does not get out of hand and that the pay-off is in proportion to the work invested is something of which we're certainly aware. Another issue (already mentioned) is how each of these complements the others, rather than repeats (or even undermines) them. While some posts and projects are better suited to certain sorts of content, others attract a distinct audience; so, while a shotgun approach to social media seems inevitable (for one never knows which post will stick with which reader), ensuring that one's efforts are well invested means trying to make sure all facets of this initiative play their own role while also contributing to an overall impression of what we're up to in the Department of Reli-

---

14 Although also available on other platforms, you can find the podcast at https://soundcloud.com/studyreligion; the seven extended, unedited interviews for the J. Z. Smith episode were then used in a book I am co-editing with Emily Crews, an Instructor on our faculty, entitled *Remembering J. Z. Smith: A Career and its Consequence* (forthcoming in 2020 from Equinox).

gious Studies at the University of Alabama—an impression no longer simply in the minds of administrators who were once asked to question whether the unit had a future and students unsure whether to do their major with us but now also in the minds of graduate students and faculty around the country and around the world, who have found us only because we've taken our presence online seriously. After all, there are many Departments in the country, let alone elsewhere in the world, yet people somehow know that we exist and are interested in what we're up to in Tuscaloosa. Hopefully, we have the substance in our individual work to back-up the presence that we have online, so that the tweet, the post, the picture, the video, or the audio file are not just digitally accessible but intellectually useful and provocative of thought.

But enough with the list of what we've tackled; what about their effects?

Institutional Ripples

These various pebbles certainly produced ripples. The challenge was in trying to get them all going in the same direction and reinforcing each other—either that or, as if often the case, figuring out whether (and if so, then how) to follow the direction in which they were moving us. While certainly not wanting to portray our work as omnisciently managing all of the above moving parts, reflecting with the benefit of hindsight on the past couple of decades, it seems possible to discern a variety of beneficial effects that our assorted digital experiments have had —whether alone or, as has increasingly become apparent over time, in concert.

First off, early on our Department became known on campus (which, given that we were possibly due to be closed just a couple years earlier, is an accomplishment on its own) for web-based projects—to such an extent that, within a couple years of arriving, I was asked to chair a three person committee that made sweeping recommendations for overhauling how technology was managed and utilized throughout the College of Arts and Sciences. By that time I'd also served on an A&S committee working to establish new relationships with the College of Continuing Studies, which built and managed all online courses (though Departments provided the content and the instructors); like many others in higher education, our Department was wary of moving to the online environment (especially because such courses competed with our regular lecture classes, let alone because no matter how well designed, they couldn't match standing in front of a student and answering their question), but we already had several old and poorly delivered distance learning courses (i.e., the once common model of a textbook and a booklet arriving for the student in the mail) and because distance learning plays an important role in some students' education, we at least saw this as an opportunity to revamp these courses and turn them into respectable and engaging pedagogical experiences for students far from campus—thus a good that we achieved through this relationship.[15]

So, ripple #1: the Department quickly developed an identity with the College, helping the College to achieve some of its aims while also working to benefit the Department and its own goals in the process. (Aside: anyone familiar with working in large institutions will recognize that not every unit's goals completely overlap or complement the institution's other units; making this happen, or at least minimizing the effects of goals that run counter to one another, sometimes takes craft and effort.)

This first effect was crucial, since achieving our primary goal—increasing demand for classes and increasing the number of majors and graduates (i.e., recruitment)—was a slow process, sometimes with only incremental gains that took years to accomplish. So, while working diligently to accomplish those necessary goals (or what we're now told to call outcomes), through a wide variety of strategies, the more instantaneous effect of paying attention to the web helped to secure the confidence of an administration that took a chance protecting what was then our little (and, at least back then, very exposed) Department. It demonstrated ingenuity and actually helped move the College of Arts and Sciences in its own new direction. (While the pay-off of such work is never guaranteed, I have learned over the years that, despite possible disagreements, it is never a bad idea to build up a small store of good karma with others in your institutional setting.)

A second effect concerned what we all now refer to as retention, i.e., the small number of majors we already had, plus any newly acquired majors, needed a reason to stay with us. Given our small size then and for many years prior to that (i.e., the Department had been just three faculty for decades, was 4.25 faculty when I arrived in 2001 [we had two cross-appointments], though now we're twelve with a full-time Instructor as well), we exerted little control over our own subject domain; a variety of courses from across our campus therefore focused on the topic of religion, offered by a variety of Departments (units that, to this day, still exert territorial control over some classes that other Departments of Re-

---

15 For a while, Departments on our campus lost the right to limit these courses to Distance Learning (DL) students, thereby allowing all students to enroll and, possibly, bypass a lecture course; that right, however, was eventually reinstated and so our small number of online courses (with limited seating) continue to play an important role in our teaching mission while serving as an opportunity for a small set of off-campus instructors to earn some extra money and continue to be involved in university instruction.

ligious Studies might reasonably see as their own).[16] So students could satisfy an interest in studying such things as the history of Christianity (Department of History), the study of myths and rituals (Department of Anthropology), or religion in ancient Rome (Classics) by taking courses in other Departments, and, perhaps, enrolling in other majors. Again, given that this was long before social networks were something that we talked about, the new Department webpage, with regular and timely updates, candid pictures from student events, temporary (and sometimes fun) fake homepages announcing events or new courses, and substantive pages on just what the study of religion was (and wasn't) functioned to unify our current students, in an almost Durkheimian fashion, by prominently featuring the logo that we settled on (which was also displayed around the Department as well), making plain who we were, who they were, and where you were when you landed on our page. (Given the standardized web template that units on campus must now use, thanks to those who work to brand the university and not just its various Departments, we have strived to ensure that the newly designed page continues to reflect this hard-won identity.) The engaging nature of our classes and personable style of our faculty played a significant role in this, of course, as did the fact that we quickly established a Department lounge, but I think that it would be shortsighted to downplay the role that the website played in providing these students with a virtual home.

Ripple #2, then? The establishment of an effective virtual totem of considerable reach.[17]

The third effect was also focused on students, for despite an early course release for a faculty member to create an initial website, and even though I made a number of the pages on the site myself (actually, before publishing an intro book in the field I even created an online version that was on the web for years, for free), it was undergraduate student workers who played a key role in building and maintaining the site. For the ability to engage a work-study student was soon followed by hiring one and sometimes two part-time students in the main office, often doing so with one throughout the summer as well—they tack-

---

[16] We see here a common (and for me problematic) model of religion, inasmuch as it is presumed to be an inner, pan-human experience, only secondarily expressed, and thereby something that virtually anyone—inasmuch as they are human themselves—can talk about authoritatively.

[17] In fact, it was so effective that, upon learning a couple years ago that our longtime site was being replaced, a group of alums (some of whom graduated over a decade ago and many of whom had worked on our site under my direction) quickly started an initiative to petition the Dean to let us keep the site. The result was that the new site tipped its hat enough to the look of the old site that everyone ended up being happy with the change.

led a wide variety of projects, of course, as any student working in a Department's main office will, from running errands on campus to making photocopies, sorting the mail and answering phones. But learning Dreamweaver for web design, using a FTP (file transfer protocol) software and Fireworks for handling images soon became a requirement for students who were working in the Department. Now, Photoshop has replaced Fireworks and WordPress has almost completely replaced Dreamweaver, but since we now make movies that means Final Cut Pro has also become a required software. And Altman's initial experiments in podcasting has even involved students as well (from reading and recording their own work to doing the intro and credits), meaning that learning to use GarageBand and other audio software will also be required. Inevitably the students we hired would learn far more about each software once we had taught them the basics about html coding and so, in some cases, they'd graduate and put those same skills to work in other degrees, with other Departments, or even in jobs. In fact, I recall one major and former main office student worker whose first job after graduation was working in a real estate office and her ability to make and manage their webpage played a role in her getting that job. While I wouldn't want to overplay this hand, realizing that students were not just helping us maintain our site but were acquiring and using these digital skills on their own has played an important role in how we most recently charted the future course of the Department.

So, ripple #3: the digital emphasis provided practical, transferable skills for our students.

Recognizing this third effect—an effect that was not initially apparent of course, since our minds were then just on the task of getting it done and not on the longer term effects of students learning how to get it done ("teach a man to fish ..." and all that)—marks a shift in the role that our various digital initiatives have played in the Department; for, at the outset, students were just the target audience for our various initiatives—we wanted them to know what courses were being offered, to see the events taking place, to learn more about each faculty member's research, so as to consider taking our classes or even entertain becoming a major. But, slowly, we realized that we were transitioning to students, always under the faculty's direction, becoming not only the readers but also the providers of the content and the producers of the sites—a shift that quickly puts their skills in focus (both for writing online to reach wide audiences and for learning the software necessary to distribute the material online). In part this was out of necessity—after all, we needed assistance to manage these sites—but in part it was also a conscious effort on our part to, for example, refashion campus-wide initiatives (e. g., the now national trend toward emphasizing undergraduate research as a way to retain students and improve graduation rates), to help us to make plain, by making public, what had been going on in our classes all along. (In our experience, this is also the key to the nation-wide assessment initiative: determine how to effectively publicize the work we're already doing rather than inventing things to satisfy credentialing agencies.) Although we're known among those in our classes for emphasizing writing, the student blog makes that clear to a far wider audience, helping us to showcase our students' research and writing and doing so by student workers preparing and posting their peer's work. And although it has detracted from the site's "cool factor," with moms and dads now on Facebook, we regularly reach them as well—and families are not an unimportant audience. (In our case, a few families have so welcomed our work with their children that very kind endowments have resulted.) The video initiative might be an even better example, inasmuch as no faculty member has ever learned the ins and outs of Final Cut Pro but, instead, a student worker, with some assistance from a media lab in our main library, learned it on her own, after transitioning from iMovie; then, because she overlapped with her successor, she passed the skills on to him. And he passed them along to another. And now, we have the fourth and fifth students in this successive tradition of student workers who pass along their video production skills—working under faculty supervisor, to be sure, but more as a producer of a film might work not only with a director but also a cinematographer, sound engineer, and editor (all roles our student workers play). The skills now largely operate well apart from faculty. So while I wouldn't claim that the Department's current emphasis on what others might characterize as skills-based education is wholly due to our digital emphases—for, from the start, focusing not simply on mastery of data domains but also on learning how to define, describe, compare, interpret, and explain was something that we emphasized across our classes (e.g., see McCutcheon 2001, chpt. 13 for my early attempt to discuss integrating the skills that, in my reading, Jonathan Z. Smith had long advocated)—seeing the practical effects of undergraduate students having learned these digital skills certainly reinforced our sense that a methodological focus would be beneficial to their education and, hopefully, careers and lives.

Ripple #4? The shift from delivering information *to* students to being collaborators *with* them.

And this brings me to the last, but perhaps the most significant and lasting, effect; for combining much of the above—e.g., grappling with nation-wide criticisms of the Humanities' relevance, our Department's emphasis on skills, the role students played in assisting our various digital initiatives, the shift many faculty came to share by studying situated acts of identification rather than identities that are simply expressed, and the practical benefits we witnessed these skills having for our students—we began seriously discussing (back in 2013–

14) developing a new M.A. degree program that might explicitly integrate much of what we had learned over the past decade or so. For while it could certainly be questioned whether the world needed another graduate degree in the study of religion (due to declining academic job opportunities and the increased expense of tuition), it has not yet been established whether scholars of religion are incapable of using their skills to train students in social analysis that's accessible to wide audiences—which implicitly makes the case for what we, in the Humanities, have to contribute. So, after several years of planning, led by Merinda Simmons (who chaired the committee, joined by Ramey and Altman), our new M.A. was approved by the University of Alabama Board of Trustees and also by the Alabama Commission on Higher Education—the latter being the very group that, back in the late 1990s, had come to the conclusion that our Department was "non-viable," due to a severe shortage of majors and thus graduates, and that we should therefore lose our undergraduate major. The degree that they approved, and which was launched in the Fall of 2017 with an incoming class of three students (two of whom were our own B.A. grads), involves two required foundations courses in the first semester, one focusing on social theory and the other on the digital skills necessary to make a contribution to the public humanities (designed and then taught, in alternating years, by Altman and Nathan Loewen—who, in 2019, began his own "big data" research project, working with one of our B.A. majors—and again we see the effect of working with students).[18] While learning of the scholarly literature and debates around such categories as agency, authorship, intension, race, gender, class, identity, etc., in the social theory class (categories of analysis that we expect our students to use in the eventual thesis they produce in their second year—a course Ramey and Simmons share) the public humanities foundation course ensures that, by the time they complete it, students have not only acquired skills to create and manage a variety of digital projects—from, yes, tackling GIS projects and big data to producing a podcast or obtaining a domain name and making a website of their own—but will also have developed an entrepreneurial attitude toward mastering new technologies and incorporating them in their own research, whether in data gather-

---

[18] After learning of the JSTOR Data for Research (DfR) site (https://www.jstor.org/dfr/), Loewen (half of whose teaching time is loaned to the College for technology projects) developed a project using this tool to analyze the contents of entire journals in the Philosophy of Religion (his specialty area); see the June 4, 2019, post on our Department blog for a description of his project and a narrative of how it developed/how he acquired the skills necessary to carry it out. See Nathan Loewen, "Reading and Writing and … R: How I Began to Study the Philosophy of Religion with Digital Tools, *Studying Religion in Culture*, 4 June 2019, https://religion.ua.edu/blog/2019/06/04/reading-writing-and-r-how-i-began-to-study-the-philosophy-of-religion-with-digital-tools/.

ing, analysis, or presentation. (For the software they'll work with in the course, all of which are widely available to the students, are hardly the only digital tools they could use to get the job done.) Also, their M.A. thesis can either be a traditional piece of writing (ideally, we've decided, a more practical article-length piece of original work that can be submitted to a journal) or an original and substantial digital project (one that, though presenting their findings in a novel manner, still requires the same sort of research that scholars have done for ages). Given that so many of our undergraduate majors have not gone into the study of religion for a career but yet continue to report to us how important the skills were that they learned in our classes, we've reasoned that, for it to succeed, this M.A. should be of interest both to students hoping to pursue a Ph.D. in the study of religion immediately after as well as to those who wish to work in any number of other fields where critical thinking and creative but effective communication skills are valued[19]—making clear how, in a way that was surely unplanned, our early and ongoing digital initiatives have fed back into the degree programs of the Department itself, not just helping to rejuvenate it but transforming it in the process, opening a variety of futures we'd not anticipated whatsoever but which we're now pursuing. Five students enrolled in the Fall of 2018 and four more in Fall 2019 (with all but one of the enrollees in the most recent class coming from outside the Department), with 9 more new students beginning 2020 – 21 (none of whom are own own B.A. graduates).

And so we come to ripple #5: The unforeseen feedback loop whereby what started as a variety of discrete digital initiatives coalesced into a refined identity for the Department and a new graduate degree program that further sets us apart in the field today and thereby continues to model for others possible futures for the academic study of religion. In fact, it's now so central to who we see ourselves to be that contributing to what we're calling our public/digital humanities initiative not just shaped a request for a new tenure-track line in the Department (with Jeri Wieringa, a trained digital historian, starting in Fall 2020), but it has also inspired Altman to propose a new B.A. crash course on digital tools for undergrads (being offered for the first time in Fall 2019—the only such course on campus, as far as we can tell), as well as putting one of our current M.A. students in the position of being an appealing summer hire to work full-time in one of our campus's tech offices and another interning with *Alabama Heritage* magazine and working with a local museum in northeast Alabama for the summer. If we

---

[19] At present, we have graduated seven M.A. students, with three enrolling in a Ph.D. in Religious Studies, with others, for example, working in museum curation or seeking further degrees in other fields.

add to this that one of three annual early career workshops, to be offered over the coming four years to four different groups of applicants, will be modeled after our M.A. degree's course on digital tools as part of a $350,000 Luce Foundation grant that Department received in the summer of 2019 (with Altman as the PI but with six or our faculty acting as workshop mentors)[20] then it should be obvious that a set of specific digital tools initially used to help reinvent the Department's online presence has turned out to be the major theme driving that reinvention's success all across the unit—a success that includes an enhanced entrepreneurial attitude among the faculty as well, evidenced in the increasingly novel and bold research projects that they're tackling as well as their pedagogical experiments in the classroom, with or without the use of technology.[21]

Still Learning to Fly

And so we return to the epigraph, which, by now, hopefully makes a little more sense to readers; for what has become a series of inter-related and collective initiatives that help to keep our Department central to discussions of how to renew the study of religion in the current higher ed environment (i.e., moving away from a descriptive, world religions model and, instead, investigating so-called religion-making, and even identity-making, processes and strategies) first started out as a variety of isolated digital experiments, usually linked to a discrete need and a lone professor with time on his or her hands. Whether the eventual overlaps between these projects were planned or (as was sometimes the case) sheer happenstance, we've tried to put our own social theory to good use in understanding the conditions of our own institution—conditions that, as with any social group, change and which sometimes present unanticipated opportunities to the social actor who is paying attention. And these are the skills that we've tried to convey to undergraduates for the past 20 years and which we hope to refine with the graduate students who now decide to join us in Manly Hall— with an emphasis on the qualifications implied by "tried to" and "hope to,"

---

[20] The workshop, initially piloted with a small group of participants in the spring of 2019 (with matching funding from the College and the Department) is entitled American Examples. Learn more at: https://blogs.religion.ua.edu/ae/; see the June 2019 Luce announcement here: "Luce Awards $14 Million in New Grants," *Luce Foundation*, 28 June 2019, https://www.hluce.org/news/articles/luce-foundation-awards-14-million-new-grants/.

[21] My thanks to Mike Altman for this final observation, as part of his comments after reading an earlier draft of this chapter.

since the experiment that we call the Department of Religious Studies at the University of Alabama is ongoing and we're still learning much.

Selected References

Bayart, Jean-Francois. *The Illusion of Cultural Identity*, trans. Steven Rendall, Janet Roitman, Cynthia Schoch, and Jonathan Derrick. Chicago: University of Chicago Press, 2005.

Lease, Gary. "The History of 'Religious' Consciousness and the Diffusion of Culture: Strategies for Surviving Dissolution." *Historical Reflections/Reflexions Historique* 20 (1994): 453–79.

McCutcheon, Russell T. *Critics Not Caretakers: Redescribing the Public Study of Religion*. Albany, NY: State University of New York Press, 2001.

McCutcheon, Russell T. "Afterword: Reinventing the Study of Religion in Alabama." In *Writing Religion: The Case for the Critical Study of Religion*, edited by Steven Ramey, 208–222. Tuscaloosa, AL: University of Alabama Press, 2015.

McCutcheon, Russell T. "Why I Blog." In Russell T. McCutcheon, *"Religion" in Theory and in Practice: Demystifying the Field for Burgeoning Academics*. Sheffield, UK: Equinox Publishing, 2018.

Touna, Vaia. "Preface." In *The Problem of Nostalgia in the Study of Identity: Towards a Dynamic Theory of People and Place*, edited by Vaia Touna. Sheffield, UK: Equinox Publishing, 2018.

Index

accessibility 20, 146, 149, 153, 177f., 209, 217, 219, 250, 255, 284
activism 262
algorithm 25, 46, 58, 61, 63–66
American Academy of Religion (AAR) 24, 122, 146, 150, 320
architecture 63, 75, 97, 123, 145, 160, 172, 182, 228, 242, 245, 303
archive 7, 11f., 14f., 21, 25, 27, 36f., 75, 98, 122f., 143–145, 148, 151–154, 156, 168, 178, 189, 191, 199f., 202–204, 206f., 209–213, 228f., 235, 239–246, 257, 288
artifacts 13, 116, 167, 171, 174, 182–185, 187f., 190f., 200, 202, 206, 235, 243, 296, 315
assignment 211, 249, 298, 300
audience 77–79, 81–83, 87, 104, 115f., 137, 139, 143, 145, 190, 206f., 211, 251, 253–255, 262, 265, 273, 282f., 288, 301, 316, 324, 326–328, 332–334
audio 12, 125, 129, 131, 133, 137, 139, 149, 151f., 162, 210, 231, 237, 273–275, 279–282, 288f., 291f., 295, 297, 299, 329, 332

Bible 31f., 34, 38, 40, 42–48, 85, 89, 119, 121–125, 132f., 137, 139, 169, 173, 207, 232
big data 24, 216, 334
blogs 213, 285, 288, 336
Buddhism 3f., 8–11, 13f., 17–19, 21f., 24f., 27, 56, 97–99, 206, 217, 310, 322

canon 4–13, 15f., 18–20, 23f., 27, 44, 54, 56, 59–61, 66, 104, 200
China 3, 13f., 24, 26f., 99, 121, 154, 215–219, 221, 223–225, 293
Chinese 3f., 6–10, 13f., 16–26, 99, 112, 215–221, 224
Christianity 33, 44, 71f., 78–80, 86, 299, 309f., 331
classification 12, 46f., 241, 310, 312, 326

clustering 25, 43
codicology 72, 77
coding 243, 252, 256–258, 260, 263f., 266, 269, 312, 326, 332
collaboration 5, 73, 75f., 104, 120, 140, 143, 148, 152, 156f., 160, 162, 165, 169, 177, 186, 199, 204, 212, 252, 255, 258, 265, 267, 274, 289, 304–308, 313, 315f.
computation 23–24, 31–32, 37–38, 42–46, 48–49, 55, 59, 61, 65, 71, 73–74, 77, 83, 85–86, 217
Coptic 59, 71–83, 85–89, 91
corpora 3, 6, 8, 16, 19, 23f., 27, 37, 43, 55, 60, 64, 66, 72–74, 76f., 79, 88, 91
curation 123, 125, 144, 164, 167, 203, 207, 335

3-D 125, 129–131, 133, 139, 199f., 203, 209–211, 228, 233, 243, 245, 247
data 3, 8, 11–16, 18–20, 22–25, 27, 31f., 43–47, 49f., 55, 57–60, 63, 66f., 71, 74–76, 85, 89, 91, 98, 104, 106f., 112, 116, 119, 121–123, 125f., 129–132, 137–139, 162, 164, 168f., 171f., 175, 177–183, 188–192, 199, 201–209, 211, 215f., 218–225, 230, 243, 250–266, 268f., 275, 280, 284, 298, 311f., 315f., 333f.
databases 14, 17, 20, 36f., 73, 105, 148, 152, 223, 257, 304, 309, 315f.
data mining (see also text mining) 37, 63
dictionary 8, 12, 17–20, 22, 59, 74, 76, 224
digital humanities VII-VIII, XI-XIV, 23–25, 43, 49, 53–54, 75, 111, 114, 116, 144, 146, 149–154, 162, 164=165, 177–178, 211, 249–250, 252, 254–255, 258f., 265–269, 287, 291, 303–304, 309–310, 313, 315–316, 319, 321

340 — Index

digitization 3–6, 13f., 17, 20, 23f., 26f., 37, 53, 60, 73, 88, 145, 158, 168, 177, 183, 205
distant reading 24, 77

encoding 6, 59, 75, 77
ethics 145
ethnography 119, 122, 126–132, 137–139

funding 53, 71, 103, 114–116, 146, 167, 228, 269, 274f., 282, 293, 326, 336

gender 48, 57, 59, 72, 85, 256, 258f., 285, 306, 334
Geographical Information Systems (GIS) 178–181, 202–206, 216

Hinduism 201, 310
historiography 200
HTML (HyperText Markup Language) 11, 113, 219, 298

India 3, 5f., 8–11, 13, 17f., 21–23, 27, 31, 53, 60, 99, 105, 121, 199–201, 204, 210f., 293, 297f.
institution 26f., 76, 79, 99f., 103, 144, 192, 215–217, 225, 267, 269, 275–277, 282, 286, 296, 301, 321, 323, 330, 336
interactive 38f., 48, 122, 125, 132, 139, 178, 182, 185, 188, 209f., 232, 240f., 243, 250, 252, 255, 257, 262, 296, 304–308, 316
interface 4, 7–9, 11–14, 17–19, 22, 32, 38, 42, 45, 104, 115, 189, 208, 219f., 222f., 280, 289
International Association for the History of Religions (IAHR) 273
Islam 152, 217, 249, 251, 255, 261, 287, 309f.

Judaism 310

machine learning 31, 43f., 46–49, 58
mapping 106, 206, 215f., 227–236, 238, 245f., 249–255, 260–266, 268f.
markup 4, 25f., 59, 75, 77, 104, 106, 112

material 10–13, 15, 19f., 22, 37, 44, 63, 97f., 100, 102–105, 107, 111, 114–117, 119–123, 129, 131f., 137–139, 143–149, 151–154, 156, 160, 162–165, 167–169, 177, 182–184, 188f., 191f., 199–202, 205f., 211–213, 228f., 235, 240f., 276, 284, 286, 291, 293, 297f., 326, 332
media 3, 37, 49, 98, 115, 119, 122–125, 127, 131f., 139, 145f., 149–151, 160, 163, 191, 199, 209, 228f., 235, 250f., 257f., 261, 264, 278, 280, 283, 299, 303–306, 308–313, 315f., 327, 333
metadata 5, 9–12, 14, 45, 74f., 111, 114, 116, 172, 189, 191, 205–209, 211, 242–244
modeling XVI, 38, 43, 47–48, 55, 58, 63–64, 85, 129–133, 181–182, 185–186, 187–192, 203, 209–211, 228, 230–233, 236, 243–244
MOOC (massive open online course) 291–300, 302
multimodal 98

Named Entity Recognition 14, 25
network analysis 25f., 31, 53–58, 60f., 63f., 66
North American Association for the Study of Religion (NAASR) 273

open source 14, 25, 67, 73–75, 130, 208f., 279, 281
Optical Character Recognition (OCR) 36
organization 6, 53f., 58, 114f., 129, 144, 150, 153f., 203, 206, 209, 213, 239, 257f., 260f., 286, 306

Pali 4–6, 18, 200
parsing 256
pedagogy 114, 124, 211–213, 291f., 295f.
peer review 126, 138, 150, 283f., 288, 326
photography 106, 122f., 129, 152, 154f., 157–162, 190
place 10, 14f., 25, 33, 38, 40, 54f., 58, 80, 89, 99, 104, 106, 111, 116, 119–122, 125–127, 132f., 137, 144, 146, 157, 159, 162f., 171, 177, 190, 199f., 207, 210,

216–218, 221, 228, 231f., 237–240, 243, 252f., 258, 266, 269, 275, 286, 298, 300, 303, 311, 322, 324–326, 328, 332
podcast VIII, XVII, 273–289, 294, 328, 332, 334
preservation 13, 148, 208
privacy XIII
public humanities 334
publishing 25, 31, 104, 131, 138, 145, 150–153, 188–190, 199, 202, 223, 245, 268, 279, 282f., 289, 293, 323, 328, 331
Python (programming language) 24, 49, 218, 224, 321

RAW (file format) 105

sacred 27, 34, 44, 83, 120f., 124, 133, 153, 160, 170–172, 175f., 178–182, 199–201, 203, 210, 297, 310–312
Sanskrit 10f., 15–19, 26, 60, 63
scale 24, 37, 49, 73, 99, 103, 112, 127, 131, 144, 159, 173, 177–180, 182, 184, 186f., 192, 209, 230, 236, 241
scholarship 17, 21, 23, 37f., 44, 49, 72, 74, 86, 97, 99, 119, 122f., 125f., 132, 137–140, 144f., 149f., 152f., 162f., 165, 190f., 199, 204, 209, 212f., 251, 255, 260, 262f., 265, 287, 296, 301, 307, 315f.
social media 122, 215f., 274, 285, 311, 321, 323–328
sound 45, 122, 125, 132, 134, 136f., 143, 162, 170, 210, 282, 333
students 24, 27, 49, 76, 115, 119, 144, 148, 154, 159, 161, 199, 203–205, 209, 211, 213, 216, 236, 251, 254f., 258, 260, 266–268, 273, 276, 278, 291–293, 295, 297–301, 305, 310, 313, 315f., 321–336
surrogate 103f., 128
sustainability 15, 204, 209, 235, 266–269

tagging 74, 207, 306, 326

teaching 27, 78–81, 83, 97f., 100, 103, 105, 114–116, 122, 124, 143f., 150, 159, 165, 199, 249, 266, 268, 278, 291, 294, 296–299, 301, 321, 325–328, 330, 334
technology 38, 53, 73f., 76, 97, 105f., 114, 116f., 122, 126, 133, 139, 151, 165, 177, 182, 186f., 204, 212, 230, 250f., 275f., 294, 296, 305, 309, 325, 329, 334, 336
TEI (Text Encoding Initiative) 7f., 14, 16, 59, 66, 75
tenure 149f., 289, 321f., 325, 335
text analysis 31, 38, 43, 45, 48–50, 117
text mining 37, 42, 49
Tibetan 4, 6, 9f., 12f., 16–19, 25f., 97–100, 102, 105–109, 113, 117, 215, 322
tokenize 45, 74
translation 3, 6, 8, 12, 15–20, 23, 25, 42, 44, 66, 71, 74, 78, 104f., 107f., 111f., 117, 155, 219, 221, 224, 282

Unicode 6, 73f.
Unix 205

video 123–125, 129, 131, 137, 151f., 162, 210f., 236, 241, 274, 288, 292, 294–297, 299f., 305, 320, 324, 326–329, 333
virtual reality (VR) 125, 160, 182, 200, 210, 227–229, 240f., 246
visualization 14, 26, 31, 38, 42, 74f., 79, 81f., 125, 138, 173–175, 182f., 187, 189–191, 221, 230, 243, 250f., 257, 260–264, 266, 268f.

website 5–13, 15f., 22, 31, 64, 102f., 105, 112f., 115, 117, 144, 146, 152f., 162f., 191, 209, 235f., 241, 250, 256, 258, 261, 266, 273f., 278f., 281, 283, 285, 288, 305, 309, 314, 316, 319–325, 331, 334

XML (extensible Markup Language) 5, 7f., 14f., 66, 75, 306

www.ingramcontent.com/pod-product-compliance
Lightning Source LLC
Chambersburg PA
CBHW070806300426
44111CB00014B/2443